THIRD EDITION

FDA
Regulatory Affairs

THIRD EDITION

FDA
Regulatory Affairs

Edited by
David Mantus
Douglas J. Pisano

CRC Press
Taylor & Francis Group
Boca Raton London New York

CRC Press is an imprint of the
Taylor & Francis Group, an **informa** business

CRC Press
Taylor & Francis Group
6000 Broken Sound Parkway NW, Suite 300
Boca Raton, FL 33487-2742

© 2014 by Taylor & Francis Group, LLC
CRC Press is an imprint of Taylor & Francis Group, an Informa business

No claim to original U.S. Government works

Printed on acid-free paper
Version Date: 20131202

International Standard Book Number-13: 978-1-84184-919-5 (Hardback)

This book contains information obtained from authentic and highly regarded sources. While all reasonable efforts have been made to publish reliable data and information, neither the author[s] nor the publisher can accept any legal responsibility or liability for any errors or omissions that may be made. The publishers wish to make clear that any views or opinions expressed in this book by individual editors, authors or contributors are personal to them and do not necessarily reflect the views/opinions of the publishers. The information or guidance contained in this book is intended for use by medical, scientific or health-care professionals and is provided strictly as a supplement to the medical or other professional's own judgement, their knowledge of the patient's medical history, relevant manufacturer's instructions and the appropriate best practice guidelines. Because of the rapid advances in medical science, any information or advice on dosages, procedures or diagnoses should be independently verified. The reader is strongly urged to consult the drug companies' printed instructions, and their websites, before administering any of the drugs recommended in this book. This book does not indicate whether a particular treatment is appropriate or suitable for a particular individual. Ultimately it is the sole responsibility of the medical professional to make his or her own professional judgements, so as to advise and treat patients appropriately. The authors and publishers have also attempted to trace the copyright holders of all material reproduced in this publication and apologize to copyright holders if permission to publish in this form has not been obtained. If any copyright material has not been acknowledged please write and let us know so we may rectify in any future reprint.

Except as permitted under U.S. Copyright Law, no part of this book may be reprinted, reproduced, transmitted, or utilized in any form by any electronic, mechanical, or other means, now known or hereafter invented, including photocopying, microfilming, and recording, or in any information storage or retrieval system, without written permission from the publishers.

For permission to photocopy or use material electronically from this work, please access www.copyright.com (http://www.copyright.com/) or contact the Copyright Clearance Center, Inc. (CCC), 222 Rosewood Drive, Danvers, MA 01923, 978-750-8400. CCC is a not-for-profit organization that provides licenses and registration for a variety of users. For organizations that have been granted a photocopy license by the CCC, a separate system of payment has been arranged.

Trademark Notice: Product or corporate names may be trademarks or registered trademarks, and are used only for identification and explanation without intent to infringe.

Library of Congress Cataloging-in-Publication Data

FDA regulatory affairs / editors, Douglas J. Pisano, David Mantus. -- Third edition.
   p. ; cm.
   Includes bibliographical references and index.
   ISBN 978-1-84184-919-5 (hardcover : alk. paper)
   I. Pisano, Douglas J., editor of compilation. II. Mantus, David, editor of compilation.
   [DNLM: 1. United States. Food and Drug Administration. 2. Drug Approval--United States. 3. United States Government Agencies--United States. 4. Biological Products--standards--United States. 5. Equipment and Supplies--standards--United States. 6. Government Regulation--United States. QV 771]
   RM301.25
   615'.19--dc23   2013047346

Visit the Taylor & Francis Website at
http://www.taylorandfrancis.com

and the CRC Press Website at
http://www.crcpress.com

Contents

Preface vii
Editors ix
Contributors xi

Chapter 1 Overview of FDA and Drug Development 1
Josephine C. Babiarz and Douglas Pisano

Chapter 2 What Is an IND? 41
Michael R. Hamrell

Chapter 3 The New Drug Application 77
Charles Monahan and Josephine C. Babiarz

Chapter 4 Meetings with the FDA 105
Alberto Grignolo and Sally Choe

Chapter 5 FDA Medical Device Regulation 125
Barry Sall

Chapter 6 A Primer of Drug/Device Law: What Is the Law and How Do I Find It? 169
Josephine C. Babiarz

Chapter 7 The Development of Orphan Drugs 189
Scott N. Freeman

Chapter 8 CMC Sections of Regulatory Filings and CMC Regulatory Compliance during Investigational and Postapproval Stages 199
Prabu Nambiar, Steven R. Koepke, and Kevin Swiss

Chapter 9 Overview of the GxPs for the Regulatory Professional 235

Bob Buckley, Robert Blanks, Kimberly J. White, and Tonya White-Salters

Chapter 10 FDA Regulation of the Advertising and Promotion of Prescription Drugs, Biologics, and Medical Devices 285

Karen L. Drake, Esq.

Chapter 11 The Practice of Regulatory Affairs 309

David S. Mantus

Chapter 12 FDA Advisory Committees 327

Christina A. McCarthy and David S. Mantus

Chapter 13 Biologics 347

Florence Kaltovich

Chapter 14 Regulation of Combination Products in the United States 361

John Barlow Weiner, Esq.

Index 379

Preface

This book is a roadmap to the U.S. Food and Drug Administration (FDA) and drug, biologic, and medical device development. The book is written in plain English, with an emphasis on easy access to understanding how this Agency operates with respect to the practical aspects of U.S. product approval. It is meant to be a concise reference that offers current, real-time information. It has been written as a handy reference for use by students, staff, and professionals at corporations, organizations, and schools and colleges across the United States in need of a simple, concise text from which to learn and teach. The topics in *FDA Regulatory Affairs: A Guide for Prescription Drugs, Medical Devices, and Biologics*, Second Edition, are covered in a straightforward format. It is a compilation and commentary of selected laws and regulations pertaining to the development and approval of drugs, biologics, and medical devices in the United States. It is not intended to take the place of an actual reading of the laws of the United States of America or the regulations of the United States, its agencies or anybody that regulates the development or approval of drugs, biologics, and medical devices in the United States.

David S. Mantus
C after D Inc.

Douglas J. Pisano
Massachusetts College of Pharmacy and Health Sciences

Editors

Dr. David S. Mantus, PhD, worked in the biotechnology and pharmaceutical industry for more than 20 years. He served as vice president, Regulatory Affairs at Cubist Pharmaceuticals until 2011. Prior to joining Cubist, Dr. Mantus served as vice president, Regulatory Affairs at Sention, Inc., and held various regulatory roles at Shire Biologics, PAREXEL, the Massachusetts Public Health Laboratory, and Procter & Gamble Pharmaceuticals. Dr. Mantus received his BS in chemistry from the College of William and Mary, and his MS and PhD in chemistry from Cornell University. He was a postdoctoral research fellow in biomedical engineering at the University of Washington. He currently resides in Bolton, MA.

Douglas J. Pisano is a professor of social and administrative sciences at the Massachusetts College of Pharmacy and Health Sciences (MCPHS) University. He began his career at the University in the Fall of 1984 and has maintained a full-time faculty appointment and is a member of the graduate faculty. In 1998, he became the founder and director of the master of science degree in regulatory affairs and health policy and served in that role until 2010. In the year 2000, he became the dean of the newly established School of Pharmacy—Worcester. Its cutting-edge accelerated Doctor of Pharmacy program was instrumental in establishing a companion program using distance education technology at the college's Manchester campus. In June 2005, Dr. Pisano became the associate provost for Pharmacy Education and dean of the School of Pharmacy at Boston where he was charged with blending, coordinating, and overseeing many of the aspects of the college's three distinct and accredited Doctor of Pharmacy programs, which extend over three campuses into a single face of pharmacy for the college overall. His pharmacy deanship also included the oversight of the 12 programs in the School of Graduate Studies and the Office of Pharmacy Experiential Education. He has also served as interim dean for the School of Medical Imaging and Therapeutics and Physician Assistant Studies. Dr. Pisano currently serves as vice president of academic affairs and provost for the university.

Contributors

Josephine C. Babiarz, Esq.
School of Pharmacy
Massachusetts College of Pharmacy
  and Health Sciences
Boston, Massachusetts

Robert Blanks
Izumi Biosciences
Lexington, Massachusetts

Bob Buckley
Aegerion Pharmaceuticals
Cambridge, Massachusetts

Sally Choe
PAREXEL Consulting
Waltham, Massachusetts

Karen L. Drake
Grafton, Massachusetts

Scott N. Freeman
Food and Drug Administration
Office of Orphan Products
  Development
Silver Spring, Maryland

Alberto Grignolo
PAREXEL Consulting
Waltham, Massachusetts

Michael R. Hamrell
MORIAH Consultants
Yorba Linda, California

Florence Kaltovich
Vaccine Research Center
National Institutes of Health
Gaithersburg, Maryland

Steven R. Koepke
SRK Consulting, LLC
Walkersville, Maryland

David S. Mantus
C after D Inc.
Bolton, Massachusetts

Christina A. McCarthy
Dyax Corporation
Burlington, Massachusetts

Charles Monahan
AVEO Pharmaceuticals, Inc.
Cambridge, Massachusetts

Prabu Nambiar
Syner-G Pharma CMC Consulting,
  LLC
Newton, Massachusetts

Douglas J. Pisano
Massachusetts College of Pharmacy
  and Health Sciences
Boston, Massachusetts

Barry Sall
PAREXEL Consulting
Waltham, Massachusetts

Kevin Swiss
Nautilus Neurosciences, Inc.
Blaine, Washington

John Barlow Weiner, Esq.
Food and Drug Administration
Silver Spring, Maryland

Kimberly J. White
White-Salters Consulting, LLC
Ayer, Massachusetts

Tonya White-Salters
White-Salters Consulting, LLC
Ayer, Massachusetts

1 Overview of FDA and Drug Development

Josephine C. Babiarz and Douglas Pisano

CONTENTS

Introduction	2
The Evolution of the FDA—A Century of Laws and Policies	3
The History (1848–1979)	3
1980 to the Present: AIDS, Orphans, Terrorism, Economic Incentives, and the Food and Drug Administration Safety and Innovation Act	6
Biologics Price Competition and Innovation Act of 2009	13
FDASIA	14
Summary	17
Agency Role and Organization	17
New Drug Development and Approval Summary	19
Preclinical Investigation	20
Investigational New Drug Application	20
Phase I	22
Phase II	22
Phase III	22
New Drug Application	22
Fast Track, Accelerated Approval, Priority Review, and "Breakthrough Therapy" Designations	24
Phase IV and Postmarketing Surveillance	26
Orphan Drugs	26
Abbreviated New Drug Applications	26
OTC Regulations	28
Biologics	29
Devices	31
Regulating Drug and Device Marketing	33
Violations and Enforcement	33
Conclusion	35
Notes	35
References	36

INTRODUCTION

The Food and Drug Administration (FDA), a single agency, regulates a trillion dollars of products, roughly 25 cents in every dollar spent, ranging from 80% of the US food supply to all human health-care products, prescription drugs and devices, electronic products that emit radiation, animal products, cosmetics, and even manufacture of tobacco products. FDA is responsible for the quality of shellfish, safety of cardiac stents, Over the Counter (OTC) cough syrups, HIV drug regimens, tetanus shots, artificial sweeteners, mammography standards, toothpaste, pediatric medicines, pet food, vitamins, and lipsticks, not to mention the readability of calorie and trans fat information on a bag of potato chips. The scope of FDA's activities is so broad, and the number of decisions, approvals, clearances, recalls, hearings, and regulatory reports generated is too widespread to easily summarize. The FDA has created a portal on its Web page to present links to information and reports. That portal, titled "FDA-Track: Agency-Wide Program Performance,"[1] organizes noteworthy developments and performance reports for each center and also links to those reports by month and year. The economic impact of the FDA is difficult to calculate, the scientific challenges and increasing medical needs overwhelming, and the expectations contradictory.

In the regulation of pharmaceuticals, a main focus of this book, the economics of bringing a product to market reaches a scale difficult to comprehend. Recent estimates of the costs of drug development and approval from the industry's perspective are enormously expensive. As noted by Matthew Herper,[2] the average drug developed by a major pharmaceutical company costs between US$4 billion and US$11 billion. When research and development costs of failed formulations are taken into account, Herper estimates that the industry spends US$4 billion in research alone for every drug that is approved. He further estimates that fewer than 1 in 10 medicines that start clinical trials succeed. While some have called Herper out, pointing out that he may not have included favorable tax treatment of these costs or the value of these medications over time, the core point is accurate—this is a very expensive industry, with few large successes. These costs are spent in the name of safety and efficacy. The evaluation of the scientific data collected to support these claims remains the FDA's primary responsibility as a governmental agency, but the public expects much more from it.

The FDA is expected to protect us and our pets from harm, but also to allow us rapid access to unproven therapies that show promise to either cure or benefit us. The FDA is to act quickly to get products to market, but must be right the first time, and is criticized as being too permissive or lax if a drug or device must be recalled later for safety concerns or unpredicted adverse events (AEs). Given the pace of scientific advancement, this is no small demand.

The FDA Mission Statement reflects the scope of these responsibilities:

The FDA's Mission Statement

FDA is responsible for protecting the public health by assuring the safety, efficacy and security of human and veterinary drugs, biological products, medical devices, our nation's food supply, cosmetics, and products that emit radiation.

FDA is also responsible for advancing the public health by helping to speed innovations that make medicines more effective, safer, and more affordable and by helping the public get the accurate, science-based information they need to use medicines and foods to maintain and improve their health. FDA also has responsibility for regulating the manufacturing, marketing and distribution of tobacco products to protect the public health and to reduce tobacco use by minors.

Finally, FDA plays a significant role in the Nation's counterterrorism capability. FDA fulfills this responsibility by ensuring the security of the food supply and by fostering development of medical products to respond to deliberate and naturally emerging public health threats.[3]

The FDA's authority and influence are the product of compromise, evolving over time. It is an agency that is governed as much by law as by science. History shows us that the FDA's authority has grown commensurate with the magnitude of harm suffered by the public because of the food and drugs consumed, as well as the devices used. The Agency and its statutory framework remain a work in progress. To better understand the FDA, its controlling laws, and its role in public health, a brief summary is in order.

In the United States, all food, drugs, cosmetics, and medical devices for both humans and animals are regulated under the authority of the Food, Drug, and Cosmetic Act (FDCA) of 1938, which in turn establishes the FDA. The FDA and all of its regulations were created by the government in response to the pressing need to address the safety of the public with respect to food and medicinals. The purpose of this chapter is to describe and explain the nature and extent of these regulations as they apply to medical products in the United States. A historical perspective is offered as a foundation for regulatory context. In addition, this chapter will discuss the FDA's regulatory oversight and that of other agencies, the drug approval and development process, the mechanisms used to regulate manufacturing and marketing, as well as various violation and enforcement schemas.

THE EVOLUTION OF THE FDA—A CENTURY OF LAWS AND POLICIES

THE HISTORY (1848–1979)

Prior to 1902, the US government took a hands-off approach to the regulation of drugs. Many of the drugs available were so-called patent medicines, which were so named because each had a more or less descriptive or patent name. No laws, regulations or standards existed to any noticeable extent, even though the US Pharmacopoeia (USP) became a reality in 1820 as the first official compendium of the United States. The USP set standards for strength and purity, which could be used by physicians and pharmacists who needed centralized guidelines to extract, compound, and otherwise utilize drug components that existed at that time.[2,4]

However, in 1848, the first American drug law, the Drug Importation Act, was enacted when American troops serving in Mexico became seriously affected when adulterated quinine, an antimalarial drug, was discovered. This law required laboratory inspection, detention, and even destruction of drugs that did not meet acceptable standards. Later, in 1902, the Virus–Serum–Toxin Act (Biologics Control Act) was passed in response to tetanus-infected diphtheria antitoxin, which was manufactured by a small laboratory in St. Louis, Missouri. Ten school children died as a result of the tainted serum. No national standards were as yet in place for purity or potency of medical products. The Act authorized the Public Health Service to license and regulate the interstate sale of serum, vaccines, and related biological products used to prevent or treat disease.

This Act also spurred Dr. Harvey W. Wiley, chief chemist for the Bureau of Chemistry, a branch of the US Department of Agriculture (USDA) and the forerunner of today's FDA, to investigate the country's foods and drugs. He established the "Hygienic Table," where a group of young men who volunteered to serve as human guinea pigs allowed Dr. Wiley to feed them a controlled diet laced with a variety of preservatives and artificial colors. More popularly known as the "Poison Squad," they helped Dr. Wiley gather enough data to prove that many of America's foods and drugs were "adulterated," the products' strength or purity was suspect or "misbranded," or the products had inadequate or inaccurate labeling. Dr. Wiley's efforts, along with publication of Upton Sinclair's *The Jungle* (a novel depicting the putrid conditions in America's meat industry), were rewarded when Congress passed America's first food and drug law, in 1906, the United States Pure Food and Drug Act (PFDA; also known as the Wiley Act). The Wiley Act prohibited interstate commerce of misbranded foods or drugs based on their labeling. It did not affect unsafe drugs in that its legal authority would only come to bear when a product's ingredients were falsely labeled. Even intentionally false therapeutic claims were not prohibited.

This began to change in 1911 with the enactment of the Sherley Amendment, which prohibited the labeling of medications with false therapeutic claims that were intended to defraud the purchaser. These amendments, however, required the government to find proof of intentional labeling fraud. Later, in 1937, a sentinel event occurred that changed the entire regulatory picture. Sulfa became the miracle drug of the time and was used to treat many life-threatening infections. It tasted bad and was hard to swallow, which led entrepreneurs to seek a palatable solution. S.E. Massengill Company of Bristol, Tennessee, developed what it thought was a palatable, raspberry-flavored liquid product. However, it used diethylene glycol to solubilize the sulfa. Six gallons of this dangerous mixture, elixir of sulfanilamide, killed some 107 people, mostly children.

The result was the passage of one of the most comprehensive statutes in the history of American health law. The federal FDCA repealed the Sherley Amendment and required that all new drugs be tested by their manufacturers for safety and that those tests be submitted to the government for marketing approval via a new drug application (NDA). The FDCA also mandated that drugs be labeled with adequate directions if they were shown to have had harmful effects. In addition, the FDCA authorized the FDA to conduct unannounced inspections of drug manufacturing facilities. Though amended many times since 1938, the FDCA is still the broad foundation for statutory authority for the FDA as it exists today.

However, a new crisis loomed. Throughout the late 1950s, European and Canadian physicians began to encounter a number of infants born with a curious birth defect called "phocomelia," a defect that resulted in limbs that resembled "flippers," similar to those found on seals. These birth defects were traced back to mothers who had been prescribed the drug thalidomide in an effort to relieve morning sickness during pregnancy. The manufacturer of this drug applied for US marketing approval as a sleep aid. However, because of the efforts of Dr. Frances O. Kelsey, the FDA's chief medical officer at that time, a case was made that the drug was not safe, and therefore not effective for release in the US marketplace.

Dr. Kelsey's efforts and decisive work pushed the US Congress to pass another necessary amendment to the FDCA, in 1962, the Kefauver–Harris Act. This Act essentially closed many of the loopholes regarding drug safety in American law. These "Drug Efficacy Amendments" now required drug manufacturers to prove safety and efficacy of their drug products, register with the FDA and be inspected at least every two years, have their prescription drug advertising approved by the FDA (this authority being transferred from the Federal Trade Commission), provide and obtain documented "informed consent" from research subjects prior to human trials, and increase controls over manufacturing and testing to determine drug effectiveness.

In an effort to address these new provisions of the Act, the FDA contracted with the National Academy of Sciences along with the National Research Council to examine some 3400 drug products approved between the years 1938 and 1962 on the basis of safety alone. Called the Drug Efficacy Study Implementation (DESI) Review of 1966, it charged these organizations to determine whether post-1938 drug products were "effective" for the indications claimed in their labeling, or "probably effective," "possibly effective," or "ineffective." Those products not deemed effective were removed from the marketplace, reformulated, or sold with a clear warning to prescribers that the product was not deemed effective.

Later in 1972, the FDA began to examine OTC drug products. Phase II of the Drug Efficacy Amendments required the FDA to determine the efficacy of OTC drug products. This project was much larger in scope than the analysis of prescription drugs. In the America of the 1970s, consumers could choose from more than 300,000 OTC drug products. The FDA soon realized that it did not have the resources to evaluate each and every OTC drug product. Hence, the FDA created advisory panels of scientists, medical professionals, and consumers who were charged with evaluating active ingredients used in OTC products within 80 defined therapeutic categories. After examining both the scientific and medical literature of the day, the advisory panels made decisions regarding active ingredients and their labeling. The result was a "monograph" that described in detail acceptable active ingredients and labeling for products within a therapeutic class. Products that complied with monograph guidelines were deemed category I: safe and effective, not misbranded. However, products not in compliance with monograph guidelines were deemed category II: not safe and effective or misbranded. Category II products were removed from the marketplace or reformulated. Products for which data were insufficient for classification were deemed category III and were allowed to continue in the market until substantive data could be established or until they were reformulated and were in compliance with the monograph. The OTC Drug Review took approximately 20 years to complete.

Although there were numerous other federal laws and regulations that were passed throughout the 1970s, many were based on regulating the professional practice of medical professionals or for the direct protection of consumers. For example, the federal Controlled Substances Act (CSA), part of the Comprehensive Drug Abuse Prevention and Control Act of 1970, placed drugs with a relatively high potential for abuse into five federal schedules along with a "closed record-keeping system," designed to track federally controlled substances via a definite paper trail, as they were ordered, prescribed, dispensed, and utilized throughout the health-care system.

1980 to the Present: AIDS, Orphans, Terrorism, Economic Incentives, and the Food and Drug Administration Safety and Innovation Act

The 1980s brought significant regulatory changes. Biotechnology had begun on a grand scale and the pharmaceutical industry was on its cutting edge. Many of the medicinal compounds being discovered were shown to be very expensive and have limited use in the general US population. However, these compounds could prove lifesaving to demographically small patient populations who suffered from diseases and conditions that were considered rare. In an effort to encourage these biotech pharmaceutical companies to continue to develop these and other products, Congress passed the Orphan Drug Act in 1983.[5] The Act continues to allow manufacturers incentives for research, development, and marketing of drug products used to treat rare diseases or conditions that would otherwise be unprofitable via a system of market exclusivity, and substantial breaks and deductions in a manufacturer's corporate taxes. Though the success of the Orphan Drug Act proved to be of great medical benefit for a few, a scandal was looming in other parts of the pharmaceutical industry.

The generic pharmaceutical industry experienced steady growth as many of the exclusive patents enjoyed by major pharmaceutical companies for brand named products were beginning to expire. Generic versions of these now freely copied products were appearing much more frequently in the marketplace. However, these generic copies were required to undergo the same rigorous testing that brand name, pioneer, or innovator products did, thereby increasing costs, duplicating test results, and substantially slowing the availability of less expensive but equivalent drugs. To speed access to cheaper therapies, Congress passed the Drug Price Competition and Patent Restoration Act in 1984.[6] This Act, also called the Hatch–Waxman Act after its sponsors, was designed to level the playing field in the prescription drug industry with regard to patent-protected prescription drug products and their generic copies.

The Hatch–Waxman Act was composed of two distinct parts or "titles." Title I was for the benefit of the generic pharmaceutical industry. It extended the scope of the Abbreviated NDA (ANDA) to cover generic versions of post-1962-approved drug products. It required that generic versions have the same

Overview of FDA and Drug Development

relevant aspects as the pioneer or innovator drugs with regard to bioequivalence (rate and extent of absorption of the active drug in the human body) and pharmaceutical equivalence (same dosage form as the pioneer drug to which it is compared). Though somewhat simplified, the Hatch–Waxman Act permitted easier market access to generic copies of pioneer drugs, provided they were not significantly different from the pioneer drug in their absorption, action, and dosage form. In addition, Title II of the Act was designed to aid and encourage research-based or innovator pharmaceutical companies in continuing their search for new and useful medicinal compounds by extending the patent life of pioneer drug products to compensate for marketing time lost during the FDA "review period."[i]

While the patent extension benefit has become somewhat less valuable today due to an overall reduction in the FDA review time as a result of management improvements implemented with the prescription drug user fees, the value of patent-protected drugs has skyrocketed, with the so-called blockbuster drugs garnering millions of dollars in sales in less than a year. Market exclusivity and patent extensions remain powerful motivators used to encourage orphan drug development and pediatric applications, as discussed below.

Congress recognized that counterfeit drugs, as well as improper control over drug samples, and sales and marketing materials posed serious health hazards. Accordingly, the Prescription Drug Marketing Act (PDMA) of 1987[7] requires that all drugs be distributed through legitimate commercial channels, that pharmaceutical sales representatives maintain detailed accounts of drug samples (giving birth to the term "detailer"), and that importation of drugs from foreign countries be restricted.

Congress focused on devices and nutrition in the year 1990. The Safe Medical Devices Act of 1990[8] established a user reporting system to improve device safety. If a medical device probably caused or contributed to death, serious injury, or illness, representatives of the institution or facility where the incident occurred were required to file a report with the FDA. In turn, the device manufacturers were required to address or respond to the incident. The statute also gives FDA the power and authority to recall devices, which it does not have in the case of drugs (drug recalls are voluntary actions by the manufacturers; using US Marshals, the FDA can and will seize drug lots, however). This Act also addressed combination products, establishing that the jurisdiction of the FDA centers over combination products would be based on the primary indication of the product (see Chapter 14 for a fuller discussion of combination products). The year 1990 also brought regulation to food; the Nutrition Labeling and Education Act of 1990 requires nutrition labeling and health claims to be consistent with the format and rules established by the FDA. This law brought new—and uniform—meaning to the words "low fat" and "light."[9]

The year 1992 saw three major laws enacted. An unintended side effect of the Hatch–Waxman Act was a very public scandal in which a few unscrupulous generic pharmaceutical companies took shortcuts in reporting data, submitted fraudulent samples, and offered bribes to FDA officials to gain easy and rapid market approval for their products. The Generic Drug Enforcement Act of 1992 provided for debarment and other serious penalties for bribery, fraud, or misconduct, among other deterrents.[ii] Congress also strengthened device oversight; the Medical Device Amendments of 1992 added penalties if a manufacturer did not comply with postmarketing surveillance testing and reporting.[10]

The most significant change of that year came in the form of the first Prescription Drug User Fee Act (PDUFA[11]; in hindsight, it is labeled PDUFA I; PDUFA V was passed in 2012). The Act was intended to help the FDA generate additional funds to upgrade and modernize its operations and to accelerate drug approval. It authorized the FDA to charge pharmaceutical manufacturers a "user fee" to accelerate drug review. These funds in turn are used by the FDA to hire additional staff for reviews and modernize the review process. The PDUFA framework requires the FDA to develop and implement performance goals, primarily directed at reducing review time and making the drug development process more efficient. Critics and supporters alike quickly point out that the user fee is fully paid only when the FDA approves a product for market—and not if the final clinical results do not prove that the benefit outweighs the risk. This fee assessment has sparked a great deal of debate about the real conflict of interest present when the FDA reviewers are examining a product whose approval fees go directly to fund the reviewers' employment. As will be discussed later, there are checks and balances in this system, as Congress appropriates funds to cover FDA administration, including reviewers' salaries.

As a result of PDUFA I, the FDA hired more personnel and reduced the approval time of new pharmaceutical products, which was greater than 30 months in 1992. However, the first Act had a "sunset" provision, which ended the FDA's authority to charge user fees in 1997. The Act was so successful that PDUFA has been reauthorized and extended five additional times, and the fee concept has been expanded to include medical devices and biologics [Medical Device User Fee and Modernization Act (MDUFMA) of 2002],[12] and with the most recent reauthorization, PDUFA V part of the FDA Safety and Innovation Act (FDASIA), passed in 2012, Generics and Biosimilars.[13]

Congress relaxed the regulation of certain industries. The Dietary Supplement Health and Education Act (DSHEA) of 1994 shifted the burden of proof from the industry to the FDA. For drugs, devices, and biologics, a sponsor or manufacturer must prove that the product is safe and effective for the indication claimed. The opposite is true of dietary supplements; thanks to this law, the FDA "bears the burden of proof... to show that a dietary supplement is adulterated."[14]

Congress continued to expand and enhance the scope and powers of the FDA. One example is the FDA Modernization Act (FDAMA) of 1997.[15] FDAMA gave FDA authority to conduct "fast-track" product reviews to speedily launch lifesaving drug therapies to market, permitted an additional six-month patent exclusivity for pediatric prescription drug products, and required the National Institutes of Health (NIH) to build a publicly accessible database on clinical studies of investigational drugs or life-threatening diseases.

FDAMA addressed the real dilemma of terminally ill patients, who were routinely denied access to experimental drugs because of the lack of safety and efficacy data on the drugs; the FDA had no authority to allow the use of such drugs outside enrollment in a controlled clinical investigation. However, pressures from acquired immunodeficiency syndrome (AIDS) activists in particular moved Congress to change the rules. Under FDAMA, there was expanded access to unapproved drug and devices, specifically therapies and diagnostics for serious diseases or life-threatening conditions.[16] Individuals not enrolled in a formal clinical trial could obtain unapproved products—that is, products covered by an investigational new drug (IND) or investigational device exemption (IDE)—during emergencies or for personal use. Unapproved drugs are available under "expanded access" or "compassionate use,"[17] and experimental devices are available under the Humanitarian Device Exemptions (HDEs).[18]

FDAMA addressed, albeit briefly, issues of "off-label" promotion. Generally, a manufacturer may only advertise those claims and indications that are stated in the label; any deviation can be prosecuted as misbranding.[19] The FDA took the position that certain publications directed at prescribers were in fact off-label promotion. Various critics, including the Washington Legal Foundation, felt that this position violated the right to freedom of speech, guaranteed by the First Amendment to the US Constitution. A federal court agreed with the foundation,[20] and the FDA conceded that round. The core of the debate centers on public health—should a manufacturer be able to promote therapeutic claims that have not been subject to scientific review by the Agency, aka "off-label" claims? The FDA has traditionally taken the view that off-label use of a drug subjects the patient with certainty to the risks associated with that formulation, but will afford no possible proven benefit. The FDCA does not give the FDA any authority over the practice of medicine,[21] and consequently, the FDA cannot prevent the prescribing of products for off-label use, hence the risk to public health.

This debate continues, reinvigorated by the recent US Supreme Court decision in *Citizens United v. the Federal Election Commission*,[22] which decided that corporations had a right to free speech. While the *Citizens United* case is not directly applicable to the FDCA, the decision presents additional complexity in the ongoing debate about the type and quality of scientific or "off-label" information which a manufacturer can publish. The fundamental question was and remains—how can the public be adequately protected if a manufacturer is allowed to promote all the uses of a product and not only those that the FDA has determined to be supported by scientific evidence?

The compromise Congress crafted was to allow dissemination of information on unapproved uses of products to a limited group of professionals—that is, physicians, insurance companies, and other health-care practitioners. This particular provision has expired,[23] but newer laws reflect the battlefront has shifted to the arena of postmarketing surveillance and drug database registries, and with the passage of the FDASIA in 2012,[24] to the Internet as well.

Imitation is the sincerest compliment—the success of PDUFA gave birth to the MDUFMA in 2002. MDUFMA was enacted in order to provide the FDA with the resources necessary to better review medical devices, enact needed regulatory reforms so that medical device manufacturers can bring their safe and effective devices to the American people at an earlier time, and ensure that reprocessed medical devices are as safe and effective as original devices.[25]

MDUFMA continues to be strengthened and is now reauthorized through 2017. This same fee scheme has been adapted by veterinary medicines with the Animal Drug User Fee Act (ADUFA) of 2003.[26]

In 2002 and again in 2003, Congress began addressing the untested use of adult drugs for pediatric indications, using a carrot and stick approach. The Best Pharmaceuticals for Children Act of 2002,[27] the carrot, provided a six-month patent exclusivity, as a reward for manufacturers who tested the formulation in pediatric indications. This law was followed in 2003 by the Pediatric Research Equity Act,[28] the stick, which mandates that new drugs and biologics be tested in children if they can be used by children.

Under Section 505A of the Best Pharmaceuticals for Children Act, the FDA may request a sponsor to conduct ethnically and racially diverse pediatric studies of both approved and unapproved uses of a drug. If a sponsor disagrees with this request on the basis that it is not possible to develop a pediatric formulation, but gives no other reason, then the drug labeling must clearly state that it is unsafe/untested for pediatric use. If the sponsor agrees to test, the sponsor must provide not only all postmarket AE reports regarding the drug but also those generated during the pediatric trial. Additionally, the law requires the publication of pediatric labeling changes, including both on- and off-label indications and the results of the study of both. If the FDA and sponsor cannot agree on the labeling change, the matter will be referred to the Pediatric Advisory Committee, which can make recommendations but not bind the Agency. Changes required by the FDA but not made by the manufacturer means the product is misbranded. If the FDA does not determine that a drug is safe and effective, or the data are inconclusive, the labeling must reflect that information. While the FDA can grant a waiver for this mandatory pediatric testing, the FDA takes the position that prescribing adult formulations for children without adequate studies is off-label and in effect, constitutes unapproved and unmonitored ongoing drug trial. These initiatives were permanently enacted—that is, the expiration dates were removed—by FDASIA, passed in 2012. FDASIA also strengthens the FDA's powers in the pediatric area. The age range for pediatrics now includes neonates and the FDA's reasons for granting written requests to exclude pediatric testing will be posted on the FDA Website. Additionally, if there is a deferral, under FDASIA, the FDA must track the time frames and issue noncompliance letters to those who fail to comply; again, this information would be publicly available.

Under the Pediatric Research Equity Act, all marketed drugs must have a pediatric label unless a waiver is granted. However, there are "workarounds" in the law. Section 505B allows a sponsor to extrapolate pediatric effectiveness from adequate and well-controlled studies in adults, supplemented with other

Overview of FDA and Drug Development 11

information obtained in pediatric patients, for example, pharmacokinetic studies, and also allows the extrapolation of data from one pediatric age group to another pediatric age group. Pediatric studies themselves can be deferred if the adult form is ready before pediatric studies are complete or until additional safety and effectiveness data have been collected. The law provides for full and partial waivers from pediatric studies if they are impossible or impractical, if there is strong evidence that the product is ineffective or unsafe in all pediatric age groups, or if the new product does not have a meaningful benefit over existing pediatric treatments and is not likely to be used in a substantial number of pediatric patients, provided, however, that the product label reflects that the product would be ineffective or unsafe in pediatric populations.

Devices may also have pediatric indications; due to the physical differences, for example, size between adults and children, adult devices can simply be unusable in pediatric populations. Earlier laws established the device equivalent of the drug's compassionate use program for orphan indications, which is the HDE. FDASIA now allows manufacturers who provide the HDE to pediatric patients to charge patients for the device and profit from the transaction. (Readers are referred to Chapter 5 for a complete discussion of device approvals.)

The events of 9/11 also impacted the FDA. The Public Health Security and Bioterrorism Preparedness and Response Act of 2002[29] requires the stockpiling of certain drugs and enhances protection of the food supply, among the measures that address national emergency situations. The Project Bioshield Act of 2004[30] will "...provide protections and countermeasures against chemical, radiological, or nuclear agents that may be used in a terrorist attack against the United States... [by] streamlining the FDA approval process of countermeasures."

In 2004, the FDA launched the "Critical Path Initiative" (CPI), which outlines the FDA's national strategy to address the gap between scientific discovery and successful development of novel drugs; this initiative is an important tool in expediting overall drug development; the CPI continues to be updated.[31]

The Food and Drug Administration Amendments Act (FDAAA) of 2007[32] is broad and detailed legislation that impacted each pressure point, not only in drug and device development, but also throughout the FDA's purview. The law established new and robust requirements for postmarketing surveillance and improved food safety. Specifically, FDAAA expanded the FDA's implementation of guidance for product review, in particular, guidelines for industry on clinical trial design, provided funding for electronic data processing, and improved clinical trial databases, a responsibility shared between the FDA and NIH and finally, placed additional emphasis on postmarketing surveillance across all fields.

Specific sections of FDAAA merit review. Clinical trial databases are the subject of Section 80. The statute generally exempts inclusion of information for drugs in Phase I and devices in feasibility or prototype trials. However, the NIH clinical trial registry has been expanded to include most types of clinical trial results, in which a clinical trial is the primary basis of an efficacy claim or if the trial is conducted after the drug is approved or the device is cleared. This means that devices studied under a 501(k), a premarket approval application, or a HDE, are now reportable, as are pediatric postmarketing studies. The reportable elements are extensive and include the product, the indications being studied, eligibility criteria, and links to existing results. The FDA has implemented this mandate, in part, by launching the ClinicalTrials.gov Website[33] which itemizes the requirements of a "responsible party" (either the sponsor or the principal investigator) must meet, along with numerous other requirements.[34] The results of these clinical trials must be disclosed when a drug or device is initially approved, when there is a new previously off-label use, and where the product is not approved or cleared. Serious AEs and frequent AEs are also reportable. The law is clear that this information is to be available on the Internet.

The public demanded even greater changes to the FDCA. Section 901 greatly strengthens the FDA's authority over the resources in conducting postmarket surveillance. This law applies to drugs and biologics but not to veterinary drugs. Under the law now, a responsible person may not introduce or deliver a new drug if that person has not (1) conducted postapproval studies or postapproval clinical trials on the basis of scientific data selected by the FDA or (2) not made safety labeling changes requested by the FDA.

The purposes of the study or clinical trial are to (1) assess a known, serious risk related to the use of the drug involved; (2) to assess signals of serious risk related to the use of the drug; and (3) to identify an unexpected serious risk when available data indicate the potential for serious risks. Regarding the safety labeling changes, there is a discussion and dispute resolution procedure, but the Secretary of Health and Human Services (HHS) wields considerable power in the outcome. The guidance, "Postmarketing Studies and Clinical Trials—Implementation of Section 505(o)(3) of the Federal Food Drug and Cosmetic Act,"[35] details the responsibilities of the applicants subject to this law.

The Act also requires a risk evaluation and mitigation strategy (REMS) before certain new drugs are marketed; failure to conduct the aforementioned postmarket studies when required constitutes a violation of this law.[36]

A "risk determination" is necessary to "ensure that the benefits of the drug outweigh the risks of the drug." The criteria to be used to determine when the evaluation is necessary are as follows:

- The estimated size of the population that is likely to use the drug involved,
- The seriousness of the disease or condition to be treated with the drug,
- The expected benefit of the drug,
- The expected or actual duration of treatment,
- The seriousness of any known or potential AEs that may be related to the drug and the background incidence of such events in the population that is likely to use the drug, and
- Whether the drug is a new molecular entity (NME).

The law specifically extends the postapproval requirements to drugs approved before the effective date of the Act of 2007 if the FDA "becomes aware of new safety information and makes a determination that such a strategy is necessary to

Overview of FDA and Drug Development

ensure that the benefits of the drug outweigh the risks of the drug." The law brings its own definitions of "adverse events" and "new safety information," among other terms. These terms are now well-developed, and the FDA has created a sentinel system to provide insight into postmarketing problems.[35]

An adverse drug experience occurs when an AE is associated with drug use, whether or not drug related, including instances of overdose, abuse, withdrawal, or failure of pharmacological action. The term "new safety information" is broadly defined to include information obtained from a clinical trial, AE reports, postapproval studies, or notably, peer-reviewed biomedical literature and data derived from the postmarket risk identification and analysis system or other scientific data that reveal a serious risk or an unexpected serious risk associated with the use of the drug that has been unearthed since the drug was approved, the risk evaluation was filed, or the last assessment. Other terms are redefined for these purposes to capture negative experiences or harmful side effects, whether due directly to the drug or attributable to the treatment experience.

The risk assessment strategies are keyed to timetables—the first assessment is due 18 months after initial approval of the strategy, the second at three years, and the third at seven years. It can be basically eliminated if the Agency is satisfied that all risks have been identified and managed. The strategy also contemplates new medication guides for patients, package inserts, and enhanced health-care professional communications.

The Act ratifies earlier risk management programs of the FDA, allowing enhanced patient monitoring, patient registries, and increased training and licensing for prescribing or dispensing professionals. These risk evaluation and mitigation rules are truncated for generic drugs undergoing the ANDA process, in that a generic must only comply with the revised medication guide requirements, and may use a single, shared system, such as a registry.

The 2012 enactment of FDASIA has updated REMS; under the new law, the FDA is required to develop an assessment strategy to determine if the REMS is effective or should be modified to achieve its purpose. The purpose of REMS remains to assure that the benefits of a drug continue to outweigh the risks. This provision will help alleviate the burden on the US health-care system.

BIOLOGICS PRICE COMPETITION AND INNOVATION ACT OF 2009

The Biologics Price Competition and Innovation (BPCI) Act of 2009[iii] was enacted as part of the Patient Protection and Affordable Care Act (also known as "Health Care Reform" or "Obamacare"). This law formalizes the concept of "generic" or follow-on biologic, by establishing a category of product, a "biosimilar" and gives the FDA the authority to approve Biologics License Applications (BLAs) for "biosimilar" products which are highly similar to a previously licensed biologic (the "reference product"), and has no clinically meaningful difference. (Please refer to the Biologics paragraph later in this chapter for a discussion of biosimilar.) The BPCI is intended to increase competition among biologics, by allowing the approval, on a theoretically streamlined basis, of a "biosimilar" product that is "interchangeable" with the reference product. This Act is conceptually the equivalent of the generic drug procedures, which are designed to allow efficient and timely approval of follow-on products, reducing the need to duplicate testing for essentially the same therapy; however, due to the nature of biologics and the standards defining "biosimilars" and "interchangeable," questions remain as to the nature and extent of testing which will be required for approval.

At the time of publication, the FDA is in the process of developing regulations and guidances defining biosimilar and the various methods used to demonstrate interchangeability with the reference biologic. The leading concept is that there are "no clinically meaningful differences" between the biosimilar and the innovator product, especially in terms of safety, purity, and potency. Industry does expect the Agency to strike a balance in testing and costs, to afford manufacturers and the public, the advantages of the elimination of redundant testing. These regulatory activities are ongoing.

Also of note in the BPCI Act is enhanced exclusivity. The BPCI Act included market exclusivity for biologics. There is now a 12-year provision for biologics, in contrast to five years for NDA drugs. Readers are directed to Chapter 13 for a fuller discussion of biologics.

FDASIA

Signed into law on July 9, 2012, the FDASIA[37] continues the modernization of the FDA begun in 1997. FDASIA has multiple sections.

FDASIA extends the user fee legislation for drugs and devices, and now adds user fees for generics and biosimilars. The underlying concept of all user fee systems is that the FDA would receive additional funds paid by the applicants and not the taxpayers, which the FDA would use to improve management, especially the various review processes and decision-making timetables. In what is the fifth reiteration of this legislation, FDASIA brings new initiatives as well as continuing and substantial management improvements. The new fees are significant. The fee for prescription drug applications with clinical data for fiscal year 2013 is approximately US$2 million; the generic and biologics fees are keyed to the drug fee.[iv] The Medical Device User Fee for PMA applications for fiscal year 2013 is approximately US$250,000; the fee for 510(k) clearances is set at 2% of the PMA application fee.[v]

An integral part of the user fee legislative concept is the requirement that the FDA improve its performance in reviewing applications, to make the process more efficient without sacrificing safety. The law establishes performance goals that the FDA must meet; as a practical matter, these goals are negotiated between the FDA and its stakeholders, that is, with industry and with the public/patient health advocates, and formalized by Congress at the time the laws are passed. FDASIA, through PDUFA V, brings the "Program" to drug approvals.[38] The "Program" applies to all NME, NDAs, and original BLAs; it provides for

Overview of FDA and Drug Development

a presubmission meeting between the manufacturer and the FDA, a mid-cycle communication, and a late-cycle meeting. At the presubmission meeting, the

> FDA and the applicant will agree on the content of a complete application for the proposed indication(s), including preliminary discussions on the need for REMS or other risk management actions... FDA and the applicant may also reach agreement on submission of a limited number of application components not later than 30 calendar days after the submission of the original application.

If the applicant does not have this presubmission meeting, the FDA considers that the submission is complete when first submitted.[38] The FDA expects to act on 90% of standard NME NDAs and original BLAs within 10 months of the filing date and 90% of priority NME NDAs and original BLAs within six months of the filing date.[38] Note the use of the phrase "filing date." The submission date is the day that the FDA physically or electronically receives a document; the filing date is the date that the FDA has made a threshold determination that the application is sufficiently complete to begin an in-depth review.[39] This provision intends that the applicant and FDA will reach substantial agreement on the contents of the NDA at the time of initial filing. If a major amendment should be necessary, the review times are extended by three months. This extension also applies to manufacturing amendments. The PDUFA agreement defines "major amendment" to include submission of an REMS or a significant change to an REMS.

The MDUFAs included in FDASIA bring substantial changes to medical devices. Fees must now be paid by any establishment "engaged in the manufacture, preparation, propagation, compounding or processing of a device." However, fees may be waived or reduced if it is in the interest of public health. There are corollary performance goals here as well.[40] However, the most substantial changes in devices occur later in the law, under Title VI, Medical Device Regulatory Improvements. The IDE has been amended to allow broader clinical use of unapproved or uncleared devices. As a balance to this new use, the FDA has clearer authority to put a clinical hold on an IDE at any time, if it is determined that the device involves unreasonable risk to the participants in the study. The FDA must provide a rationale for any significant decision made regarding a 510(k) or application for exemption. There are improvements to the device recall system geared toward identifying trends and mitigating health risks related to defective or unsafe medical devices. There are new procedures for determining device classifications. Devices are also to be included in the FDA's sentinel system, a database used for postmarket risk identification and analysis. There is also a requirement that the FDA publish a risk-based strategy for regulating health information technology, including mobile medical applications that promote innovation, protect patient safety, and avoid regulatory duplication. Postmarket surveillance of devices is also strengthened in several ways; most significantly, a manufacturer must begin surveillance within 15 months of an FDA order, and this can last 36 months or longer. These significant changes are discussed more fully in Chapter 5.

FDASIA significantly enhances the FDA's inspection authority and the drug supply chain. Under FDASIA, the FDA will create a Unique Facility Identifier (UFI) system and drug facilities will be registered and listed in the database. The significant change here is that any drug or device is misbranded if it is imported from an unregistered foreign facility. Additionally, any drug will be misbranded if it has been manufactured, processed, packed, or held in any factory, warehouse, or establishment by an owner or operator who has delayed, denied, or limited an inspection or has refused to permit entry or inspection. Additionally, manufacturers are required to implement quality oversight over their suppliers and may be required to start this oversight earlier in the drug development process. While the inspection schedule for devices remains unchanged at biennial intervals, the drug inspection schedule is moving to a "risk-based" schedule. Additionally, commercial importers of drugs must now register with the FDA and register a principal place of business; failure to do so means that the drugs imported or offered for import are misbranded. There is also a new standard of admission for drug products intended to be imported into the United States.

FDASIA adds provisions to encourage the development of new antibiotics and antifungals that address serious and life-threatening infections. This new category of drug, formalized in Section 505E of the FDCA, receives an additional five years of market exclusivity upon the approval of an NDA for a Qualified Infectious Disease Product (QIDP). A QIDP is an antibacterial or antifungal drug for human use intended to treat serious or life-threatening infections or certain pathogens. When combined with existing exclusivity provisions under new clinical investigation and orphan drug, the QIDP exclusivity can range from 8 to 12 years. These provisions must be carefully scrutinized, because the exclusivity grants are very precise, and compliance with the various notice periods, filings, and conditions is essential.[41]

Under FDASIA, there is more patient access to new medical treatments. The law expands the types of products that qualify for expedited development, and significantly, the range of endpoints that may be used for approval. A significant change is to expand the definition of fast-track products to include those products that treat a serious or life-threatening disease or condition. The product may be approved by the use of a surrogate endpoint that is reasonably likely to predict clinical benefit, or on a clinical endpoint that can be reasonably measured earlier than irreversible morbidity or mortality, which is reasonably likely to predict an effect on irreversible morbidity or mortality or other clinical benefit, taking into account the severity, rarity or prevalence of the condition and the availability or lack of alternative treatments, as given in the FDCA, Section 506(c) (1)(A), as amended by FDASIA, Section 901(b). The sponsor must still conduct post-approval studies. The FDA has the authority to expedite withdrawal of product approval if the follow-up studies do not verify the predicted benefits.

FDASIA adds a new tool in the accelerated approval arsenal, with the creation of "breakthrough therapies" FDCA Section 506 as amended by FDASIA Section 902. A "breakthrough therapy" is a drug that[42]

... is intended, alone or in combination with 1 or more other drugs, to treat a serious or life-threatening disease or condition and preliminary clinical evidence indicates that the drug may demonstrate substantial improvement over existing therapies on 1 or more clinically significant endpoints, such as substantial treatment effects observed early in clinical development.

The designation of "breakthrough therapy" is determined at submission or any time after submission of an IND. If the FDA determines that the product qualifies as a breakthrough therapy, the FDA must take action to expedite the development and review of the product.

FDASIA also addresses drug shortages. Manufacturers of all drugs that are life-supporting, life-sustaining, or intended for use in the prevention or treatment of a debilitating disease or condition are required to notify the FDA six months in advance when the product is to be permanently discontinued or its production would be interrupted. This encompasses any disruption in supply that would affect the ability of the manufacturer to fill orders or meet expected demand. Routine or insignificant interruptions are not reportable.

SUMMARY

The body of laws regulating food, drugs, devices, and biologics has changed considerably over the past century, reflecting evolving health priorities and scientific advancement. These laws have at their core the balance struck between government regulation to protect public health and private enterprise to develop new medical products for patients to use. The laws are regularly amended and adjusted, most recently to expedite new therapies, increase patient access to treatments, and ease drug shortages and continue postmarketing surveillance and reporting. Each of these laws is implemented by a single agency, with nationwide field offices, centers which concentrate in specific areas, and numerous offices and programs. The structure of the FDA itself is the focus of the following section.

AGENCY ROLE AND ORGANIZATION

The FDA is organized into a number of offices and centers headed by a commissioner who is appointed by the President with the consent of the Senate. It is a scientifically based law enforcement agency whose mission is to safeguard the public health and ensure honesty and fairness between health-regulated industries, that is, pharmaceutical, device, biologic and the consumer.[43] It licenses and inspects manufacturing facilities; tests products; evaluates product submissions; assesses postmarket safety and effectiveness; evaluates claims and prescription drug advertising; monitors research; and creates regulations, guidelines, standards, and policies.

The most familiar entities are the five centers: the Center for Drug Evaluation and Research (CDER), Center for Biologics Evaluation and Research (CBER), Center for Devices and Radiological Health (CDRH), Center for Food Safety and Applied Nutrition (CFSAN), and the Center for Veterinary Medicine (CVM). The

Center for Tobacco Products is responsible for the implementation of the Family Smoking Prevention and Tobacco Control Act.[44] The Office of Regulatory Affairs (ORA) is the lead for FDA inspections, imports, and enforcement policy; of particular note is a publication of the ORA, the Investigations Operation Manual (IOM), the primary guidance document on FDA inspection policy and procedures for field investigators and inspectors.[45] The IOM describes in considerable detail the procedures field inspectors follow, including which documents they will be looking for and which types of evidence to demonstrate compliance are acceptable. The IOM is an invaluable guide to prepare for an inspection.

Other frequently noted resources include the Office of Combination Products; products which combine a drug/device, drug/biologic, or biologic/device are assigned to a center based on the combination product's "primary mode of action" (PMOA).[46] The primary mission of the Office of Pediatric Therapeutics is to assure access for children to innovative, safe, and effective medical products.[47] The work of the Office of Orphan Products Development is to advance the evaluation and development of products (drugs, biologics, devices, or medical foods) that demonstrate promise for the diagnosis and/or treatment of rare diseases or conditions.[48] The Office of Good Clinical Practice (GCP) is an essential portal for any applicant involved in clinical trials—this office "is the focal point within the FDA for GCP and Human Subject Protection (HSP) issues arising in human research trials regulated by FDA."[49] Related to the Office of GCP is the Bioresearch Monitoring Program (BIMO) where one can find Compliance Program Guidance Manuals (CPGMs); these are the written manuals the FDA uses to direct its field personnel on the conduct of inspectional and investigational activities.[50] The division of Drug Marketing, Advertising, and Communications (DDMACs) has been reorganized, and it is now the Office of Prescription Drug Promotion (OPDP). The mission of OPDP is "To protect the public health by assuring prescription drug information is truthful, balanced and accurately communicated. This is accomplished through a comprehensive surveillance, enforcement and education program, and by fostering better communication of labeling and promotional information to both healthcare professionals and consumers."[51] Finally, there is the Office of the Ombudsman, which "is the agency's focal point for addressing complaints and assisting in resolving disputes between companies or individuals and FDA offices concerning fair and even handed application of FDA policy and procedures."[52]

Each of these entities has a defined role, although sometimes their authorities overlap. For example, if a pharmaceutical company submits a drug that is contained and delivered to a patient during therapy by a device not comparable to any other; CDER and CDRH may need to coordinate that product's approval. Though most prescription drugs are evaluated by CDER, any other center or office may become involved with its review. One of the most significant resources to industry and consumers is the FDA's Website, www.fda.gov. Easily accessible and navigable, each center and office has its own link within the site.

Having reviewed the organization of the FDA, we turn to a summary of approval and clearance procedures for medical products. This section is intended

Overview of FDA and Drug Development   **19**

as an introductory summary; the details and nuances are addressed in the individual chapters in this book.

NEW DRUG DEVELOPMENT AND APPROVAL SUMMARY

Prior to any discussion of how pharmaceuticals make their way through the FDA for market approval, one must understand what a "drug" is. A drug is a substance that exerts an action on the structure or function of the body by chemical action or metabolism and is intended for use in the diagnosis, cure, mitigation, treatment, or prevention of disease.[53] Many different elements are metabolized, but it is that intended use—diagnosis, treatment, and mitigation—which is the core of the drug "indication" and the basis of the claims and labeling, once proved. The concept of "new drug" stems from the Kefauver–Harris Amendments to the FDCA. A new drug is defined as one that is not generally recognized as safe and effective (GRASE) for the indications proposed.[54] However, this definition has much greater reach than simply a "new" chemical entity. The term "new drug" also refers to a drug product already in existence, though never approved by the FDA for marketing in the United States: new therapeutic indications for an approved drug; a new dosage form; a new route of administration; a new dosing schedule; or, any significant clinical differences other than those approved.[43] Therefore, any chemical substance intended for use in humans or animals for medicinal purposes, or any existing chemical substance that has some significant change associated with it, is considered a new drug and is neither safe nor effective until proper testing is done and the FDA's approval is obtained.

As noted earlier, the FDA's approval is generally a lengthy and, almost always, an expensive process. While laws and regulations are designed to make the process more responsive to patient needs, the approval process essentially involves the precise identification of an active ingredient or NME, the proper manufacture of that product, the conduct of scientific tests—on a preclinical and clinical basis, and the marshaling of data collected from those tests to provide the evidence needed to prove that the drug is effective for its indicated uses, that the risks associated with that drug are known on the basis of reasonable testing, and that the nature of the known risks is appropriate with the anticipated benefit for the indication.

For a pharmaceutical manufacturer to place a product on the market for human use, a multiphase procedure must be followed. It must be remembered that the mission of the FDA is to protect the public and the Agency is a scientific organization. Therefore, all drug approvals must satisfy certain scientific benchmarks as part of the review process.

These steps begin with the identification of a substance that is going to be tested, and the development of a process which ensures that definite quantities of this exact substance can be produced. Scientific data collection requires that the doses of a substance being tested are identical, and that each dose has the same properties as any other. These requirements are fulfilled by following the FDA's procedures on current Good Manufacturing Practices (cGMPs)[55] and the FDA's Good Laboratory Practices (GLPs).[56] Next follow a number of preclinical or "before human" studies. These studies, done *in vitro* and *in vivo*, examine the pharmacology, toxicology, and possible carcinogenicity of the substance, and answer the question whether the drug is reasonably safe for human consumption.

The information from these studies, along with the documentation of the manufacturing and lab controls, is used to create a protocol to test the substance in humans. This protocol, along with the results of the earlier tests, is organized into an IND that is submitted to the FDA. Once approved, the applicant can begin to test the product on a human subject. This is usually done in three phases. Approved drugs are also subject to a fourth phase of testing, known as postmarket surveillance, which may include additional formal trials. The nature and scope of postmarket surveillance have been dramatically increased by the provisions in the FDAAA of 2007, as discussed earlier.

The FDA has published extensive information on the Drug Development Process; readers should refer to the FDA Web page, http://www.fda.gov/drugs/developmentapprovalprocess/default.htm, for detailed information.

PRECLINICAL INVESTIGATION

The testing of new drugs in humans cannot begin until there is solid evidence that the drug product can be used with reasonable safety in humans. This phase is called "preclinical investigation." The basic goal of preclinical investigation is to assess potential therapeutic effects of the substance on living organisms and to gather sufficient data to determine reasonable safety of the substance in humans through laboratory experimentation and animal investigation.[43] The FDA requires no prior approval for investigators or pharmaceutical industry sponsors to begin a *preclinical* investigation on a potential drug substance. Investigators and sponsors are, as noted earlier, required to follow GLPs.[57] GLPs govern laboratory facilities, personnel, equipment, and operations. Compliance with GLPs requires procedures and documentation of training, study schedules, processes, and status reports, which are submitted to facility management and included in the final study report to the FDA. A preclinical investigation usually takes one to three years to complete. If at that time enough data are gathered to reach the goal of potential therapeutic effect and reasonable safety, the product sponsor must formally notify the FDA of its wishes to test the potential new drug on humans.

INVESTIGATIONAL NEW DRUG APPLICATION

Unlike the preclinical investigation stage, the IND phase has much more direct FDA activity throughout the process. Because a preclinical investigation is designed to gather significant evidence of reasonable safety and efficacy of the compound in live organisms, the IND phase is the clinical phase where all activity is designed to gather significant evidence of reasonable safety and efficacy data about the potential drug compound in humans. Clinical trials in humans are carefully scrutinized and regulated by the FDA to protect the health and safety of human test subjects and to ensure the integrity and usefulness of the clinical study data.[58] Numerous meetings between both the Agency and the sponsor occur during this time. As a result, the clinical investigation phase may take as long as 12 years to complete. Only one in five compounds tested may actually demonstrate clinical effectiveness and safety and reach the US marketplace. (Note that the total development time is not the same as actual FDA review time. PDUFA and other initiatives, including fast track and accelerated approval, have shortened the drug approval and, consequently, the drug development cycle, with the result that some products are approved within a year of commencement of testing in humans.)

The sponsor will submit the IND to the FDA. The IND must contain information on the compound itself and information about the study. All INDs must have the same basic components: a detailed cover sheet; a table of contents; an introductory statement and basic investigative plan; an investigator's brochure, comprehensive investigation protocols; the compound's actual or proposed chemistry, manufacturing, and controls; any pharmacology and toxicology information; any previous human experience with the compound; and any other pertinent information the FDA deems necessary. After submission, the sponsor company must wait for 30 days to commence clinical trials. The FDA does not "approve" an IND; rather, if the FDA does not object within that period, the trials may begin.

Prior to the actual commencement of the clinical investigations, however, a few ground rules must be established. For example, a clinical study protocol must be developed, proposed by the sponsor, and reviewed by an Institutional Review Board (IRB). An IRB is required by regulation[59] and is a committee of medical and ethical experts designated by an institution such as a university medical center, in which the clinical trial will take place. The charge of the IRB is to oversee the research to ensure that the rights of human test subjects are protected and that rigorous medical and scientific standards are maintained.[60]

An IRB must approve the proposed clinical study and monitor the research as it progresses. Each IRB must develop written procedures of its own regarding its study review process and its reporting of any changes to the ongoing study as they occur. In addition, an IRB must also review and approve documents for informed consent prior to commencement of the proposed clinical study. Regulations require that potential participants be informed adequately about the risks, benefits, and treatment alternatives before participating in experimental research.[61] An IRB's membership must be sufficiently diverse to review the study in terms of the specific research issue, community and legal standards, and professional conduct and practice norms. All of its activities must be well documented and open to the FDA inspection at any time.

Once the IRB is satisfied that the proposed trial is ethical and proper, the testing may begin. The clinical trial phase has three steps or phases. Each has a purpose, requires numerous patients, and can take more than one year to complete.

Phase I

A Phase I study is relatively small, has less than 100 subjects, and is brief (one year or less). Its purpose is to determine toxicology, metabolism, pharmacologic actions, and, if possible, any early evidence of effectiveness. The results of the Phase I study are used to develop the next step, which is Phase II.

Note that there are certain indications and drugs which Phase I testing in healthy volunteers is determined to be simply unethical. The examples include oncologic indications, where even minimal dosing can have a toxic effect, or AIDS/HIV vaccines into healthy individuals. In these instances, the applicant usually only tests subjects with the disease condition, and the procedure would start with Phase IIa, which is more fully described in the next section.

Phase II

Phase II studies are the first controlled clinical studies using several hundred subjects who were afflicted with the disease or condition being studied. The purpose of Phase II is to determine the compound's possible effectiveness against the targeted disease or condition and its safety in humans at that dosing. Phase II may be divided into two subparts: Phase IIa is a pilot study, which is used to determine initial efficacy, and Phase IIb, uses controlled studies on several hundred patients. At the end of Phase II studies, the sponsor and the FDA will usually confer to discuss the data and plans for Phase III.

Phase III

Phase III studies are considered "pivotal" trials, which are designed to collect all of the necessary data to meet the safety and efficacy standards that the FDA requires to approve the compound for the US marketplace. Phase III studies are usually very large, consisting of several thousand patients in numerous study centers with a large number of investigators who conduct long-term trials over several months or years. In addition, Phase III studies establish final formulation, indications, labeling, marketing claims and product stability, and packaging and storage conditions. On completion of Phase III, all clinical studies are complete, all safety and efficacy data have been analyzed, and the sponsor is ready to submit the compound to the FDA for market approval. This process begins with the submission of an NDA.

New Drug Application

An NDA is a regulatory mechanism that is designed to give the FDA sufficient information to make a meaningful evaluation of a new drug.[62] Although the quantity of information and data contained in an NDA is dependent on the drug testing, all NDAs contain essentially the same information, organized and delivered in a very precise way. The goals of the NDA are to provide enough information to permit the FDA reviewer to reach the following key decisions[63]:

- Whether the drug is safe and effective in its proposed use(s), and whether the benefits of the drug outweigh the risks.
- Whether the drug's proposed labeling (package insert) is appropriate, and what it should contain.
- Whether the methods used in manufacturing the drug and the controls used to maintain the drug's quality are adequate to preserve the drug's identity, strength, quality, and purity.

The NDA is supposed to tell the drug's whole story, including what happened during the clinical tests, identify the ingredients of the drug, results of animal studies, how the drug behaves in the body, and how it is manufactured, processed, and packaged. The NDA starts with an index and summary; moves on to the chemistry, manufacturing, and controls; and then to preclinical laboratory and animal data, human pharmacokinetic and bioavailability data, clinical data, including tabulations of individual subject case report forms, safety data, packaging, a description of the drug product and substance, a list of relevant patents for the drug, its manufacture or claims, any proposed labeling, and any additional information the FDA considers relevant.

Traditionally, NDAs consisted of hundreds of volumes of information, in triplicate, all cross-referenced. Since 1999, the FDA has continued to move toward electronic filings; today, electronic submissions are encouraged but not required. These electronic submissions facilitate ease of review and possible approval and are submitted via the FDA Electronic Submissions Gateway.[64]

The NDA must be submitted complete, in the proper form, and with all critical data. On receipt, the FDA first determines whether an application is "filable." The FDA screens the document to determine if the application is complete, justifying the time it will take to review the application. The FDA must notify the sponsor within 60 days of its "refuse-to-file" decision. Otherwise, the review process begins.

The next steps require in-depth review, and the sponsor may be required to submit additional information. The purpose of an NDA from the FDA's perspective is to ensure that the new drug meets the criteria to be "safe and effective." The FDA makes the safety and effectiveness and risk versus benefit determinations on the basis of the data; the data from the Phase III pivotal studies are given most weight.

In addition, the NDA must be very clear about the manufacture and marketing of the proposed drug product. The application must define and describe manufacturing processes; validate cGMPs; and provide evidence of quality, purity, strength, identity, and bioavailability (a preinspection of the manufacturing facility is conducted by the FDA). Finally, the FDA will review all product packaging and labeling for content and clarity. Statements on a product's package label, package insert, media advertising, or professional literature must be reviewed. Of note, "labeling" refers to all of the above and not just the label on the product container.

FDASIA has brought new performance standards to the review, especially as part of the "Program" as discussed earlier. Applicants must avail themselves of the meetings, which are intended to clarify the interpretations and requirements of each party. There are three possible results of a review, each reported through an "action letter." An "approval letter" signifies that all substantive requirements for approval are met and that the sponsor company can begin marketing the drug as of the date on the letter.

An "approvable letter" signifies that the application substantially complies with the requirements but has some minor deficiencies, which must be addressed before an approval letter is sent. Generally, these deficiencies are minor in nature, and the product sponsor must respond within 10 days of receipt. At this point, the sponsor may amend the application and address the Agency's concerns, request a hearing with the Agency, or withdraw the application entirely.

A "nonapprovable letter" signifies that the FDA has major concern with the application and will not approve the proposed drug product for marketing as submitted. The remedies available to a sponsor for this type of action letter are similar to those in the approvable letter.

FAST TRACK, ACCELERATED APPROVAL, PRIORITY REVIEW, AND "BREAKTHROUGH THERAPY" DESIGNATIONS

There are several procedures available to applicants, which speed the development and availability of drugs. They are fast track, accelerated approval, priority review and, with the passage of FDASIA, the "breakthrough therapy" designation. Drugs that qualify for these procedures still follow the basic outline of IND/NDA, but these designations and processes are intended to streamline the process. The law and the FDA Website provide the following definitions[vi]:

Fast Track

Fast track is a process designed to facilitate the development, and expedite the review of drugs to treat serious diseases and fill an unmet medical need. The purpose is to get important new drugs to the patient earlier. *Fast Track* addresses a broad range of serious diseases.

Determining whether a disease is *serious* is a matter of judgment, but generally is based on whether the drug will have an impact on such factors as survival, day-to-day functioning, or the likelihood that the disease, if left untreated, will progress from a less severe condition to a more serious one. AIDS, Alzheimer's, heart failure, and cancer are obvious examples of serious diseases. However, diseases such as epilepsy, depression, and diabetes are also considered to be serious diseases.

Filling an *unmet medical need* is defined as providing a therapy where none exists or providing a therapy which may be potentially superior to existing therapy.

Any drug being developed to treat or prevent a disease with no current therapy obviously is directed at an unmet need. If there are existing therapies, a fast track drug must show some advantage over available treatment. If FDA determines that the drug qualifies for fast track, the drug is eligible for more frequent meetings with FDA, and possibly accelerated approval, priority review, and possibly, rolling review.

Overview of FDA and Drug Development

Accelerated Approval

When studying a new drug, it can take a long time—sometimes many years—to learn whether a drug actually provides real improvement for patients—such as living longer or feeling better. This real improvement is known as a "clinical outcome." Mindful of the fact that obtaining data on clinical outcomes can take a long time, in 1992 FDA instituted the *Accelerated Approval* regulation, allowing earlier approval of drugs to treat serious diseases, and that fill an unmet medical need based on a surrogate endpoint.

A surrogate endpoint is a marker—a laboratory measurement, or physical sign—that is used in clinical trials as an indirect or substitute measurement that represents a clinically meaningful outcome, such as survival or symptom improvement. The use of a surrogate endpoint can considerably shorten the time required prior to receiving FDA approval.

Approval of a drug based on such endpoints is given on the condition that post-marketing clinical trials verify the anticipated clinical benefit.

The FDA bases its decision on whether to accept the proposed surrogate endpoint on the scientific support for that endpoint. The studies that demonstrate the effect of the drug on the surrogate endpoint must be "adequate and well controlled" studies, the only basis under law, for a finding that a drug is effective.

Priority Review

Prior to approval, each drug marketed in the United States must go through a detailed FDA review process. In 1992, under the Prescription Drug User Act (PDUFA), FDA agreed to specific goals for improving the drug review time and created a two-tiered system of review times—*Standard Review* and *Priority Review*.

Standard Review is applied to a drug that offers at most only minor improvement over existing marketed therapies. The 2002 amendments to PDUFA set a goal that a *Standard Review* of a new drug application be accomplished within a *ten-month* time frame.

A *Priority Review* designation is given to drugs that offer major advances in treatment, or provide a treatment where no adequate therapy exists. A *Priority Review* means that the time it takes FDA to review a new drug application is reduced. The goal for completing a Priority Review is *six months*.

Priority Review status can apply both to drugs that are used to treat serious diseases and drugs for less serious illnesses. The FDA goal for reviewing a drug with Priority Review status is six months.

The distinction between priority and standard review times is that additional FDA attention and resources will be directed to drugs that have the potential to provide significant advances in treatment.

"Breakthrough Therapy" Designation

A "breakthrough therapy" is one that is intended "to treat a serious or life-threatening disease or condition, and preliminary clinical evidence indicates that the drug may demonstrate substantial improvement over existing therapies on one or more clinically significant endpoints."[42] FDA has issued a Fact Sheet and Draft Guidance for Industry, on June 25, 2013; the information can be found at http://www.fda.gov/RegulatoryInformation/Legislation/FederalFoodDrugandCosmeticActFDCAct/SignificantAmendmentstotheFDCAct/FDASIA/ucm329491.htm

PHASE IV AND POSTMARKETING SURVEILLANCE

Pharmaceutical companies that successfully gain marketing approval for their products are not exempted from further regulatory requirements. In addition to the extensive postmarketing changes made by the Acts of 2007, compliance efforts take center stage. All producers must be registered and inspected, file various safety reports, meet import and export requirements, and maintain cGMPs. Many products are approved for market on the basis of a continued submission of clinical research data to the FDA. These data may be required to further validate efficacy or safety, detect new uses or abuses for the product, or determine the effectiveness of labeled indications under conditions of widespread usage.[65] The FDA may also require a Phase IV study for drugs approved under FDAMA's fast-track provisions.

Any change to the approved product's indications, active ingredients, manufacture, or labeling requires the manufacturer to submit a supplemental NDA (SNDA) for agency approval. Also, as emphasized in the relevant section of *The Food and Drug Administration Amendments Act of 2007*, "adverse drug reports" must be reported to the Agency. All reports must be reviewed by the manufacturer promptly; and if found to be serious, life threatening, or unexpected (not listed in the product's labeling), the manufacturer is required to submit an "alert report" within 15 working days of receipt of the information.

ORPHAN DRUGS

Orphan drugs are approved using many of the same processes as any other application. However, there are several significant differences. An orphan drug as defined under the Orphan Drug Act of 1983 is a drug used to treat a "rare disease," which would not normally be of interest to commercial manufacturers in the ordinary course of business. A rare disease is defined in the law as any disease that affects fewer than 200,000 persons in the United States, or one in which a manufacturer has no reasonable expectation of recovering the cost of its development and availability (e.g., manufacturing and marketing) in the United States. The Act creates a series of financial incentives for manufacturers. For example, the Act permits grant assistance for clinical research, tax credits for research and development, and a seven-year market exclusivity to the first applicant who, to obtain market approval for a drug, is designated as an orphan. This means that if the sponsor gains approval for an orphan drug, the FDA will not approve any application by any other sponsor for the same drug, for the same disease or condition for seven years from the date of the first applicant's approval, provided certain conditions are met, such as an assurance of sufficient availability of drug to those in need or a revocation of the drugs' orphan status.[66]

ABBREVIATED NEW DRUG APPLICATIONS

ANDAs are used when a patent has expired for a product that has been in the US marketplace and a company wishes to market a generic. In the United States, a drug patent is valid for 20 years. Subsequently, a manufacturer is able to submit an abbreviated application for that product, provided that it certifies that the product patent in question has already expired, is invalid, or will not be infringed.

The generic copy must meet certain other criteria as well. The drug's active ingredient must have already been approved for the conditions of use proposed in the ANDA, and nothing should have changed to call into question the basis for approval of the original drug's NDA.[65] Sponsors of ANDAs are required to prove that their version meets the standards of bio- and pharmaceutical equivalence. The FDA publishes a list of all approved drugs called *Approved Drug Products with Therapeutic Equivalence Evaluations* commonly known as the *Orange Book* because of its orange-colored cover. It lists marketed drug products that are considered by the FDA to be safe and effective and provides information on therapeutic equivalence evaluations for approved multisource prescription drug products[67] monthly. The *Orange Book* rates drugs on the basis of their therapeutic equivalence. For a product to be considered therapeutically equivalent, it must be both pharmaceutically equivalent (i.e., the same dose, dosage form, strength, etc.) and bioequivalent (i.e., rate and extent of its absorption are not significantly different from the rate and extent of absorption of the drug with which it is to be interchanged).

Realizing that there may be some degree of variability in patients, the FDA allows pharmaceuticals to be considered bioequivalent through either of the two methods. The first method studies the rate and extent of absorption of a test drug, which may or may not be a generic variation, and a reference or brand name drug under similar experimental conditions and in similar dosing schedules in which the test results do not show significant differences. The second approach uses the same method and the results determine that there is a difference in the test drug's rate and extent of absorption, except for the difference that is considered medically insignificant for the proper clinical outcome of that drug.

The FDA provided the following information:

> Bioequivalence of different formulations of the same drug substance involves equivalence with respect to the rate and extent of drug absorption. Two formulations whose rate and extent of absorption differ by 20% or less are generally considered bioequivalent. The use of the 20% rule is based on a medical decision that, for most drugs, a 20% difference in the concentration of the active ingredient in blood will not be clinically significant.[68]

The FDA's *Orange Book* uses a two-letter coding system, which is helpful in determining which drug products are considered therapeutically equivalent. The first letter, either an "A" or a "B," indicates a drug product's therapeutic equivalence rating. The second letter describes dose forms and can be any one of a number of different letters.

The A codes are described in the FDA's *Orange Book* as follows:

> Drug products that the FDA considers to be therapeutically equivalent to other pharmaceutically equivalent products, that is, drug products for which

1. There are no known or suspected bioequivalence problems. These are designated AA, AN, AO, AP, or AT, depending on the dose form; or
2. Actual or potential bioequivalence problems have been resolved with adequate *in vivo* and/or *in vitro* evidence supporting bioequivalence. These are designated AB.[69]

The B codes are a much less desirable rating when compared with a rating of A. Products that are rated B may still be commercially marketed; however, they may not be considered therapeutically equivalent. The FDA's *Orange Book* describes B codes as follows:

> Drug products that the FDA at this time does not consider to be therapeutically equivalent to other pharmaceutically equivalent products, i.e., drug products for which actual or potential bioequivalence problems have not been resolved by adequate evidence of bioequivalence. Often the problem is with specific dosage forms rather than with the active ingredients. These are designated BC, BD, BE, BN, BP, BR, BS, BT, or BX.[70]

The FDA has adopted an additional subcategory of B codes. The designation "B*" is assigned to former A minus-rated drugs "if the FDA receives new information that raises a significant question regarding therapeutic equivalence."[71] Not all drugs are listed in the *Orange Book*. Drugs obtainable only from a single manufacturing source, DESI drugs, or drugs manufactured prior to 1938 are not included. Those that do appear are listed by their generic names.

OTC REGULATIONS

The 1951 Durham–Humphrey Amendments of the FDCA specified three criteria to justify prescription-only status. If the compound is shown to be habit forming, requires a prescriber's supervision, or has an NDA prescription-only limitation, it will require a prescription. The principles used to establish OTC status (nonprescription required) are a wide margin of safety, method of use, benefit to risk ratio, and adequacy of labeling for self-medication. For example, injectable drugs may not be used OTC, with certain exceptions such as insulin. OTC market entry is less restrictive than that for ℞ drugs and does not require premarket clearance; these pose fewer safety hazards than ℞ drugs because they are designed to alleviate symptoms rather than disease. Easier access far outweighs the risks of side effects, which can be adequately addressed through proper labeling.

As previously discussed, OTC products underwent a review in 1972. Because agency review of the 300,000+ OTC drug products in existence at that time would be virtually impossible, the FDA created OTC advisory panels to review data based on some 26 therapeutic categories. OTC drugs would only be examined by active ingredients within a therapeutic category. Inactive ingredients would only be examined, provided they were shown to be safe and suitable for the product and not interfering with effectiveness and quality.

This review of active ingredients would result in the promulgation of a regulation or a monograph, which is a "recipe" or set of guidelines applicable to all

OTC products within a therapeutic category. OTC monographs are general and require that OTC products show "general recognition of the safety and effectiveness of the active ingredient." OTC products do not fall under prescription status if their active ingredients (or combinations) are deemed by the FDA to be GRASE. The monograph system is a public system with a public comment component included after each phase of the process. Any products for which a final monograph has not been established may remain on the market until one is determined.

There are four phases in the OTC monograph system. In Phase I, an expert panel was selected to review data for each active ingredient in each therapeutic category for safety, efficacy, and labeling. Its recommendations were made in the Federal Register. A public comment period of 30–60 days was permitted and supporting or contesting data accepted for review. Then the panel reevaluated the data and published a "proposed monograph" in the Federal Register, which publicly announced the conditions for which the panel believes that OTC products in a particular therapeutic class are GRASE and not misbranded. A tentative "final monograph" was then developed and published stating the FDA's position on safety and efficacy of a particular ingredient within a therapeutic category and acceptable labeling for indications, warnings, and directions for use. Active ingredients were deemed: Category I—GRASE for claimed therapeutic indications and not misbranded, Category II—not GRASE and/or misbranded, or Category III—insufficient data for determination.

After public comment, the final monograph was established and published with the FDA's final criteria for which all drug products in a therapeutic class become GRASE and not misbranded. Following the effective date of the final monograph, all covered drug products that fail to conform to the FDA's requirements are considered misbranded and/or an unapproved new drug.[43]

However, since monograph panels are no longer convened, many current products are switched from prescription status. A company that wishes to make this switch and offer a product to the US marketplace can submit an amendment to a monograph to the FDA that will act as the sole reviewer. The company may also file an SNDA, provided that it has three years of marketing experience as a prescription product, can demonstrate a relatively high use during that period, and can validate that the product has a mild profile of adverse reactions. The last method involves a "citizen's petition," which is rarely used.[43]

BIOLOGICS

Biologics are defined as substances derived from or made with the aid of living organisms, which include vaccines, antitoxins, serums, blood, blood products, therapeutic protein drugs derived from natural sources (i.e., antithrombin III) or biotechnology (i.e., recombinantly derived proteins), and gene or somatic cell therapies, which are all individually intended for use in the diagnosis, cure, mitigation, treatment, or prevention of disease. This legal definition is not part of the FDCA, but found in the Public Health Act (PHA).[72] The PHA requires that there must be an approved biologics license before it is marketed; this document, the BLA however, is submitted to and the license is issued by the FDA, which classifies biologics as a subset of drugs. Applicants for BLAs will use form 356h,[73] which is identical to the form that is used for drugs, but the BLA itself has some different emphases from the NDA. The FDA deals with biologics primarily through the CBER. CBER regulates cellular and gene therapy products, allergenics, and xenotransplantation, and has a large regulatory role in vaccine development, tissue safety, and blood. However, CDER has certain responsibilities for certain therapeutic biological products that were transferred from CBER. CDER's duties include premarket review and oversight; however, despite this transfer, the products continue to be regulated as licensed biologics. Some of the transferred products include growth factors and enzymes.[vii]

The core concept here is that biologics are large molecules, and cannot be described efficiently with a precise formula. As the FDA notes[74]

> ...in contrast to chemically synthesized small molecular weight drugs, which have a well-defined structure and can be thoroughly characterized, biological products are generally derived from living material—human, animal, or microorganism—are complex in structure, and thus are usually not fully characterized.

As noted further,

> Changes in the manufacturing process, equipment or facilities could result in changes in the biological product itself and sometimes require additional clinical studies to demonstrate the product's safety, identity, purity and potency. Traditional drug products usually consist of pure chemical substances that are easily analyzed after manufacture.

This difference has come to mean that the biologic is defined by its manufacturing process, and has also redefined standards of "safety," "purity," and "potency" as applied to biologics. As defined by the FDA on that page,[74] "safety" means the relative freedom from harmful effects, direct or indirect, when a product is prudently administered, taking into consideration the character of the product in relation to the condition of the recipient at the time. The word "purity" means relative freedom from extraneous matter in the finished product, whether or not harmful to the recipient or deleterious to the product. Purity includes, but is not limited to, relative freedom from residual moisture or other volatile and pyrogenic substances. The word "potency" is interpreted to mean the specific ability or capacity of the product, as indicated by appropriate laboratory tests, to yield a given result. Additional information is found in the regulations at 21 Code of Federal Regulations (CFR) parts 600, 601, and 610.

Biologics follow the broad regulatory steps to approval, with appropriate adjustments made for their composition. There are preclinical stages, clinical stages, and finally manufacturing hurdles, using the same basic stepping stones as a drug—for example, a biologic must have an IND, and approval culminates with the BLA. Again, biologics are treated and tested differently than drugs because of their composition.

There is a category of biologics known as "innovator" or "brand," just as with drugs. There is a concept of "generic biologic," and as one would expect with biologics, this concept required adjustment to take into account the complex nature of the biologic product, the manufacturing considerations, and the different standards of safety, purity, and potency noted earlier. As discussed before, the BPCI Act formalizes the concept of "biologic–generic" or follow-on biologic, by establishing a category of product, a "biosimilar." This new category is necessary again because of the generally recognized differences between a biologic and a "small-molecule" drug as noted earlier. A brand name or innovator drug is relatively easily identified and identification of a generic is clear-cut as well. According to the FDA, a generic drug product is one that is comparable to an innovator drug product in dosage form, strength, route of administration, quality, performance characteristics, and intended use.[viii] A biosimilar is not the bioequivalent of the innovator biologic in the same sense that a generic drug is the bioequivalent of an innovator drug, but rather a biosimilar is similar to the innovator product and can be expected to produce the same result as the innovator in a patient and presents no greater risk to a patient. The FDA Website declares "A biological product may be demonstrated to be 'biosimilar' to an already-licensed FDA biological product (the 'reference product') if data show that the product is highly similar to the reference product notwithstanding minor differences in clinically inactive components and there are no clinically meaningful differences between the biological product and the reference product in terms of safety, purity and potency of the product."[75] A biosimilar meeting these criteria is considered "interchangeable" with the innovator or reference product. This Act is conceptually the equivalent of the generic drug procedures, which are designed to allow efficient and timely approval of follow-on products, reducing the need to duplicate testing for essentially the same therapy.

At the time of publication, the FDA is in the process of developing regulations and guidelines defining biosimilar and the various methods used to demonstrate interchangeability with the reference biologic. The leading concept is that there are "no clinically meaningful differences" between the biosimilar and the innovator product, especially in terms of safety, purity, and potency. Industry does expect the Agency to strike a balance in testing and costs, to afford manufacturers and the public, the advantages of the elimination of redundant testing. These regulatory activities are ongoing.

DEVICES

The FDA regulates an extraordinary range of devices. At the simplest level, devices include toothbrushes and band-aids, which are under the FDA jurisdiction because they are sold with a therapeutic claim, such as prevention of tooth decay and a barrier to infection. At the most complex level, devices are life-saving, life-supporting, or implantable devices, such as cardiac stents, which are considered to pose the greatest risk to patients. The statutory definition of device is based on that of a drug, except that the device "... does not achieve its primary intended purposes through chemical action within or on the body of man or other animals and which is not dependent upon being metabolized for the achievement of its primary intended purposes."[76] Additionally, CDRH has jurisdiction over all radiation-emitting products and processors that sterilize and prepare single-use devices for reuse (reprocessors of single-use devices).

Devices are classified by the risk they pose to users. Class I is the simplest and subject to the least oversight and regulation; Class III includes all devices of the highest risk, which support or sustain human life, and are subjected to the strictest of testing standards. Class II devices are everything that are not Class I or III. An example of a Class II device would be a home pregnancy test; the reasoning is that this test must be subject to some level of oversight, because of patient reliance on the test results. CDRH has several Web pages to assist device manufacturers; Device Advice is a straight-forward portal, easy to use, which is an excellent starting point.[77]

The FDA has different regulatory standards for different classes of products.[ix] Class I devices are the least regulated, and are subject to general controls. All devices, including those in Class I, must be manufactured in compliance with the Quality System Regulations (QSRs; 21 CFR Part 820), be properly labeled and packaged (21 CFR Part 801), and the manufacturer or distributor must meet the establishment registration requirements.

Class II devices are generally cleared for sale, using the 501(k) premarket notification procedure. This procedure is named after the FDCA section which describes it; it requires that the manufacturer submit an application which demonstrates that the subject device is substantially similar to an existing device which is on the market, called the predicate device. If preclinical tests demonstrate that the new device is as safe and effective as the predicate device, clinical tests are not required. If clinical studies are required, the manufacturer must seek an IDE (21 CFR Part 812). Device clinical studies have the same phases as drug trials, Phases I–III, but typically involve much smaller cohorts, a few hundred rather than thousands of subjects. After the 510(k) is submitted, the FDA reviews the submission and determines whether the device is substantially equivalent; marketing begins with the receipt of this letter.

Class III devices are the most serious, and are approved with the premarket approval process (PMA; 21 CFR Part 814). This is a much more rigorous process than the 510(k) and approval depends upon the FDA's determination that the PMA contains "valid scientific evidence" that the device is safe and effective for its indications. A PMA always requires clinical testing. A device that has undergone PMA approval is not a predicate device and cannot be used to support a 510(k); each Class III device must undergo the PMA. This "regulatory patent" is discussed in more detail in Chapter. Some PMA devices are given "conditions of approval" letters; these letters allow the clinical use of PMA devices, subject to manufacturer compliance with each condition in the letter. These conditions usually require additional postmarketing follow-up.

Overview of FDA and Drug Development 33

Devices are subject to quality system requirements rather than cGMPs. Devices are also subject to stringent AE reporting and postmarket surveillance. All device manufacturing facilities should expect to be inspected every two years.

Finally, FDASIA has changed some aspects of the device laws; there is a mandate to improve the device recall system, expanded authority of the FDA to issue a clinical hold over IDEs, and requirements for the FDA to actively regulate health information technology, especially mobile medical applications, to name a few.

REGULATING DRUG AND DEVICE MARKETING

The FDA has jurisdiction over prescription drug advertising and promotion. The basis for these regulations lies within the 1962 Kefauver–Harris Amendments. Essentially, any promotional information, in any form, must be truthful, fairly balanced, and fully disclosed. The FDA views this information as either "advertising" or "labeling." Advertising includes all traditional outlets in which a company places an advertisement. Labeling includes everything else such as brochures, booklets, lectures, slide kits, letters to physicians, and company-sponsored magazine articles. All information must be truthful and not misleading. All material facts must be disclosed in a manner that is fairly balanced and accurate. If any of these requirements are violated, the product is considered misbranded for the indications for which it was approved under its NDA. The FDA is also sensitive to the promotion of a product for "off-label use." Off-label use occurs when a product is in some way presented in a manner that does not agree with or is not addressed in its approved labeling. In addition, provisions of the PDMA of 1987 apply. The Act prohibits company representatives from directly distributing or reselling prescription drug samples. Companies are required to establish a closed system of record keeping, which will be able to track a sample from their control to that of a prescriber in order to prevent diversion. Prescribers are required to receive these samples and record and store them appropriately.[78]

Additionally, television, the Internet, and print advertisements must comply with the standards set by law and regulation. The OPDP [formerly DDMACs], within CDER, reviews such advertisements; the advertisements may not be false or misleading, must present a "fair balance" between side effects, contraindications, and effectiveness information, and presumably, neutrality under the Acts of 2007, reveal material facts, and include established, scientific, and brand names in specified font ratios.[79]

Although there are fewer regulations and guidances regarding device and biologic marketing and advertisement, these are still regulated activities, and the same concepts apply and are more fully addressed in Chapter 10.

VIOLATIONS AND ENFORCEMENT

The FDA has the power to enforce the regulations for any product as defined under the FDCA. It has the jurisdiction to inspect a manufacturer's premises and its records. After a facilities inspection, an agency inspector will issue an

FDA form 483, which describes observable violations. Response to the finding as described in this form must be made promptly. A "warning letter" may be used when the Agency determines that one or more of a company's practices, products, and procedures are in violation of the FDCA. The FDA district office has 15 days to issue a warning letter after an inspection. The company has got 15 days to respond. If the company response is satisfactory to the Agency, no other action is warranted. If the response is not satisfactory, the Agency may request a "recall" of the violated products. However, the FDA has no authority to force a company to recall a drug product. But, it may force removal of a product through the initiation of a seizure.

Recalls can fall into one of the three classes. A Class I recall exists when there is a reasonable possibility that the use of a product will cause either serious adverse effects on health or death. A Class II recall exists when the use of a product may cause temporary or medically reversible adverse effects on health, or where the probability of serious adverse effects on health is remote. A Class III recall exists when the use of a product is not likely to cause adverse health consequences. Recalls are also categorized as consumer level, where the products are requested to be recalled from the consumers' homes or control; retail level, where the products are to be removed from the retail shelves or control; and wholesale level, where the products are to be removed from wholesale distribution. Companies that conduct a recall of their products are required to conduct "effectiveness checks" to determine the effectiveness of recalling the product from the marketplace.

If a company refuses to recall the product, the FDA will seek an injunction against the company.[80] An injunction is recommended to the Department of Justice (DOJ) by the FDA. The DOJ takes the request to the federal court which issues an order that forbids a company from carrying out a particular illegal act, such as marketing a product that the FDA considers is a violation of the FDCA. Companies can either comply with the order or sign a "consent agreement" that will specify changes required by the FDA for the company to continue operations or to litigate.

The FDA may also initiate a seizure of violative products.[81] A seizure is ordered by the federal court in the district that the products are located. The seizure order specifies products, their batch numbers, and any records as determined by the FDA as violative. US Marshals carry out this action. The FDA institutes a seizure to prevent a company from selling, distributing, moving, or otherwise tampering with the product.

The FDA may also debar individuals or firms from assisting or submitting an ANDA or directly providing services to any firm with an existing or pending drug product application.

However, one of the more powerful deterrents that the FDA uses is adverse publicity. The Agency has no authority to require a company to advertise adverse publicity. It does publish administrative actions against a company in any number of federal publications such as the *Federal Register*, the *FDA Enforcement Report*, the *FDA Medical Bulletin*, and the *FDA Consumer*. Additionally, letters detailing a company's or a person's violation of regulation can be found at the warning letters link on the FDA home page, www.fda.gov.

CONCLUSION

The laws and regulations that govern US pharmaceutical and medical product industry are both vast and complicated; the Agency that implements these mandates regulates one-quarter of the country's gross national product. The mission of the FDA, to protect human health through the assessment and evaluation of scientific data, is important and extremely challenging, given the pace of scientific discovery, the increasing number of people dependening upon medications, and the significant diseases that obdurately refuse to be cured. Over the past century, we have seen the laws strengthen, and also accommodate the needs of private enterprise to develop and market these medications and devices. There is growing transparency and increased communication between the Agency and its stakeholders.

These Acts have paved the way for meaningful change as we continue in our battle against disease. The US system of investigating new drugs, devices, and biologics is one that continues to have merit by balancing the accumulation of scientific evidence with the public's need for access to therapies. The risk–benefit analysis that guides the use of medical products is focused on providing the patient with the necessary information to determine whether or not to use a therapeutic. This regulatory system, albeit complex and at many times frustrating, remains the gold standard throughout the world. The American public can look forward to great advances from the industry and should be comfortable that the FDA is watching.

NOTES

i. The FDA cannot compel the manufacture of generic drugs once patent rights have expired. In recent times, the Hatch–Waxman Act has come under criticism for that reason. Under a free market system, companies that hold expired patents may, and some do, make "reverse payments" to potential competitors to keep generic drugs off the market. This practice clearly frustrates the spirit of the law; however, it can be legal. See the Drug Price Competition and Patent Restoration Act containing its legislative history at http://www.fda.gov/ohrms/DOCKETS/dockets/06p0242/06p-0242-cp00001-13-Tab-12-vol1.pdf, accessed September 22, 2012. The United States Supreme Court decided that reverse payments in patent settlement agreements are not presumptively unlawful, but these agreements are subject to the 'rule of reason' [*Federal Trade Commission v. Acatavis, Inc.* 570 US _____ (20130)].

ii. http://www.ehow.com/about_5630912_generic-drug-enforcement-act.html, accessed September 22, 2012. See generally, Generic Drug Enforcement Act of 1992 (21 USC 335a), Public Law 102-282. Note that Section 306(k) of the FDCA [21 USC 335a(k)] requires that drug product applicants certify that they did not and will not use in any capacity the services of any debarred persons in connection with a drug product application. See http://www.fda.gov/downloads/Drugs/GuidanceComplianceRegulatoryInformation/Guidances/ucm080584.pdf, accessed September 22, 2012. Note that this certification applies to combination products that include any drug component; this certificate is commonly used by device manufacturers as well.

iii. See the Patient Protection and Affordable Care Act; Public Law 111-148; see specifically Title VII, Section 7001FF for the Biologics Price Competition and Innovation Act, http://www.gpo.gov/fdsys/pkg/PLAW-111publ148/pdf/PLAW-111publ148.pdf.
iv. See http://www.fdalawblog.net/fda_law_blog_hyman_phelps/2012/07/pdufa-v-begins-with-relativelymodest-changes-to-user-fee-rates.html, accessed September 28, 2016; for the actual PDUFA Fees and FR, Vol. 77, No. 148, Wednesday, August 1, 2012/ Notices for background.
v. See http://www.fdalawblog.net/fda_law_blog_hyman_phelps/2012/08/fda-publishes-medicaldevice-user-fee-rates-for-fy-2013-includes-rate-for-expanded-establishment-reg.html, accessed September 28, 2012; for the FY 2013 MDUFA fees.
vi. Fast Track, Breakthrough Therapy, Accelerated Approval and Priority Review, http://www.fda.gov/forconsumers/byaudience/forpatientadvocates/speedingaccesstoimportantnewtherapies/ucm128291.htm, accessed June 26, 2013.
vii. Readers are referred to the FDA Web page, *Transfer of Therapeutic Products to the Center for Drug Evaluation and Research*, at http://www.fda.gov/AboutFDA/CentersOffices/OfficeofMedicalProductsandTobacco/CBER/ucm133463.htm, for information on specific products; the bottom of the page lists the center with authority over the diverse products.
viii. Definition of generic at http://www.fda.gov/Drugs/DevelopmentApprovalProcess/HowDrugsareDevelopedandApproved/ApprovalApplications/AbbreviatedNewDrugApplicationANDAGenerics/, accessed September 24, 2012.
ix. The requirements for premarket notification and premarket approval are simplified for this introduction; please check with CDRH for the specific rules for any device.

REFERENCES

1. FDA-TRACK, http://www.fda.gov/AboutFDA/Transparency/track/default.htm, accessed September 22, 2012.
2. Herper, M. W., "The truly staggering costs of inventing new drugs." *Forbes Magazine*, see http://www.forbes.com/sites/matthewherper/2012/02/10/the-truly-staggering-cost-of-inventing-new-drugs/, accessed September 26, 2012.
3. FDA's Regulatory Responsibilities, http://www.fda.gov/AboutFDA/WhatWeDo/default.htm
4. Valentino, J. "Practical uses for the USP: A legal perspective." In: *Strauss's Federal Drug Laws and Examination Review*. 5th edn., Strauss, S. (ed.), Lancaster, PA: Technomic Publishing Co., 1999, p. 38.
5. Public Law 97-414 (1983).
6. Public Law 98-417 (1984).
7. Public Law 100-292; 102 Stat. 95. Prescription Drug Marketing Act of 1987, http://www.fda.gov/RegulatoryInformation/Legislation/FederalFoodDrugandCosmeticActFDCAct/SignificantAmendmentstotheFDCAct/PrescriptionDrugMarketingActof1987/default.htm, accessed September 22, 2012.
8. Safe Medical Devices Act of 1990, http://thomas.loc.gov/cgi-bin/bdquery/z?d101:HR03095:@@@D&summ2=1&|TOM:/bss/d101query.html, accessed September 22, 2012.
9. Nutrition Labeling and Education Act, http://thomas.loc.gov/cgi-bin/bdquery/z?d101:HR03562:@@@D&summ2=3&|TOM:/bss/d101query.html, accessed September 22, 2012.

10. http://thomas.loc.gov/cgi-bin/bdquery/z?d102:SN02783:@@@D&summ2=m&|TOM:/bss/d102query.html
11. Public Law 102-571; 21 USC 379g and ff.
12. Public Law 107-250; 21 USC 379f et seq.
13. The Food and Drug Administration Safety and Innovation Act (FDASIA), http://www.fda.gov/RegulatoryInformation/Legislation/FederalFoodDrugandCosmeticActFDCAct/SignificantAmendmentstotheFDCAct/FDASIA/default.htm
14. Dietary Supplement Health and Education Act of 1994, Public Law 103-417, http://www.fda.gov/RegulatoryInformation/Legislation/FederalFoodDrugandCosmeticActFDCAct/SignificantAmendmentstotheFDCAct/ucm148003.htm
15. Public Law 105-115; 21 USC 301 et seq., http://www.fda.gov/RegulatoryInformation/Legislation/FederalFoodDrugandCosmeticActFDCAct/SignificantAmendmentstotheFDCAct/FoodandDrugAdministrationAmendmentsActof2007/default.htm
16. 21 USC Section 360bbb and following, expanded access to unapproved therapies and diagnostics.
17. The final rules on expanded access to investigational drugs for treatment use, http://www.fda.gov/Drugs/DevelopmentApprovalProcess/HowDrugsareDevelopedandApproved/ApprovalApplications/InvestigationalNewDrugINDApplication/ucm172492.htm, accessed September 22, 2012.
18. The final rules on expanded access to humanitarian devices, http://www.fda.gov/MedicalDevices/DeviceRegulationandGuidance/GuidanceDocuments/ucm110194.htm, accessed September 22, 2012.
19. 21 USC Section 331 and what follows.
20. *Washington Legal Foundation v. Henney*; Federal Appellate District DC US Court of Appeals, decided February 11, 2000; No. 99-5304.
21. 21 USC 396, http://www.law.cornell.edu/uscode/text/21/396
22. 558 US 310 (2010).
23. 21 USC Section 360aaa and what follows.
24. Public Law 112-144; http://www.fda.gov/RegulatoryInformation/Legislation/FederalFoodDrugandCosmeticActFDCAct/SignificantAmendmentstotheFDCAct/FDASIA/default.htm
25. Public Law 107-250; http://www.fda.gov/RegulatoryInformation/Legislation/FederalFoodDrugandCosmeticActFDCAct/SignificantAmendmentstotheFDCAct/MedicalDeviceUserFeeandModernizationActMDUFMAof2002/default.htm, accessed September 22, 2012.
26. Public Law 108-130; http://www.fda.gov/RegulatoryInformation/Legislation/FederalFoodDrugandCosmeticActFDCAct/SignificantAmendmentstotheFDCAct/AnimalDrugUserFeeActof2003/default.htm, accessed September 22, 2012.
27. Public Law 107-109; http://www.fda.gov/RegulatoryInformation/Legislation/FederalFoodDrugandCosmeticActFDCAct/SignificantAmendmentstotheFDCAct/ucm148011.htm, accessed September 22, 2012.
28. Public Law 108-155; http://www.gpo.gov/fdsys/pkg/PLAW-108publ155/html/PLAW-108publ155.htm, accessed September 22, 2012.
29. Public Health Security and Bioterrorism Preparedness and Response Act of 2002 (Public Law 107-188), http://www.fda.gov/Food/FoodDefense/Bioterrorism/ucm111086.htm, accessed September 22, 2012.
30. Public Law 108-276; http://olpa.od.nih.gov/legislation/108/publiclaws/bioshield.asp, accessed September 22, 2012.
31. http://www.fda.gov/ScienceResearch/SpecialTopics/CriticalPathInitiative/default.htm, accessed September 26, 2012.

32. Public Law 110-85; http://www.fda.gov/RegulatoryInformation/Legislation/Federal FoodDrugandCosmeticActFDCAct/SignificantAmendmentstotheFDCAct/Food andDrugAdministrationAmendmentsActof2007/default.htm, accessed September 22, 2012.
33. http://clinicaltrials.gov/ct2/home, accessed September 26, 2012.
34. http://clinicaltrials.gov/ct2/manage-recs/fdaaa, accessed September 26, 2012.
35. http://www.fda.gov/downloads/Drugs/GuidanceComplianceRegulatoryInformation/Guidances/UCM172001.pdf, accessed September 28, 2012.
36. 21 USC Section 355-1; http://www.gpo.gov/fdsys/pkg/USCODE-2010-title21/pdf/USCODE-2010-title21-chap9-subchapV-partA-sec355-1.pdf and the 2009 Guidance on the Format and Content of Proposed Risk Evaluation and Mitigation Strategies (REMS), REMS Assessments, and Proposed REMS Modifications, http://www.fda.gov/downloads/Drugs/GuidanceComplianceRegulatoryInformation/Guidances/UCM184128.pdf, accessed September 26, 2012.
37. FDASIA: Public Law 112-144; http://www.fda.gov/RegulatoryInformation/Legislation/FederalFoodDrugandCosmeticActFDCAct/SignificantAmendmentstotheFDCAct/FDASIA/default.htm
38. PDUFA Reauthorization Performance Goals and Procedures Fiscal Years 2013 through 2017, http://www.fda.gov/downloads/ForIndustry/UserFees/PrescriptionDrugUserFee/UCM270412.pdf, accessed September 23, 2012; see also http://www.fda.gov/downloads/ForIndustry/UserFees/PrescriptionDrugUserFee/UCM304793.pdf.
39. Definition of filing date, http://www.fda.gov/MedicalDevices/DeviceRegulationandGuidance/HowtoMarketYourDevice/PremarketSubmissions/PremarketApprovalPMA/ucm047991.htm
40. MDUFA law and performance goals, http://www.fda.gov/MedicalDevices/DeviceRegulationandGuidance/Overview/MDUFAIII/default.htm; http://www.fda.gov/downloads/MedicalDevices/NewsEvents/WorkshopsConferences/UCM295454.pdf.
41. Phelps H. and McNamara, P.C. "Food and Drug Administration Safety and Innovation Act," July 11, 2012; http://www.hpm.com/pdf/blog/FDASIA-HP&M Summary&Analysis.pdf, accessed September 26, 2012.
42. FDASIA, Section 902(a)(3).
43. Strauss, S. (ed.), "Food and drug administration: An overview." In: *Strauss's Federal Drug Laws and Examination Review*, 5th edn., Lancaster, PA: Technomic Publishing Co., 1999, p. 176, 186, 285, 323.
44. Public Law 111-31.
45. http://www.fda.gov/ICECI/Inspections/IOM/default.htm
46. http://www.fda.gov/downloads/CombinationProducts/MeetingsConferencesWorkshops/UCM116740.pdf.
47. http://www.fda.gov/AboutFDA/CentersOffices/OfficeofMedicalProductsandTobacco/OfficeofScienceandHealthCoordination/ucm2018186.htm
48. http://www.fda.gov/AboutFDA/CentersOffices/OfficeofMedicalProductsandTobacco/OfficeofScienceandHealthCoordination/ucm2018190.htm
49. http://www.fda.gov/AboutFDA/CentersOffices/OfficeofMedicalProductsandTobacco/OfficeofScienceandHealthCoordination/ucm2018191.htm
50. http://www.fda.gov/ScienceResearch/SpecialTopics/RunningClinicalTrials/ucm160670.htm
51. http://www.fda.gov/AboutFDA/CentersOffices/OfficeofMedicalProductsandTobacco/CDER/ucm090142.htm
52. http://www.fda.gov/AboutFDA/CentersOffices/OC/OfficeofScientificandMedicalPrograms/ucm197508.htm
53. 21 USC Section 321(g)(1).

Overview of FDA and Drug Development 39

54. 21 USC Section 321(p).
55. http://www.fda.gov/drugs/developmentapprovalprocess/manufacturing/ucm090016.htm, accessed September 27, 2012.
56. http://www.fda.gov/ICECI/EnforcementActions/BioresearchMonitoring/NonclinicalLaboratoriesInspectedunderGoodLaboratoryPractices/ucm072738.htm, accessed September 27, 2012.
57. 21 CFR Part 58.
58. Pinna, K. and Pines, W. (eds.), "The drugs/biologics approval process." In: *A Practical Guide to Food and Drug Law and Regulation*. Washington, DC: FDLI, 1998, pp. 98, 119.
59. 21 CFR Part 56; 21 CFR Part 312.66.
60. 21 CFR Part 50, Human Subject Protection.
61. Institutional Review Boards (IRBs) and Protection of Human Subjects in Clinical Trials, http://www.fda.gov/AboutFDA/CentersOffices/OfficeofMedicalProductsandTobacco/CDER/ucm164171.htm, accessed September 28, 2012.
62. 21 CFR Part 314.
63. http://www.fda.gov/Drugs/DevelopmentApprovalProcess/HowDrugsareDevelopedandApproved/ApprovalApplications/NewDrugApplicationNDA/ucm2007029.htm, accessed September 28, 2012.
64. http://www.fda.gov/ForIndustry/ElectronicSubmissionsGateway/default.htm
65. Pinna, K. et al. p. 111.
66. The Orphan Drug Act of 1982, Public Law 97–414; The Orphan Drug Amendments of 1985, Public Law 99–91.
67. USP/DI, Volume III, 13th Edition, Preface, v.
68. USP/DI, p. I/7.
69. USP/DI, p. I/9.
70. USP/DI, p. I/10.
71. USP/DI, p. I/12.
72. 42 USC Section 262(i).
73. http://www.fda.gov/downloads/AboutFDA/ReportsManualsForms/Forms/ucm082348.pdf.
74. http://www.fda.gov/Drugs/DevelopmentApprovalProcess/HowDrugsareDevelopedandApproved/ApprovalApplications/TherapeuticBiologicApplications/ucm113522.htm
75. http://www.fda.gov/Drugs/DevelopmentApprovalProcess/HowDrugsareDevelopedandApproved/ApprovalApplications/TherapeuticBiologicApplications/Biosimilars/ucm291186.htm; http://www.fda.gov/Drugs/DevelopmentApprovalProcess/HowDrugsareDevelopedandApproved/ApprovalApplications/TherapeuticBiologicApplications/Biosimilars/default.htm, accessed September 24, 2012.
76. 21 USC Section 321(h).
77. http://www.fda.gov/MedicalDevices/DeviceRegulationandGuidance/default.htm, accessed September 27, 2012.
78. 21 USC Section 301 and ff.
79. See 21 CFR 202.
80. 21 USC 302 et seq.
81. 21 USC 304 et seq.

2 What Is an IND?

Michael R. Hamrell

CONTENTS

What Is an IND? 42
   When Do I Need an IND? 42
   When Don't I Need an IND? 43
Pre-IND Meeting 45
The Content and Format of an IND Application 46
   Cover Sheet—Section 312.23(a)(1) Form FDA 1571—IND 46
   Table of Contents—Section 312.23(a)(2) 50
   Introductory Statement and General Investigational
   Plan—Section 312.23(a)(3) 50
   Investigator's Brochure—Section 312.23(a)(5) 51
   Clinical Protocol—Section 312.23(a)(6) 52
   CMC Information—Section 312.23(a)(7) 53
   Pharmacology and Toxicology Information—Section 312.23(a)(8) 59
   Previous Human Experience—Section 312.23(a)(9) 61
   Additional Information—Section 312.23(a)(10) 62
   Relevant Information—Section 312.23(a)(11) 62
   Other Important Information about the Submission of an IND 62
   Electronic Submissions 65
   FDA Review of the IND 66
Maintaining an IND—IND Amendments and Other Required Reports ... 66
   The IND Safety Report 67
   The Protocol Amendment 69
   Information Amendments 71
   IND Annual Reports 71
Other Types of INDs 72
Promotion and Charging for Investigational Drugs 73
   Promotion of Investigational Drug Products 73
   Charging for Investigational Drugs 74
More Information about INDs 74
   The FDA Websites 74
   Other Websites 75
References 75

WHAT IS AN IND?

The Federal Food, Drug, and Cosmetic (FD&C) Act requires that all new drugs have an approved marketing application [new drug application (NDA) or biologics license application (BLA)] before they can be shipped in interstate commerce. An IND, or investigational new drug application, is a submission to the US Food and Drug Administration (FDA) requesting permission to initiate a clinical study of a new drug product in the United States. From a legal perspective, the IND is a request for exemption from the Act that prohibits from introducing any new drug into interstate commerce without an approved application. The IND allows you to legally ship an unapproved drug, or import a new drug from a foreign country.

In reality, the IND is much more than a legal tool allowing a company to ship an IND. The IND application allows a company to initiate and conduct clinical studies of their investigational drug product. The IND application provides the FDA with the data necessary to decide whether the new drug and the proposed clinical trial pose a reasonable risk to the human subjects participating in the study. The Act directs the FDA to place investigations on *clinical hold* if the drug involved presents unreasonable risk to the safety of the subjects. The safety of the clinical trial subjects is always the primary concern of the FDA when reviewing an IND, regardless of the phase of the clinical investigation. In later phases (phases 2 and 3), the FDA also evaluates the study design in terms of demonstrating efficacy; but safety of the subjects is critical throughout the drug development process. When preparing an IND, and throughout the drug development process, the primary goal of the sponsor should be to demonstrate to the FDA that the new drug, the proposed trial, and the entire clinical development plan described in the IND is designed to minimize risk to the trial subjects.

IND TERM

Clinical hold—an order issued by the FDA to the sponsor of an IND to delay a proposed clinical investigation or to suspend an ongoing investigation. Subjects may not be given the investigational drug or the hold may require that no new subjects be enrolled into an ongoing study. The clinical hold can be issued before the end of the 30-day IND review period to prevent a sponsor from initiating a proposed protocol or at any time during the life of an IND.

WHEN DO I NEED AN IND?

Simply put, an IND is required anytime you want to conduct a clinical trial of an unapproved drug in the United States. The term "drug" in this context also applies to an investigational biologic drug product and the exact same regulations for an IND apply equally to an investigational study for a new biologic product. However, what is actually considered a new or unapproved drug and how the Act defines a drug often make the decision about filing an IND more complicated. The Act

What Is an IND?

defines a drug, in part, as "articles intended for use in the diagnosis, cure, mitigation, treatment, or prevention of disease in man or other animals; and articles (other than food) intended to affect the structure or any function of the body of man or other animals."[1] The Act further defines a new drug, in part, as "any drug the composition of which is such that such drug is not generally recognized as safe and effective for use under the conditions prescribed, recommended, or suggested in the labeling."[1] Because of these legal definitions, an approved drug can be considered a new drug and would require an IND to conduct a study if it is for an unapproved indication or use. An IND would be required to conduct a clinical trial if the drug is

- A new chemical entity.
- Not approved for the indication under investigation.
- In a new dosage form.
- Being administered at a new dosage level.
- In combination with another drug and the combination is not approved.

A less obvious situation in which a clinical study must be conducted under the authority of an IND is when the chemical compound is being used as a "clinical research tool" but will not be developed for therapeutic use. Sometimes, these "tools" are administered to human subjects to elicit specific physiologic responses that are being studied. In this context, these compounds are considered drugs because the Act states that compounds intended to affect the structure or any function of the body of man or other animals are drugs. There is no exemption from the IND requirements in the Act or regulations for studies conducted with compounds that are considered drugs not being developed for a therapeutic use. All clinical studies where an unapproved drug is administered to human subjects, regardless of whether the drug will be commercially developed, require an IND.

WHEN DON'T I NEED AN IND?

An IND is not required to conduct a study if the drug

- Is not intended for human subjects, but is intended for *in vitro* testing or laboratory research animals (nonclinical studies).
- Is an approved drug and the study is within its approved indication for use.

The regulations also exempt studies of *approved* drugs if *all* of the following criteria are satisfied[2]:

1. The study will not be reported to the FDA in support of a new indication or other change in labeling or advertising for the product.
2. The study will not involve a route of administration, dose level, or patient population that increases the risks associated with the use of the drug.
3. The studies will be conducted in compliance with the Institutional Review Board (IRB) and informed consent regulations.
4. The studies will not be used to promote unapproved indications.

The FDA will generally not accept an IND application for investigations that meet these exemption criteria. The IND regulations also provide an exemption for studies that use placebos,[3] as long as the study would not otherwise require submission of an IND. The use of a placebo in a clinical study does not automatically necessitate an IND.

Finally, an IND is not required for bioequivalence studies in humans to support a generic drug application (ANDA, Abbreviated NDA), if the test and reference product will be dosed at or below the recommended dose and there is no additional risk associated with the conduct of the study. The considerations of when an IND is needed for a bioequivalence study are defined in 21 Code of Federal Regulation (CFR) Section 320.31.

In January 2004, the FDA published a final guidance document clarifying under what circumstances an IND would not be required for the study of marketed cancer drugs.[4] The guidance specifically discusses how investigators assess increased risk to cancer patients when there is scientific literature or other clinical experience available to support the proposed uses. The guidance states that studies may be considered exempt from the IND requirements if the studies involve a new use, dosage, schedule, route of administration, or new combination of marketed cancer drugs in a patient population with cancer if the four exemption criteria for approved products listed earlier are met. They also clarified that as a basis for assessing whether there is an increased risk associated with the proposed use, the investigators and the IRBs must determine that on the basis of scientific literature and generally known clinical experience, there is no significant increase in the risk associated with the use of the drug product. The guidance also provides a clarification for drug manufacturers who provide approved cancer drugs to sponsor investigators for clinical study. Providing an approved cancer drug for an investigator-sponsored trial would not, in and by itself, be considered a promotional activity on the part of the manufacturer if it were for a bona fide clinical investigation.

Whenever a sponsor or investigator considers conducting a clinical study, careful consideration should be given to the need for an IND. Companies should consult with their regulatory affairs staff to determine if an IND is required, and investigators can also consult with the IRB at their institution. If after consultation it is still unclear whether an IND is required, potential sponsors should contact the FDA for advice. Conducting a study without an IND when one is required can lead to regulatory action by the FDA.

IND TERM

Institutional Review Board—A board or committee formally designated by an institution to review and approve the initiation of biomedical research involving human subjects. The primary purpose of the IND is to protect the rights and welfare of human subjects.

> **IND FACTS**
>
> In 2008, the Center for Drug Evaluation and Research (CDER) at the FDA received 2039 original INDs (including therapeutic biologics in CDER). Of these, 883 were commercial INDs and 1156 were noncommercial INDs. At the close of the 2008 calendar year, there were 15,892 active INDs (5962 commercial and 9930 noncommercial).

PRE-IND MEETING

Frequent meetings between the sponsor and the FDA are useful in resolving questions and issues raised during the preparation of an IND. The FDA encourages such meetings to the extent that they aid in the solution of scientific problems and to the extent that the FDA has available resources. To promote efficiency, all issues related to the submission of the IND should be included to the extent practical, since the FDA generally expects to grant only one pre-IND meeting. On occasion, when there are complex manufacturing issues, a separate chemistry, manufacturing, and controls (CMC) meeting can be granted. Meetings at this stage regarding the CMC information are often unnecessary when the project is straightforward. A pre-IND meeting is considered a type B meeting. It is a "formal" meeting requiring a written request that includes, among other things, a list of specific objectives and outcomes and a list of specific questions, grouped by discipline. Most issues and questions are usually related to the design of animal studies needed to initiate clinical trials as well as the scope and design of the initial study in humans. Type B meetings should be scheduled to occur within 60 days of the FDA's receipt of the written request for the meeting. A briefing document is required at least four weeks before the meeting. The briefing document should provide summary information relevant to the product and supplementary information that the FDA can use to provide responses to the questions that have been identified by the sponsor for the submission of the IND. There should be full and open communication about the scientific or medical issue to be discussed during the meeting. The meeting may be a face-to-face one or the FDA may prefer to have a telephonic conference call to serve as the meeting. Typically, the FDA will have a premeeting to address the issues that have been raised and may provide initial feedback before the meeting. The attendance at the pre-IND meeting is multidisciplinary, involving FDA personnel in clinical, pharmacology/toxicology, biopharmaceutics, chemistry, statistics, microbiology, and other disciplines as needed. At the conclusion of the meeting, there should be a review of all the issues, responses, and agreements. An assigned individual from the FDA, typically the project manager, will prepare the minutes of the meeting, and the FDA's version of the minutes are considered the official version, so they should be reviewed carefully to assure that all discussion points and agreements were captured properly. In general, they should be available to the sponsor within 30 days after the meeting but are often made available just before the meeting in draft to form the basis for any discussion. It is most important that all issues and agreements be addressed in the IND submission. There are other meetings that can be held during the IND phases of development and include an end-of-phase 1 meeting (generally for fast-track products), an end-of-phase 2 meeting, and a pre-NDA or pre-BLA meeting.

THE CONTENT AND FORMAT OF AN IND APPLICATION

The content and format of an initial IND is laid out in 21 CFR Part 312 and in numerous guidance documents published by the FDA. This section outlines the required content of an initial IND based on CFR requirements and the published guidance. In addition, since the FDA has adopted the common technical document (CTD) format for NDAs, it is also possible and strongly encouraged that sponsors consider submitting the IND in the CTD format. The FDA is also strongly encouraging companies to submit the CTD document in electronic format. This provides better consistency and ease of review for reviewers and facilitates the later preparation of the NDA, since everything is already in the proper format. Electronic submissions are discussed in more detail later in this chapter. The initial IND application to the FDA can be for a phase 1 first in human study, or it can be for a later-phase study where clinical studies of the compound have already been conducted in the United States or in another country and some clinical data already exist. Although the basic content is the same, the expected level of detail is different. The information expected in later-phase studies is based on the phase of investigation, the amount of human experience with the drug, the drug substance, and the dosage form of the drug. In the outline, requirements will be addressed both for INDs for phase 1 studies as well as initial INDs for later-stage studies. This section is not intended to be a recitation of 21 CFR Section 312.23 or the guidance documents, but an overview of the key elements of the initial IND, regardless of the phase of the proposed study. The specific references to Section 312.23 for each of the sections of an IND are included for reference.

COVER SHEET—SECTION 312.23(a)(1) FORM FDA 1571—IND

The Form 1571 (Figure 2.1) is a required part of the initial IND and every subsequent submission related to the IND application. Each *IND Amendment, IND Safety Report, IND Annual Report*, or general correspondence with the FDA regarding the IND must include a Form 1571. The Form 1571 serves as a cover sheet for IND submissions and provides the FDA with the basic information about the submission—name of the sponsor, IND number, name of the drug, type of submission, serial number, and the contents of the application. Each submission to the IND must be consecutively numbered, starting with the initial IND application, which is numbered 0000. The next submission (response to clinical hold, correspondence, amendment, etc.) should be numbered 0001, with subsequent submissions numbered consecutively in the order of submission. It is important to note that the FDA expects that every submission, even the most routine correspondence, be made with a completed Form 1571 and have a serial number. The FDA tracks all IND submissions on the basis of serial numbers and files

Next Page | Export Data | Import Data | Reset Form

DEPARTMENT OF HEALTH AND HUMAN SERVICES
Food and Drug Administration
INVESTIGATIONAL NEW DRUG APPLICATION (IND)
(Title 21, Code of Federal Regulations (CFR) Part 312)

Form Approved: OMB No. 0910-0014
Expiration Date: April 30, 2015
See PRA Statement on page 2.

NOTE: No drug/biologic may be shipped or clinical investigation begun until an IND for that investigation is in effect (21 CFR 312.40)

1. Name of Sponsor
2. Date of Submission *(mm/dd/yyyy)*
3. Sponsor Address
   Address 1 *(Street address, P.O. box, company name c/o)*
   Address 2 *(Apartment, suite, unit, building, floor, etc.)*
   City | State/Province/Region
   Country | ZIP or Postal Code
4. Telephone Number *(Include country code if applicable and area code)*
5. Name(s) of Drug *(Include all available names: Trade, Generic, Chemical, or Code)*
   Continuation Page for #5
6. IND Number *(If previously assigned)*
7. (Proposed) Indication for Use
   Is this indication for a rare disease (prevalence <200,000 in U.S.)? ☐ Yes ☐ No
   Does this product have an FDA Orphan Designation for this indication? ☐ Yes ☐ No
   If yes, provide the Orphan Designation number for this indication:
   Continuation Page for #7
8. Phase(s) of Clinical Investigation to be conducted ☐ Phase 1 ☐ Phase 2 ☐ Phase 3 ☐ Other *(Specify)*:
9. List numbers of all Investigational New Drug Applications (21 CFR Part 312), New Drug Applications (21 CFR Part 314), Drug Master Files (21 CFR Part 314.420), and Biologics License Applications (21 CFR Part 601) referred to in this application.
10. IND submission should be consecutively numbered. The initial IND should be numbered "Serial number: 0000." The next submission (e.g., amendment, report, or correspondence) should be numbered "Serial Number: 0001." Subsequent submissions should be numbered consecutively in the order in which they are submitted.
    Serial Number
11. This submission contains the following *(Select all that apply)*
    ☐ Initial Investigational New Drug Application (IND) ☐ Response to Clinical Hold ☐ Response To FDA Request For Information
    ☐ Request For Reactivation Or Reinstatement ☐ Annual Report ☐ General Correspondence
    ☐ Development Safety Update Report (DSUR) ☐ Other *(Specify)*:

Protocol Amendment(s)	**Information Amendment(s)**	**Request for**	**IND Safety Report(s)**
☐ New Protocol	☐ Chemistry/Microbiology	☐ Meeting	☐ Initial Written Report
☐ Change in Protocol	☐ Pharmacology/Toxicology	☐ Proprietary Name Review	☐ Follow-up to a Written Report
☐ New Investigator	☐ Clinical ☐ Statistics	☐ Special Protocol Assessment	
☐ PMR/PMC Protocol	☐ Clinical Pharmacology	☐ Formal Dispute Resolution	

12. Select the following only if applicable. *(Justification statement must be submitted with application for any items selected below. Refer to the cited CFR section for further information.)*
    ☐ Emergency Research Exception From Informed Consent Requirements, 21 CFR 312.23 (f)
    ☐ Charge Request, 21 CFR 312.8
    Expanded Access Use, 21 CFR 312.300
    ☐ Individual Patient, Non-Emergency 21 CFR 312.310
    ☐ Individual Patient, Emergency 21 CFR 312.310(d)
    ☐ Intermediate Size Patient Population, 21 CFR 312.315
    ☐ Treatment IND or Protocol, 21 CFR 312.320

For FDA Use Only

CBER/DCC Receipt Stamp	DDR Receipt Stamp	Division Assignment
		IND Number Assigned

FORM FDA 1571 (10/12) **Page 1 of 2** PSC Publishing Services (301) 443-6740 EF

(a)

FIGURE 2.1 Form 1571—IND application form.

Previous Page | Next Page

13. Contents of Application – This application contains the following items *(Select all that apply)*

- ☐ 1. Form FDA 1571 *(21 CFR 312.23(a)(1))*
- ☐ 2. Table of Contents *(21 CFR 312.23(a)(2))*
- ☐ 3. Introductory statement *(21 CFR 312.23(a)(3))*
- ☐ 4. General investigational plan *(21 CFR 312.23(a)(3))*
- ☐ 5. Investigator's brochure *(21 CFR 312.23(a)(5))*
- ☐ 6. Protocol(s) *(21 CFR 312.23(a)(6))*
  - ☐ a. Study protocol(s) *(21 CFR 312.23(a)(6))*
  - ☐ b. Investigator data *(21 CFR 312.23(a)(6)(iii)(b))* or completed Form(s) FDA 1572
  - ☐ c. Facilities data *(21 CFR 312.23(a)(6)(iii)(b))* or completed Form(s) FDA 1572
- 6. Protocol(s) *(Continued)*
  - ☐ d. Institutional Review Board data (21 CFR 312.23(a)(6)(iii)(b)) or completed Form(s) FDA 1572
- ☐ 7. Chemistry, manufacturing, and control data *(21 CFR 312.23(a)(7))*
  - ☐ Environmental assessment or claim for exclusion *(21 CFR 312.23(a)(7)(iv)(e))*
- ☐ 8. Pharmacology and toxicology data (21 CFR 312.23(a)(8))
- ☐ 9. Previous human experience *(21 CFR 312.23(a)(9))*
- ☐ 10. Additional information *(21 CFR 312.23(a)(10))*
- ☐ 11. Biosimilar User Fee Cover Sheet *(Form FDA 3792)*
- ☐ 12. Clinical Trials Certification of Compliance *(Form FDA 3674)*

14. Is any part of the clinical study to be conducted by a contract research organization? ☐ Yes ☐ No

If Yes, will any sponsor obligations be transferred to the contract research organization? ☐ Yes ☐ No

If Yes, provide a statement containing the name and address of the contract research organization, identification of the clinical study, and a listing of the obligations transferred *(use continuation page)*.

Continuation Page for #14

15. Name and Title of the person responsible for monitoring the conduct and progress of the clinical investigations

16. Name(s) and Title(s) of the person(s) responsible for review and evaluation of information relevant to the safety of the drug

I agree not to begin clinical investigations until 30 days after FDA's receipt of the IND unless I receive earlier notification by FDA that the studies may begin. I also agree not to begin or continue clinical investigations covered by the IND if those studies are placed on clinical hold or financial hold. I agree that an Institutional Review Board (IRB) that complies with the requirements set forth in 21 CFR Part 56 will be responsible for initial and continuing review and approval of each of the studies in the proposed clinical investigation. I agree to conduct the investigation in accordance with all other applicable regulatory requirements.

17. Name Of Sponsor or Sponsor's Authorized Representative

18. Telephone Number *(Include country code if applicable and area code)*

19. Facsimile (FAX) Number *(Include country code if applicable and area code)*

20. Address

Address 1 *(Street address, P.O. box, company name c/o)*

Address 2 *(Apartment, suite, unit, building, floor, etc.)*

City | State/Province/Region

Country | ZIP or Postal Code

21. Email Address

22. Date of Signature *(mm/dd/yyyy)*

23. Name of Alternate Contact

24. Telephone Number of Alternate Contact *(Include country code if applicable and area code)*

25. Signature Of Sponsor or Sponsor's Authorized Representative

Sign

WARNING : A willfully false statement is a criminal offense (U.S.C. Title 18, Sec. 1001).

The information below applies only to requirements of the Paperwork Reduction Act of 1995.

The burden time for this collection of information is estimated to average 100 hours per response, including the time to review instructions, search existing data sources, gather and maintain the data needed and complete and review the collection of information. Send comments regarding this burden estimate or any other aspect of this information collection, including suggestions for reducing this burden to the address to the right.

"An agency may not conduct or sponsor, and a person is not required to respond to, a collection of information unless it displays a currently valid OMB number."

Department of Health and Human Services
Food and Drug Administration
Office of Chief Information Officer
Paperwork Reduction Act (PRA) Staff
1350 Piccard Drive, Room 400
Rockville, MD 20850

DO NOT SEND YOUR COMPLETED FORM TO THIS PRA STAFF ADDRESS.

FORM FDA 1571 (10/12) **Page 2 of 2**

(b)

FIGURE 2.1 (Continued) Form 1571—IND application form.

them according to the serial number on receipt. If more than one group within a company submits IND amendments (e.g., a pharmacovigilance group may submit safety reports directly to the FDA), it is essential that the serial numbers be consecutive. With electronic submissions, it is important that the file sequence number be consecutive and tracked.

The Form 1571 provides a section for the sponsor to state whether a contract research organization (CRO) will conduct any parts of the clinical study and if any sponsor obligations will be transferred to the CRO. If sponsor responsibilities are to be transferred, a list of the obligations transferred and the name and address of the CRO must be attached to the Form 1571. Although the sponsor may transfer some or all of its obligations to a CRO, the sponsor of the IND is ultimately responsible for the conduct of the clinical investigation and all the regulatory and legal requirements pertaining to a clinical trial.

When signing the Form 1571, the sponsor also makes three important commitments to the FDA, which are outlined on page 2 of the form.

1. The sponsor is committing not to initiate the clinical study until 30 days after the FDA receives the IND, unless otherwise notified by the FDA, and not to begin or continue clinical studies covered by the IND if they are placed on clinical hold.
2. The sponsor is committing to ensure that an IRB will be responsible for initial and continuing review and approval of each study in the proposed clinical investigation.
3. The sponsor is committing to conduct the investigation in accordance with all other applicable regulatory requirements.

These are significant commitments and the sponsor should be aware that signing the Form 1571 is more than a formality. Making a willfully false statement on the Form 1571 or accompanying documentation is a criminal offense. Detailed information on completing the Form 1571 can be found on the FDA Website,[5] in Section 312.23(a)(1) and from the FDA review division responsible for reviewing the IND.

IND TERM

IND amendment—A submission to the IND file that adds new or revised information. Every submission adds to, revises, or affects the body of information within the IND and is, therefore, considered an IND amendment. Protocol amendments and information amendments are two examples of information that is filed to an IND in the course of clinical development. A protocol amendment is submitted when a sponsor intends to conduct a new study, wishes to modify the design or conduct of a previously submitted study protocol, or adds a new investigator to a protocol. An information amendment is used to submit new CMC, toxicology, pharmacology, clinical, or other information that does not fall within the scope of a protocol amendment, annual report, or IND safety report.

> **IND TERM**
>
> *IND safety report*—An expedited report sent to the FDA and all participating investigators of a serious and unexpected adverse experience associated with use of the drug or findings from nonclinical studies that suggest a risk to human subjects.

> **IND TERM**
>
> *IND annual report*—A brief report to the FDA on the progress of the clinical investigations. It is submitted each year within 60 days of the anniversary date that the IND went into effect.

Table of Contents—Section 312.23(a)(2)

This should be a comprehensive listing of the contents of the IND broken down by section, volume, and page number. The table of contents (TOC) should include all required sections, appendices, attachments, reports, and other reference material. The TOC must be accurate and building the table should not be a last-minute task. An accurate, well-laid-out TOC will allow the FDA reviewers to quickly find the information they need and ultimately speed up review of the IND application. Many sponsors begin planning the IND submission by laying out the TOC first. This allows the team to clearly see what information is required for the submission and how the document will be structured, and it allows the TOC to be updated as the application is being built.

Introductory Statement and General Investigational Plan—Section 312.23(a)(3)

This section should provide a brief, three- to four-page overview of the investigational drug and the sponsor's investigational plan for the coming year. The goal of this section is simply to provide a brief description of the drug and lay out the development plan for the drug. For a phase 1 first-in-person (FIP) submission, two to three pages may be sufficient if the sponsor is attempting to determine early pharmacokinetic and pharmacodynamic properties of the drug. The sponsor should not attempt to develop and present a detailed development plan that will, in all likelihood, change considerably should the product proceed to further development.[6]

The introductory statement should begin with a description of the drug and the indication(s) to be studied and include the pharmacologic class of the compound, the name of the drug and all active ingredients, the structural formula of the drug, the dosage form, and the route of administration. This section must also describe the sponsor's plan for investigating the drug during the following year and should include a rationale for the drug and the research study proposed, the general approach to be followed in studying the drug, the indication(s) to be studied, the type of clinical studies to be conducted, the estimated number of patients receiving the drug, and the risks anticipated on the basis of nonclinical studies or prior studies in humans.

If the drug has been previously administered to humans, the introductory statement should include a brief summary of human clinical experience to date, focusing mainly on safety of the drug in previous studies and how that supports studies proposed in the IND. If the drug was withdrawn from investigation or marketing in any country for safety reasons, the name of the country and the reasons for withdrawal should also be briefly discussed in the introductory statement.

INVESTIGATOR'S BROCHURE—SECTION 312.23(a)(5)

The content and format of the investigator's brochure (IB) is described in 21 CFR Section 312.23(a)(5) and in greater detail in the "International Conference on Harmonization (ICH) E6 Good Clinical Practice" guidance document.[7] An exhaustive discussion of the IB is not presented here, preferring to focus more broadly on the purpose of the document and the general content required by the regulations.

The IB is a key document provided to each clinical investigator and the IRB at each of the clinical sites. The IB presents, in summary form, the key nonclinical (safety), clinical, and CMC (quality) data that support the proposed clinical trial. The IB provides the clinical investigators with the information necessary to understand the rationale for the proposed trial and to make an unbiased risk–benefit assessment of the appropriateness of the proposed trial.[7]

> **IND TERM**
>
> *Chemistry, manufacturing, and controls (CMC) describes the chemical structure and chemical properties of the compound, composition, manufacturing process and control of the raw materials, drug substance, and drug product that ensure the identity, quality, purity, and potency of the drug product. The ICH guidance refers to this as the quality section of the file.*

The type and extent of information provided in the IB will be dependent on the stage of development of the drug product, but the IB must contain the following information:

1. A brief summary of the CMC information, including the physical, chemical, and pharmaceutical properties of the drug, and the chemical name and chemical structure, if known. It should also include a description of the formulation and how the drug is supplied, as also the storage and handling requirements.
2. A summary of all relevant nonclinical pharmacology, toxicology, pharmacokinetic, and drug metabolism information generated to support

human clinical studies. It should include a tabular summary of each nonclinical study conducted, outlining the methodology used and the results of each study.
3. If human clinical studies have been conducted with the drug, a summary of information relating to safety and efficacy should be presented, including any information from those studies on the metabolism, pharmacokinetics, pharmacodynamics, dose response, or other pharmacologic activities.
4. A summary of data and guidance for the investigator in the management of subjects participating in the trial. An overall discussion of the nonclinical and clinical data presented in the IB and a discussion of the possible risks and adverse reactions associated with the investigational drug product and the specific tests, observations and precautions that may be needed for the clinical trial.

It is important to remember that the IB is a living document and must be updated by the sponsor as new information becomes available from ongoing clinical and nonclinical studies. Keep in mind though, that the document must be a readable and useful document, so it is recommended that the IB should ideally not exceed 75–80 pages. At a minimum, the IB should be reviewed and updated annually. However, important safety information should be communicated to the investigator, the IRB, and the FDA, if required, before it is included in the IB.

CLINICAL PROTOCOL—SECTION 312.23(a)(6)

As with the IB, the content and format of the protocol is described in 21 CFR Section 312.23 and in greater detail in the "ICH E6 Good Clinical Practice" guidance document[7] and will not be presented here.

A clinical protocol describes how a particular clinical trial is to be conducted. It describes the objectives of the study, the trial design, the selection of subjects, and the manner in which the trial is to be carried out. The initial IND is required to have a clinical protocol for the initial planned study. However, the IND regulations specifically allow phase 1 protocols to be less detailed and more flexible than protocols for phase 2 or 3 studies.[6] The regulations state that phase 1 protocols should be directed primarily at providing an outline of the investigation: an estimate of the number of subjects to be included; a description of safety exclusions; and a description of the dosing plan, including duration, dose, or method to be used in determining dose. Phase 1 protocols should specify in detail only those elements for the study that are critical to subject safety, such as necessary monitoring of vital signs and blood chemistries, toxicity-based stopping, or dose adjustment rules.[6]

Although the regulations allow phase 1 protocols to be less detailed, the sponsor cannot submit a protocol summary in lieu of a complete protocol as part of the initial IND. A protocol summary may be acceptable in some instances, but the submission of a summary should be discussed and agreed to by the reviewing division at the FDA during the pre-IND meeting. Later-phase protocols should be more detailed than a phase 1 protocol and reflect that stage of development of the drug. It should contain efficacy parameters, methods, and timing for assessing and analyzing the efficacy parameters and detailed statistical sections, describing the statistical methods to be employed and the timing of any planned interim analysis.

The regulations require any protocol submitted as part of an IND to contain the following elements.

1. A statement of the objectives and the purpose of the study
2. The name, address, and qualifications [curriculum vitae (CV)] of each investigator and each subinvestigator participating in the study; the name and address of each clinical site; and the name and address of each IRB responsible for reviewing the proposed study. The required information regarding all investigators is collected on the Form FDA 1572—statement of investigator (Figure 2.2). The Form 1572 collects basic information about the investigator, such as the name and address of the investigator, a description of the education and training of the investigator (a copy of the investigator's CV is usually attached), the name and address of the IRB at the site, and the names of any subinvestigators at the site. The Form 1572 includes a series of commitments (see box 9 in Figure 2.2) that the investigator agrees to by signing the form. These commitments include, among others, agreeing to conduct the study according to the protocol, agreeing to personally conduct or supervise the investigation, agreeing to report adverse events (AEs) to the sponsor, agreeing to maintain accurate records, and agreeing to comply with all other obligations and requirements outlined in the regulations. Investigators and sponsors should be aware that making willfully false statements on the Form 1572 is a criminal offense.
3. The criteria for study subject inclusion and exclusion and an estimate of the number of subjects to be enrolled in the study
4. A description of the study design, control groups to be used, and methods employed to minimize bias on the part of the subjects, investigators, and analysts
5. The planned maximum dose, the duration of patient exposure to the drug, and the methods used to determine the doses to be administered
6. A description of the measurements and observations to be made to achieve the study objectives
7. A description of the clinical procedures and laboratory tests planned to monitor the effects of the drug in the subjects

CMC Information—Section 312.23(a)(7)

This key section of an IND describes the quality information, comprising the composition, manufacturing process, and control of the drug substance and drug product. The CMC section must provide sufficient detail and information to demonstrate the identity, quality, purity, and potency of the drug product. The amount of information needed to accomplish this is based on the phase of the proposed study, the duration of the study, the dosage form of the investigational drug, and the amount of additional information available.[6] For a phase 1 IND, the

DEPARTMENT OF HEALTH AND HUMAN SERVICES
FOOD AND DRUG ADMINISTRATION

STATEMENT OF INVESTIGATOR
(TITLE 21, CODE OF FEDERAL REGULATIONS (CFR) PART 312)
(See instructions on reverse side.)

Form Approved: OMB No. 0910-0014
Expiration Date: April 30, 2015
See OMB Statement on Reverse.

NOTE: No investigator may participate in an investigation until he/she provides the sponsor with a completed, signed Statement of Investigator, Form FDA 1572 (21 CFR 312.53(c)).

1. NAME AND ADDRESS OF INVESTIGATOR

Name of Principal Investigator

Address 1 | Address 2

City | State/Province/Region | Country | ZIP or Postal Code

2. EDUCATION, TRAINING, AND EXPERIENCE THAT QUALIFY THE INVESTIGATOR AS AN EXPERT IN THE CLINICAL INVESTIGATION OF THE DRUG FOR THE USE UNDER INVESTIGATION. ONE OF THE FOLLOWING IS PROVIDED *(Select **one** of the following.)*

☐ Curriculum Vitae ☐ Other Statement of Qualifications

3. NAME AND ADDRESS OF ANY MEDICAL SCHOOL, HOSPITAL, OR OTHER RESEARCH FACILITY WHERE THE CLINICAL INVESTIGATION(S) WILL BE CONDUCTED

CONTINUATION PAGE for Item 3

Name of Medical School, Hospital, or Other Research Facility

Address 1 | Address 2

City | State/Province/Region | Country | ZIP or Postal Code

4. NAME AND ADDRESS OF ANY CLINICAL LABORATORY FACILITIES TO BE USED IN THE STUDY

CONTINUATION PAGE for Item 4

Name of Clinical Laboratory Facility

Address 1 | Address 2

City | State/Province/Region | Country | ZIP or Postal Code

5. NAME AND ADDRESS OF THE INSTITUTIONAL REVIEW BOARD (IRB) THAT IS RESPONSIBLE FOR REVIEW AND APPROVAL OF THE STUDY(IES)

CONTINUATION PAGE for Item 5

Name of IRB

Address 1 | Address 2

City | State/Province/Region | Country | ZIP or Postal Code

6. NAMES OF SUBINVESTIGATORS *(If not applicable, enter "None")*

CONTINUATION PAGE – for Item 6

7. NAME AND CODE NUMBER, IF ANY, OF THE PROTOCOL(S) IN THE IND FOR THE STUDY(IES) TO BE CONDUCTED BY THE INVESTIGATOR

FORM FDA 1572 (7/13) **PREVIOUS EDITION IS OBSOLETE.** **Page 1 of 2**

PSC Publishing Services (301) 443-6740 EF

(a)

FIGURE 2.2 Form 1572—Statement of investigator.

What Is an IND?

8. PROVIDE THE FOLLOWING CLINICAL PROTOCOL INFORMATION. *(Select **one** of the following.)*	
☐ For Phase 1 investigations, a general outline of the planned investigation including the estimated duration of the study and the maximum number of subjects that will be involved.	
☐ For Phase 2 or 3 investigations, an outline of the study protocol including an approximation of the number of subjects to be treated with the drug and the number to be employed as controls, if any; the clinical uses to be investigated; characteristics of subjects by age, sex, and condition; the kind of clinical observations and laboratory tests to be conducted; the estimated duration of the study; and copies or a description of case report forms to be used.	

9. COMMITMENTS

I agree to conduct the study(ies) in accordance with the relevant, current protocol(s) and will only make changes in a protocol after notifying the sponsor, except when necessary to protect the safety, rights, or welfare of subjects.

I agree to personally conduct or supervise the described investigation(s).

I agree to inform any patients, or any persons used as controls, that the drugs are being used for investigational purposes and I will ensure that the requirements relating to obtaining informed consent in 21 CFR Part 50 and institutional review board (IRB) review and approval in 21 CFR Part 56 are met.

I agree to report to the sponsor adverse experiences that occur in the course of the investigation(s) in accordance with 21 CFR 312.64. I have read and understand the information in the investigator's brochure, including the potential risks and side effects of the drug.

I agree to ensure that all associates, colleagues, and employees assisting in the conduct of the study(ies) are informed about their obligations in meeting the above commitments.

I agree to maintain adequate and accurate records in accordance with 21 CFR 312.62 and to make those records available for inspection in accordance with 21 CFR 312.68.

I will ensure that an IRB that complies with the requirements of 21 CFR Part 56 will be responsible for the initial and continuing review and approval of the clinical investigation. I also agree to promptly report to the IRB all changes in the research activity and all unanticipated problems involving risks to human subjects or others. Additionally, I will not make any changes in the research without IRB approval, except where necessary to eliminate apparent immediate hazards to human subjects.

I agree to comply with all other requirements regarding the obligations of clinical investigators and all other pertinent requirements in 21 CFR Part 312.

INSTRUCTIONS FOR COMPLETING FORM FDA 1572
STATEMENT OF INVESTIGATOR

1. Complete all sections. Provide a separate page if additional space is needed.
2. Provide curriculum vitae or other statement of qualifications as described in Section 2.
3. Provide protocol outline as described in Section 8.
4. Sign and date below.
5. FORWARD THE COMPLETED FORM AND OTHER DOCUMENTS BEING PROVIDED TO THE SPONSOR. The sponsor will incorporate this information along with other technical data into an Investigational New Drug Application (IND). INVESTIGATORS SHOULD NOT SEND THIS FORM DIRECTLY TO THE FOOD AND DRUG ADMINISTRATION.

10. DATE *(mm/dd/yyyy)*	11. SIGNATURE OF INVESTIGATOR	Sign

(**WARNING**: A willfully false statement is a criminal offense. U.S.C. Title 18, Sec. 1001.)

The information below applies only to requirements of the Paperwork Reduction Act of 1995.

The burden time for this collection of information is estimated to average 100 hours per response, including the time to review instructions, search existing data sources, gather and maintain the data needed and complete and review the collection of information. Send comments regarding this burden estimate or any other aspect of this information collection, including suggestions for reducing this burden to the address to the right:	Department of Health and Human Services Food and Drug Administration Office of Chief Information Officer Paperwork Reduction Act (PRA) Staff PRAStaff@fda.hhs.gov
"An agency may not conduct or sponsor, and a person is not required to respond to, a collection of information unless it displays a currently valid OMB number."	**DO NOT SEND YOUR COMPLETED FORM TO THIS PRA STAFF EMAIL ADDRESS.**

(b)

FIGURE 2.2 (Continued) Form 1572—Statement of investigator.

CMC information provided for the raw materials, drug substance, and drug product should be sufficiently detailed to allow the FDA to evaluate the safety of the subjects participating in the trial. A safety concern or lack of data, which makes it impossible for the FDA to conduct a safety evaluation, is the only reason for a clinical hold based on the CMC section. Safety concerns may include the following:

1. Product is made with unknown or impure components
2. Product has a chemical structure(s) of known or highly likely toxicity
3. Product does not remain chemically stable throughout the testing program
4. Product has an impurity profile indicative of a potential health hazard or an impurity profile insufficiently defined to assess potential health hazard
5. Master or working cell bank is poorly characterized[6]

A key aspect to assuring the safety of the subjects participating in clinical trials is adherence to current good manufacturing practices (cGMPs). The FDA requires that any drug product intended for administration to humans be manufactured in conformance with cGMP. Adherence to GMP provides a minimum level of control over the manufacturing process and final drug product and helps ensure the identity, quality, purity, and potency of the clinical trial material. The GMP controls used to manufacture drug products for clinical trials should be consistent with the stage of development, and they should be manufactured in suitable facilities, using appropriate production and control procedures to ensure the quality of the drug product.[8]

INDs for later-phase studies must contain the CMC information outlined in Section 312.23, but the focus should be on safety issues relating to the proposed phase and expanded scope of the investigation. The FDA expects that the CMC section for a later-phase IND will be more detailed than a phase 1 study and demonstrate a higher level of characterization of the drug substance and drug product and greater control over the raw materials and manufacturing process. For phase 2 studies, the sponsor should be able to document the manufacturing process that is controlled at predetermined points and yields products that meet the tentative acceptance criteria.[9]

The regulations require the CMC section of an IND to contain the following sections:

1. CMC introduction

    This section should provide a brief overview of the investigational drug product. In this section, the sponsor should state whether there are any signals of potential risk to human subjects because of the chemistry of the drug substance or drug product or the manufacturing process for the drug substance or drug product. If potential risks are identified, the risks should be discussed and steps to monitor for the risks should be described, or the reasons the potential risks are acceptable should be presented. In the introduction, the sponsor should also describe any differences between the drug product to be used in the proposed study and the drug product used in the nonclinical toxicology studies that support the clinical investigations. How these differences affect the safety profile should be discussed, and if there are no differences, those should be stated.

2. Information on the drug substance in the form of a summary report containing the following information:
   a. A brief description of the drug substance and evidence to support its chemical structure. INDs for later-phase trials should include a more complete description of the physical, chemical, and biological characteristics of the drug substance and provide additional supporting evidence characterizing the chemical structure.
   b. The name and address of the manufacturer.
   c. A brief description of the manufacturing process. The description should include a detailed flow diagram of the process and a list of all the reagents, solvents, and catalysts used in the process. INDs for later-phase trials will include a more detailed description of the manufacturing process and the controls. A process flow diagram that includes chemical structures and configurations and significant side products should be included, and the acceptance criteria for the product described.
   d. A brief description of the acceptable limits (specifications) and analytical methods used to assure the identity, strength, quality, potency, and purity of the drug substance. This section should include a description of the test methods used and outline the proposed acceptance criteria. The proposed acceptance criteria should be based on analytical data [e.g., IR spectrum to prove identity, and high-performance liquid chromatography (HPLC) to support purity level, and impurities profile].[6] Validation data and established specifications are not required for phase 1 studies; however, a certificate of analysis for the lot(s) of clinical trial material should be included with the initial IND. Initial INDs for later-phase studies should provide the same type of information as for earlier-phase studies, but analytical procedures and acceptance criteria should be better defined and validation data should be available if requested by the FDA.
   e. Data to support the stability of the drug substance. For a phase 1 IND, a brief description of the stability studies conducted and the methods used to monitor stability should be provided, including a table outlining stability data from representative lots of material. For later-phase studies, a stability protocol should be submitted, including a list of all tests, analytical procedures, sampling time points for each test, and the duration of the stability studies. Preliminary stability data should be submitted along with stability data from clinical material used in earlier-phase studies.
3. Information on the drug product in the form of a summary report containing the following information:
   a. A list of all components used in the manufacture of the drug product, including components intended to be in the drug product and those that may not appear, but are used in the manufacturing process. The components should be identified by their established name

(chemical name) and their compendial status [National Formulary (NF), United States Pharmacopoeia (USP)] should be listed, if it exists. Analytical procedures and acceptance criteria should be presented for noncompendial components. If applicable, the quantitative composition of the drug product should be summarized and any expected variations should be discussed. The same type of information should be presented in an IND for a later-phase study.
b. The name and address of the manufacturer of the drug product.
c. A brief, step-by-step description of the manufacturing and packaging procedures including a process flow diagram. For sterile products, a description of the sterilization process should be included. The same type of information should be included in an IND for a later-phase study.
d. A description of the proposed acceptable limits (specifications) for the drug product and the test methods used. Validation data and established specifications are not required in the phase 1 IND; however, a complete description of the analytical procedures and validation data should be available on request for later-phase studies. For sterile products, sterility and endotoxin tests should be submitted in the initial IND. A certificate of analysis for the drug product lot(s) to be used in the proposed investigation should also be provided.
e. A description of the proposed container closure system and a brief description of the stability study and test methods. Stability data on representative material should be presented in a tabular format. A copy of the stability protocol is not required for a phase 1 study. An initial IND for a later-phase study should include a copy of the stability protocol that includes a list of tests, analytical procedures, sampling time points, and the expected duration of the stability program. When applicable, stability data on the reconstituted drug product should be included in the initial IND.
4. Information on any placebo or comparator product that will be utilized in the proposed clinical study. This should include a brief written description of the composition, manufacture, and control of the placebo. Process flow diagrams and tabular summaries can be used in the description.
5. Copies of all proposed product labels and any other proposed labeling that will be provided to the investigators. Mock-ups of the proposed labeling are acceptable or actual printed labeling can be submitted. The investigational drug must be labeled with the caution statement: "Caution: New Drug—Limited by Federal (or United States) law to investigational use."[10]
6. A claim for categorical exclusion from an environmental assessment. The National Environmental Policy Act (NEPA) of 1969 requires all Federal agencies to assess the environmental impacts of their actions and to ensure that the interested and affected public is informed of environmental analyses.[11] The FDA is required to consider the environmental

impact of approving drug and biologic applications and requires all such applications to include an environmental assessment or a claim for categorical exclusion. IND applications are generally categorically excluded from the requirement to prepare and submit an environmental assessment.[12] In this section of the IND, the sponsor should state that the action requested (approval of an IND application) qualifies for categorical exclusion in accordance with 21 CFR Section 25.31(e) and that, to the sponsor's knowledge, no extraordinary circumstances exist [21 CFR Section 25.15(d)].

PHARMACOLOGY AND TOXICOLOGY INFORMATION—SECTION 312.23(a)(8)

The decision to proceed to the initial administration of the investigational drug to humans must include the careful conduct and review of the data from nonclinical *in vivo* and *in vitro* studies. These data must provide a good level of confidence that the new drug product is reasonably safe for administration to human subjects at the planned dosage levels. The goals of the nonclinical safety testing include characterization of toxic effects with respect to target organs, dose dependence, relationship to exposure, and potential reversibility. Nonclinical safety information is important for the estimation of an initial safe starting dose for human trials and the identification of parameters for clinical monitoring for potential AEs.[13]

The pharmacology and toxicology section of the IND includes the nonclinical safety data that the sponsor generated to conclude that the new drug is reasonably safe for clinical study. The amount and type of nonclinical data needed to support a new drug product depends on the class of the new drug, the duration of the proposed clinical trials, and the patient population that will be exposed to the drug. Generally, the following nonclinical safety studies are required before initiating phase 1 studies, and the results of these studies must be included in the IND:

Safety pharmacology studies (often conducted as part of the toxicity studies).

- Single dose and repeat dose toxicity studies (duration of the repeat dose studies should equal or exceed the duration of the human clinical trials).
- Genotoxicity studies (*in vitro* studies evaluating mutations and chromosomal damage).
- Reproduction toxicity studies (Nonclinical animal studies conducted to reveal any effects the investigational drug may have on mammalian reproduction). These studies are needed before including women of child-bearing potential in any clinical study and are usually not needed for an initial phase 1 study in normal male volunteers.
- Other supplementary studies may be needed if safety concerns are identified.
- The FDA and ICH are proposing that nonclinical studies evaluating the potential of the new drug to delay ventricular repolarization (QT interval prolongation) be conducted prior to initiation of phase 1 studies.[14,15]

The CDER guidance document Web page[16] provides access to all of the key guidance documents, discussing required nonclinical testing for new drugs. The pharmacology and toxicology information or safety section of the initial IND should contain the following elements.

1. A summary report describing the pharmacologic effects and mechanism of action of the drug and information on the absorption, distribution, metabolism, and excretion (ADME) of the drug. If this information is not known at the time the initial IND is submitted, it should be stated. Lack of this information should not generally be a reason for a phase 1 IND to be placed on clinical hold.[6] However, most sponsors will have at least early pharmacologic data, including exposure, half-life of the drug, and an understanding of the major factors that influence the pharmacokinetics of the drug, for example, the enzymes responsible for metabolism of the drug. Initial INDs for later-phase studies should be able to provide this pharmacology information and it may be derived from earlier-phase clinical investigations.
2. An integrated summary of the toxicologic effects of the drug in animals and *in vitro*. The summary presents the toxicologic findings from completed animal studies that support the safety of the proposed human investigation. The integrated summary is usually 10–20 pages long, includes text and tables, and should contain the following information:
   a. A brief description of the design of the trials and any deviations from the design in the conduct of the studies, including the dates the studies were conducted
   b. A "systematic" presentation of the findings from the animal toxicology and toxicokinetic studies. This data should be presented by organ system (cardiovascular, renal, hepatic, etc.); and if a particular body system was not assessed, it should be noted.
   c. The names and qualifications of the individuals who evaluated the animal safety data and concluded it to be reasonably safe to begin the proposed human studies
   d. A statement of where the studies were conducted and where the study records are stored and available for inspection
   e. A declaration that each nonclinical safety study reported in the IND was performed in full compliance with good laboratory practices (GLPs)[17] or if a study was not conducted in compliance with GLP, a brief statement of why it was not, and a discussion on how this might affect the interpretations of the findings.
   The integrated summary can be developed on the basis of unaudited draft toxicology reports of the completed animal studies. Final, fully quality-assured individual study reports are recommended, but not required for submission of an initial IND. If the integrated summary is based on unaudited draft reports, the toxicology reports should be finalized, and an update to the summary submitted to the FDA within

120 days after submission of the original integrated summary.[18] The updated summary, as well as the final study reports, should identify any differences found in the preparation of the final, fully quality-assured study reports, and the information submitted in the initial integrated summary. If there were no differences found, that should be stated in the update. The final reports must be available to the FDA upon request or by the 120-day time frame. In any case, the final reports are submitted with the NDA.

3. Full data tabulations for each animal toxicology study supporting the safety of the proposed trial. This should be a full tabulation of the data suitable for detailed review and consists of line listings of individual data points, including laboratory data for each animal in the trials and summary tabulations of the data points. This section will also include either a brief technical report or abstract for each study or a copy of the study protocol and amendments. These are provided to help the FDA reviewer interpret the data included in the line listings. Many sponsors will include copies of the final toxicology study reports in this section in lieu of the technical report or protocol. However, this is not required, and submission of the initial IND does not need to be delayed until final, fully quality-assured study reports are available.

> ### *GOOD LABORATORY PRACTICE*
>
> *A GLP is quality system that applies to the conduct of nonclinical safety studies used to support an IND, NDA, BLA, or other regulatory submission. GLP regulations set standards for the organization of the laboratory, facilities, personnel, and operating procedures. Clinical studies with human subjects, basic exploratory studies to determine potential utility of a compound, or tests to determine the chemical or physical characteristics of a compound are not subject to GLP regulations.*

Previous Human Experience—Section 312.23(a)(9)

This section should contain an integrated summary report of all previous human studies and experiences with the drug. When the planned study will be the first administration to humans, this section should be indicated as not applicable. However, if initial clinical investigations have been conducted in other countries before the US IND is filed, this section could be extensive. The summary should focus on presenting data from previous trials that are relevant to the safety of the proposed investigation (e.g., PK and PD data, the observed AE profile in previous studies or other experiences, and ADME data) and any information from previous trials on the drug's effectiveness for the proposed investigational use. Any published material relevant to the safety of the proposed investigation or assessment of the drug's effectiveness in the proposed indication should be provided in the IND. Other published material may be listed as a bibliography.

If the drug is marketed outside of the United States, or was previously, a list of those countries should be provided as well as a list of any countries where the drug was withdrawn from marketing because of safety or efficacy issues.

ADDITIONAL INFORMATION—SECTION 312.23(a)(10)

This section is used to present information on special topics. The following topics should be discussed, if relevant, in this section:

1. *Drug dependence and abuse potential*. If the drug is a psychotropic or otherwise shows potential for abuse, data from clinical studies or animal studies that may be relevant to assessment of the investigational drug.
2. *Radioactive drugs*. Data from animal or human studies that allow calculation of radiation-absorbed dose to the whole body and critical organs upon administration to human subjects
3. *Pediatric studies*. Any plans the sponsor has for assessing the safety and efficacy of the drug in the pediatric population
4. *Other information*. Any other relevant information that might aid in the evaluation of the proposed clinical investigations

RELEVANT INFORMATION—SECTION 312.23(a)(11)

Any information specifically requested by the FDA that is needed to review the IND application. It is common to place the meeting minutes from any pre-IND meeting or discussion in this section. This is especially useful if the information is referenced elsewhere in the IND.

OTHER IMPORTANT INFORMATION ABOUT THE SUBMISSION OF AN IND

- For clinical studies that will be submitted as part of an NDA or BLA, an IND sponsor must collect *financial disclosure* information from each investigator or subinvestigator who is directly involved in the treatment or evaluation of clinical trial subjects. Each investigator or subinvestigator must supply sufficient and accurate financial information that will allow the sponsor to eventually submit certification or disclosure statements in an NDA or BLA. Each investigator or subinvestigator must commit to update this information if any changes occur during the course of the investigation and for one year following completion of the study. Most phase 1 studies, large open safety studies conducted at multiple sites, treatment protocols, and parallel track protocols are exempted from financial disclosure requirements.[19–21]

> **IND TERM**
>
> *Financial disclosure*—*When submitting a marketing application for a drug, device, or biologic product, the applicant is required to include a list of all clinical investigators who conducted clinical studies and certify and/or disclose certain financial arrangements that include certification that no financial arrangements with an investigator have been made where study outcome could affect compensation, that the investigator has no proprietary interest in the product, that the investigator does not have significant equity interest in the sponsor, and that the investigator has not received significant payments of other sorts and/or disclosed specified financial arrangements and any steps taken to minimize the potential for bias. By collecting the financial disclosure information at the start of a study, the sponsor will be aware of potential conflicts and will be able to consult with the FDA early on and take steps to minimize the potential for bias. The thresholds for disclosure are defined in the regulation in 21 CFR Part 54.*

- Although not a required component of an IND, some FDA review divisions may ask the sponsor to submit a copy of the informed consent form for the study. This is often requested by Center for Biologics Evaluation and Research (CBER) for INDs for biologic products, especially for new technology such as gene or cellular therapy studies. However, if you include a consent template as part of the protocol submission, the FDA will likely look at it and may provide comments/suggestions regarding the document.
- Within the IND application, a sponsor may include references to other information pertinent to the IND that may have been previously submitted to the FDA, for instance, in another IND or in a marketing application. Another IND might be referenced if the sponsor is submitting a treatment use protocol that references the technical sections of an open IND for the same drug, or a sponsor might be conducting a clinical study of an approved drug but for a new indication. In this instance, the sponsor may reference the nonclinical and CMC sections of the NDA instead of submitting the same information in a new IND.
- The sponsor may also reference a drug master file (DMF) in the IND application that contains important information necessary to complete review of the IND. A DMF might contain proprietary information about a unique excipient, component, technology, or specialized drug delivery device that the owner of the information does not want to share with the sponsor of the IND. In this case, the company will submit a DMF to the FDA and allow the sponsor to reference it in the IND. Reference to any DMF or other information submitted by an entity other than the sponsor

must include a letter authorizing the sponsor to make the reference and giving the FDA permission to review the DMF in support of the IND.

> ### *DRUG MASTER FILE*
>
> *A DMF is a submission to the FDA that is used to provide confidential and detailed information about processes, or articles used in the manufacturing, processing, packaging, and storing of one or more human drugs. The information contained in the DMF may be used to support an IND, an NDA, an ANDA, another DMF, an export application, or amendments and supplements to any of these.*

- Reports or journal articles in a foreign language must be accompanied by a complete and accurate English translation.
- Each IND submission must include a four-digit serial number. The initial IND must be numbered 0000, and each subsequent submission (correspondence, amendment, safety report) must be numbered consecutively. This serial number is included on the Form 1571, any cover letter included with the submission and on any labels affixed to the binders containing submission. If the submission is electronic, this information is still included, along with the electronic sequence number for the submission.
- The FDA requires sponsors to submit the original and two copies of all IND paper submissions, including the initial IND application and any amendments, correspondence, or reports if submitted by paper. For electronic submissions, only a single electronic version is required. The FDA can request that a sponsor submit additional copies of a particular submission at any time.
- The initial IND and all subsequent submissions more than one page in length should be fully paginated, including all appendices and attachments.
- All paper IND submissions should be printed on good quality 8.5- by 11-inch paper with a 1.25-inch left margin to allow for binding. For electronic submissions, format the pages as PDF files that if printed will fit on standard paper with the same dimensions and margins. Individual volumes should be no more than approximately 2 inches thick and bound in pressboard-type binders. Three-ring binders are not used. The FDA requires the following types of binders for specific sections of IND submissions:
    One copy of the submission will serve as an archive copy and should be bound in a red polyethylene binder.
    The CMC section should be bound in a green pressboard binder.
    Microbiology information should be bound in an orange pressboard binder.
    The pharmacology/toxicology information should be bound in an orange pressboard binder.

Each volume should be labeled with permanent adhesive labels printed in permanent black ink. The labels should contain the volume number of the submission (vol. X of XX), the name of the drug, the IND number, and the sponsor's name. Specifications for the binders for a variety of FDA submissions, including INDs, can be found on the FDA Web page.[22]
- For complete traceability and adequate documentation, the initial IND application, and subsequent submissions to the IND should be sent to the FDA, using a delivery service that documents delivery (i.e., FedEx, UPS, or DHL). Many of these services also offer e-mail notification to the sender upon delivery and other customer service tools that make routine shipments easier. Sponsors should keep records of receipt for all IND submissions as documented proof of submission should questions arise.

ELECTRONIC SUBMISSIONS

Under the recent reauthorized Prescription Drug User Fee Act (PDUFA), the FDA now strongly encourages that all application submissions, including INDs be submitted to the FDA in electronic CTD format. The FDA has provided several guidances on preparing electronic submissions.[23] All of the files included in the electronic submission are to be included as PDF files. The TOC is also replaced by an XML file that links all of the documents to the appropriate section and heading. The ten sections of the IND are reorganized and placed in the appropriate module of the CTD format, as specified by the FDA (see Table 2.1).

TABLE 2.1
Detailed Mapping of IND Section to CTD Structure

Standard IND Item	CTD Module Location
Cover letter	1.2
Item 1: Form FDA 1571	1.1.1
Item 2: Table of contents	2.1
Item 3: Introductory statement	2.2 and 2.5
Item 4: General investigational plan	1.13.9
Item 5: Investigator's brochure	1.14.4.1
Item 6: Protocol(s)	5.3
Item 7: CMC information	1.12.14, 1.14.4.2, 2.3, and 3
Item 8: Pharmacology/toxicology	2.4, 2.6, and 4.2
Item 9: Previous human experience	2.5, 2.7, and 5.3
Item 10: Additional information	2.7.4 and 2, 4, or 5 (as appropriate)

FDA Review of the IND

When the initial IND submission is made to the FDA, it is logged in the Document Management Room and assigned an IND number. A sponsor can call in advance of the submission and receive the number, and this number can then be used within the submission document. Many companies commonly call ahead to receive this information. Once the IND is stamped as received, it is sent to the appropriate review division within CDER or CBER. If there is any question about which division the IND will reside, the ombudsman office is contacted. Once the IND arrives at the review division, it is critically evaluated by several reviewers of chemistry, biopharmaceutics, medical, statistics, microbiology, and pharmacology/toxicology sections, as appropriate. All these areas review the data submitted with the primary purpose of ensuring appropriate safety of the individuals who will be enrolled in the study.

Once an IND is submitted, the study cannot be initiated until a period of 30 calendar days has passed, or if the FDA has given agreement to start the study before the 30-day period expires. The usual practice is to contact the FDA shortly before the 30-day period has expired to see if there are any issues rather than going ahead at day 30 if nothing is heard from the FDA. If there are any major issues relating to the safety of the volunteers or patients in the proposed study, the FDA can institute a clinical hold (*Manual of Policies and Procedures* [MaPP] 6030.1). A clinical hold is an order issued by the FDA to the sponsor of an IND to delay or to suspend a clinical investigation. A clinical hold may be either a "complete clinical hold"—a delay or suspension of all clinical work requested under an IND—or a "partial clinical hold"—a delay or suspension of only part of the clinical work (e.g., a specific protocol or part of a protocol). If a clinical hold is imposed, the specific reasons for the clinical hold will be specified in the clinical hold letter to the sponsor of the IND. If the FDA concludes that there may be grounds for imposing a clinical hold, the Agency will attempt to discuss and satisfactorily resolve the matter before issuing a clinical hold letter. A sponsor must respond to *all* clinical hold issues before the FDA will review the responses. When the FDA receives all responses from the sponsor, it has another 30 calendar days to review and respond in writing. Under no circumstances can the study be initiated unless the FDA lifts the clinical hold. Review divisions differ in the frequency of clinical holds that are imposed.

MAINTAINING AN IND—IND AMENDMENTS AND OTHER REQUIRED REPORTS

Clinical development of a new drug will take a number of years and can take as many as 10 or 12 years, all the time requiring an active IND to conduct the necessary clinical studies. Because of the long development times, the IND is continuously updated with new information and new protocols as the drug moves from one phase of investigation to the next. The IND regulations discuss two types of amendments, protocol amendments and information amendments, and two types of required reports, safety reports and annual reports. Most other routine communication with

What Is an IND?

the FDA regarding an IND is referred to as general correspondence. It is important to remember, however, that the FDA considers any submission to the IND, an amendment and every submission must be labeled with the next sequential four-digit serial number. Even if the sponsor does not assign a submission the next serial number, the FDA will and this very often leads to confusion in future submissions. The Form 1571 cover sheet has an area for the sponsor to include the serial number and an area to designate specifically what type of submission it is they are submitting. Sponsors who maintain multiple INDs and other regulatory filings use electronic archiving systems that have powerful searching and cross-referencing capabilities. This allows for searching a database on the basis of key words or serial numbers.

In this section, we will discuss the most common types of amendments and reports to the IND, and review the required content and timing for the submissions.

THE IND SAFETY REPORT

The sponsor of an IND is responsible for continuously reviewing the safety of the investigational drug(s) under investigation. IND regulations require each sponsor to review and investigate all safety information obtained about the drug regardless of the source of the information. Safety information can come from a wide variety of sources, including the clinical studies being conducted under the IND, animal studies, other clinical studies, marketing experience, and reports in scientific journals and unpublished reports. These can be foreign or domestic sources and may be information that is not generated by the sponsor. The ongoing safety review is also a critical component of the sponsor's responsibility to keep all participating investigators updated on new observations regarding the investigational drug, especially any information regarding potential AEs.

The FDA regulations (21 CFR Section 312.32) and the "ICH E6" guidance[7] define an AE as any unfavorable and unintended sign (including an abnormal laboratory finding), symptom, or disease temporally associated with the use of the investigational product, whether or not related to the investigational product. The regulations further define a serious adverse drug reaction as any AE at any dose that

- Results in death.
- Is life threatening.
- Requires inpatient hospitalization or prolongation of an existing hospitalization.
- Results in persistent or significant disability/incapacity.
- Is a congenital anomaly/birth defect.

An AE that does not result in death, is not life threatening, or does not require hospitalization may still be considered serious, if in the opinion of the investigator, the event may have jeopardized the subject and medical intervention may be necessary to prevent one of the outcomes that define a serious AE. The final key definition related to IND safety reports is what constitutes an unexpected AE. The IND regulations define an unexpected AE as any adverse drug experience, the specificity or severity of which is not consistent with the current IB.[24] Essentially, what this means is an adverse experience is unexpected if that event was not listed in the IB as a possible side effect of the drug (not observed previously), or the event that occurred was listed in the brochure, but it occurred in a more severe way than was expected.

In 2010, the FDA finalized updated regulations for safety reporting during the IND phase of development. Importantly, the new regulation adds a definition for a suspected adverse reaction [312.32(a)] and revised requirements for the review and evaluation of AE information received.[25]

Much of the safety information obtained by the sponsor will relate to safety data that the sponsor was already aware of and included in the IB or is nonserious in nature and does not require immediate notification of the investigators or the FDA; however, all new safety information should be included in the sponsor's safety database regardless of the reporting requirements.

The IND regulations also require sponsors to notify all investigators and the FDA of certain types of safety events in an IND safety report. The IND regulations discuss two types of safety reports: a 15-day report and a more urgent 7-day report. When a reported adverse experience is considered related to the use of investigational drug and is a serious and unexpected event, the sponsor is required to notify all of the investigators in the study and the FDA within 15 calendar days of learning of the event. A 15-day safety report is submitted to the FDA on the Form FDA 3500A or in a narrative format, and foreign events can be submitted on a Council for International Organizations of Medical Sciences (CIOMS) I form. IND safety reports are sent to the reviewing division at the FDA with jurisdiction over the IND. The reports should be submitted in triplicate (one original and two copies) with a Form 1571 cover sheet and serial number. The more urgent safety report, the seven-day report, is required when any unexpected fatal or life-threatening event associated with the use of the drug occurs. The FDA must be notified by telephone or facsimile within seven calendar days of learning of a fatal or life-threatening event and followed up with a written report on Form 3500A (or CIOMS I) within 15 days of learning of the event. The telephone/facsimile report should be made to the FDA review division with jurisdiction over the IND. Other safety information that does not meet the requirements for expedited reporting should be submitted to the IND in the annual report.

IND TERM

CIOMS I Form—A standardized international reporting form used to report individual cases of serious, unexpected adverse drug reactions.

CIOMS is an international, nongovernmental, nonprofit organization established jointly by WHO and UNESCO in 1949. CIOMS has established a series of working groups that develop safety requirements for drugs and standardized guidelines for assessment and monitoring of adverse drug reactions.

The FDA interprets when the sponsor learns of the event to mean anyone in the employ of the sponsor or engaged by the sponsor's initial receipt of the information. If the sponsor's clinical research associate learns of a serious AE while visiting a site, the 15-day clock begins as soon as the associate learns of the event and not when the associate reports the event to the clinical affairs or pharmacovigilance groups. The sponsor must have strict procedures and time lines in place for employees to report potential AE.

It is important to remember that these events may not come strictly from the sponsor's ongoing clinical trials. The IND regulations require 15-day IND safety reports for adverse findings from nonclinical studies that may indicate a risk to human subjects in the ongoing clinical trials. These could be adverse findings from carcinogenicity studies, reproductive toxicology studies, or any other nonclinical studies being conducted to support clinical trials.

The sponsor must continue to investigate the adverse experience after the IND safety report is submitted. Any additional or follow-up information obtained as part of the investigation must be submitted to the FDA as soon as the new information becomes available. In practice, most sponsors will submit follow-up information to the FDA within a 15-day time frame, as with the original safety report.

Submission of an IND safety report does not mean that the sponsor or the FDA has concluded that the information being reported constitutes an admission that the drug caused or contributed to the event. In fact, the IND regulations state that a sponsor need not admit, and may deny, that the report or information submitted constitutes an admission that the drug caused or contributed to an AE.[26]

The Protocol Amendment

A protocol amendment is submitted to the FDA when a sponsor wants to initiate a new clinical study that is not described in the existing IND or when the sponsor makes changes to an existing protocol, including adding a new investigator to a trial. New protocols are submitted when clinical development of the drug advances to the next phase, for example, from phase 1 to phase 2, or when an additional study is needed during the same phase of development, for example, an additional phase 2 study to evaluate dosing or a clinical study to evaluate potential differences in pharmacokinetics or pharmacodynamics in response to changes in the formulation or route of administration of the investigational drug.

A protocol amendment for a new protocol must include a copy of the new protocol and a brief description of the most clinically significant differences between the new and previous protocols. Although not specified in the regulations, the FDA also expects phase 2 and phase 3 protocol submissions to include information on how the data will be collected (case report forms) to ensure that the study will achieve its intended scientific purposes. When submitting a new protocol to an active IND, the sponsor may initiate the study once the IRB has approved the protocol and it has been submitted to the FDA. There is no 30-day review period for the FDA, and a sponsor can initiate a study once the protocol is submitted, if IRB approval is in place. However, the FDA can still place the study on clinical hold if it believes there is a safety issue or the protocol design is insufficient to meet the stated objective. Sponsors may want to request feedback from the FDA or specifically request in the amendment that the FDA notify the sponsor if there are no objections to the proposed trial.

A protocol amendment is also required if a sponsor makes significant changes to an existing protocol. For phase 1 protocols, an amendment is required if the changes may affect the safety of the subjects participating in the study. Other modifications that do not affect the safety of the subjects can be submitted in the IND annual report and not in a protocol amendment. In the case of a phase 2 or phase 3 protocol, a protocol amendment should be submitted for any change that may affect the safety of the subjects, change the scope of the trial, or affect the scientific validity of the study.

When submitting a protocol amendment for a change to a protocol, the submission should include a description of the change, a brief discussion of the reason, and justification for the change and reference (date and serial number) to the submission that contained the protocol and other references to specific technical information in the IND or other amendments that support the proposed change.

The IND regulations allow a sponsor to immediately implement a change to a protocol if the change is intended to eliminate an immediate hazard to the clinical trial subjects. In this case, the FDA must be notified of the change by a protocol amendment as soon as possible and the IRB at each site must also be notified of the change.

A protocol amendment is required when a new investigator or subinvestigator is added to conduct the clinical trial at a new or an existing site. The investigator is the person with overall responsibility for the conduct of the clinical trial at a trial site, and a subinvestigator is any individual member of the clinical trial team designated and supervised by the investigator to perform trial-related procedures or make trial-related decisions (e.g., associates, residents, and research fellows).[27] The required information regarding the new investigators is collected on the FDA Form 1572, statement of investigator (Figure 2.2), and the sponsor must notify the FDA of new investigators and subinvestigators or changes to the submitted information by submitting Form 1572 or the required information as a protocol amendment within 30 days of the investigator being added to the study. An investigator may not participate in a study until he or she provides the sponsor with a completed and signed statement of investigator Form 1572.[28] Protocol amendments to add new investigators or to add additional information about an investigator or subinvestigator can be grouped and submitted at 30-day intervals.

All protocol amendments must be clearly labeled and identify specifically which type of protocol amendment is included, for example, "Protocol Amendment: New Protocol or Protocol Amendment: New Investigator," and as with all IND submissions, a Form 1571 cover sheet should be included with the submission. The appropriate box on the Form 1571 should be marked, indicating that the submission is a protocol amendment.

INFORMATION AMENDMENTS

Information amendments are used to submit important information to the IND that is not within the scope of a protocol amendment, annual report, or IND safety report. An information amendment may include new toxicology or pharmacology information, final study reports for completed nonclinical or other technical studies, new CMC information, notice of discontinuation of a clinical study, or any other information important to the IND. An information amendment can also include information that is specifically requested by the FDA. As with the protocol amendment, the FDA requests that information amendments be identified on the cover as an information amendment with the type of information being provided, for example, "Information Amendment: Toxicology" and as with all IND submissions, a Form 1571 cover sheet should be included. Information amendments should be submitted as needed but not more than once every 30 days, if possible.

Information typically submitted in an information amendment may also be required to support another type of amendment; for instance, a new protocol may require additional CMC information because of a change in formulation or change in manufacturing of the investigational drug. In these cases, it is not necessary to submit a separate protocol amendment and a separate information amendment with two different serial numbers. All of the protocol and CMC information can be submitted in the same amendment, but it should be clearly separated within the submission (by tabs or title pages); the submission should be labeled as containing a protocol amendment and an information amendment (Protocol Amendment: New Protocol and Information Amendment: CMC).

IND ANNUAL REPORTS

The IND regulations[29] require IND sponsors to submit an annual report that provides the FDA with a brief update on the progress of all investigations included in the IND. The regulations provide clear instruction as to the specific content and format of the annual report, so we will only briefly summarize the content here. The annual report must contain the following information:

- Individual study information—a brief summary of the status of each study in progress including the title of the study, the total number of subjects enrolled to date, the total number of subjects who completed the study, the number of subjects who dropped out for any reason, and a brief description of any study results if known.
- Summary Information—nonclinical and clinical information obtained during the previous year. This section will include a table summarizing the most frequent and most serious AEs, a listing of all IND safety reports submitted during the past year, a list of subjects who died during the investigation, including cause of death, a list of patients who dropped out of the study because of AEs, any new information about the mechanism of action, dose response or bioavailability of the drug,

a list of ongoing and completed nonclinical studies, and a list of any manufacturing changes made during the previous year.
- The general investigational plan for the coming year.
- A list of the changes along with a copy of the new brochure, if the IB was modified during the year.
- Any changes made to the protocol not reported in a protocol amendment, if there is a phase 1 protocol.
- A listing of any significant foreign marketing developments with the drug, for example, approval in another country or withdrawal or suspension of marketing approval.
- A log of any outstanding business for which the sponsor requests or expects a reply, comment, or meeting with the FDA.

As mentioned, the content of an annual report is well defined in the regulations, and sponsors should not use the annual report as a substitute for an information amendment. Final nonclinical or clinical study reports, major CMC changes, or other important PK or PD data should be submitted in an information amendment and not held until the annual report. Information of this nature must be submitted to the IND when it becomes available, which allows the FDA to review it in a timely fashion, not several months after the information first became available. The annual report should not be used to report new information, for example, new serious and unexpected AEs that could change the risk–benefit profile of the investigation, perhaps necessitating a clinical hold. The annual report is a summary of the progress of the study over the past year and provides the general investigational plan for the coming year. The annual report must be submitted to the FDA review division with jurisdiction over the IND within 60 days of the anniversary date that the IND went into effect.

OTHER TYPES OF INDs

In addition to the IND submitted by the commercial sponsor, there are investigator-sponsored INDs (Section 312.3). They usually involve a single investigator who is performing a clinical trial and serving as the sponsor as well as the investigator. The investigator usually seeks permission from a commercial sponsor to "cross-reference" manufacturing data and nonclinical pharmacology and toxicology data. Letters from the commercial supplier of the product are required to allow the FDA to review the data contained in the supplier's IND or DMF.

Additionally, there are Treatment INDs (Section 312.320), allowing for the use of investigational drugs under expanded access. These are reserved for investigational products for serious or immediately life-threatening diseases where no satisfactory alternative therapy is available. This IND would allow use in patients not in the formal clinical trials in accordance with a treatment protocol or treatment IND.[30] Special procedures apply for these INDs. The type and amount of information required depends on the number of patients to be exposed, from single patients to large populations (IND Regulations, Subpart I).

Another type of IND is the screening IND (MaPP 6030.4) or exploratory IND.[31] Generally, the FDA encourages separate INDs for different molecules and dosage forms. However, in the early phases of development, exploratory studies may be conducted on a number of closely related compounds to choose the preferred compound or formulation. These studies may be most efficiently conducted under a single IND. Its main benefit is the use of a single IND to avoid duplicative paperwork and to alert the FDA that the IND will be used to screen multiple compounds. The CMC and nonclinical pharmacology and toxicology data for each active moiety in the screening IND should be in accord with appropriate FDA guidances. These INDs only allow limited dosing with microdoses of drug to determine essential properties in humans before proceeding with the lead candidate for development. Once the lead compound is identified, the exploratory IND is closed and a full IND submitted for the drug candidate. This type of IND is not widely used but does exist as a possibility.

PROMOTION AND CHARGING FOR INVESTIGATIONAL DRUGS

PROMOTION OF INVESTIGATIONAL DRUG PRODUCTS

The determination of safety and efficacy is made by the FDA on the basis all of the information submitted in a marketing application, and a drug cannot be represented as safe or effective until the FDA has approved the product for sale. Therefore, IND regulations specifically prohibit a sponsor or investigator from promoting or commercializing an investigational drug or stating that an investigational drug is safe or effective for the indication(s) under investigation. This includes commercial distribution of the investigational drug or test marketing the drug.[32] Sponsors must be particularly aware of this prohibition when issuing press releases about ongoing or completed clinical trials. The sponsor is often eager to publicly release positive information from trials, particularly pivotal trials, but a press release cannot state that the drug is safe or effective for its intended use, no matter how positive the results of the trial may be. The FDA will consider statements such as these in a press release or other public statements, promotion of an unapproved drug. Sponsors can also run into trouble at professional meetings and trade shows. Company representatives cannot make claims about the safety or efficacy of an investigational drug either verbally or in writing or appear to be promoting an investigational drug in any way.

These prohibitions are not intended to restrict the dissemination of scientific information about the drug in scientific meetings, journals, or other lay media. The results of clinical studies can be published in peer-reviewed scientific journals, presented at medical or scientific meetings, and announced publicly in press releases. The information presented in these forums should be limited to scientific information and the actual results of a clinical study. Presenting the number of patients that met the primary efficacy measurements or other study outcomes is permissible, as long as there is no conclusion of safety and efficacy based on the reported results.

CHARGING FOR INVESTIGATIONAL DRUGS

Charging for an investigational drug product in a clinical trial conducted under an IND is prohibited unless the sponsor has submitted a written request to the FDA, seeking permission to charge for the drug and the FDA has issued a written approval.[33] In the request, the sponsor must justify why charging for the drug is necessary to initiate or continue the trial and why the cost of providing the investigational product to trial subjects should not be considered a normal part of the cost of developing the drug. Although the regulations provide this mechanism, it is rare that a sponsor will charge for an investigational drug.

The regulations do permit a sponsor to charge for an investigational drug being administered under a treatment protocol or treatment IND if certain conditions are met[34] and FDA approves the charge. If the FDA allows the sponsor to charge for the drug, the price must not be greater than the costs of handling, distribution, manufacture, and research and development of the drug. The FDA can withdraw authorization to charge for an investigational drug if it finds that any of the conditions of the authorization are no longer valid, for example, the price being charged is greater than costs associated with the drug.

MORE INFORMATION ABOUT INDs

There is a great deal of additional information available about the IND application, and much of it is now easily available via the Internet. The most complete source of information about the IND application is the FDA Website itself (www.fda.gov). The CDER and CBER Websites contain a wealth of important information about preparing, submitting, and maintaining INDs. The most important documents to be familiar with are the guidance documents (guidance for industry), but there is significantly more IND information available on the FDA Website than just the guidance documents. The FDA Website section below outlines a number of Web pages that provide significant information about INDs, how the FDA processes them, meeting with the FDA, and the drug development process in general.

The following list provides a selection of other IND resources found on the Web, in journal articles, and in books.

THE FDA WEBSITES

1. Compilation of laws enforced by the US FDA (http://www.fda.gov/RegulatoryInformation/Legislation/default.htm).
2. Title 21 CFRs (www.accessdata.fda.gov/scripts/cdrh/cfdocs/cfcfr/cfrsearch.cfm).
3. CDER guidance documents (http://www.fda.gov/Drugs/GuidanceComplianceRegulatoryInformation/Guidances/default.htm).
4. CBER guidance documents (http://www.fda.gov/BiologicsBloodVaccines/GuidanceComplianceRegulatoryInformation/default.htm).

What Is an IND? 75

5. Information for sponsor-investigators submitting INDs (http://www.fda.gov/Drugs/DevelopmentApprovalProcess/HowDrugsareDevelopedandApproved/ApprovalApplications/InvestigationalNewDrugINDApplication/ucm071098.htm).
6. Office of Drug Evaluation IV (ODE IV)—pre-IND consultation program. A program offered by the ODE IV designed to facilitate early informal communications between ODE IV and sponsors of new therapeutics for the treatment of bacterial infections, HIV, opportunistic infections, transplant rejection, and other diseases (http://www.fda.gov/Drugs/DevelopmentApprovalProcess/HowDrugsareDevelopedandApproved/ApprovalApplications/InvestigationalNewDrugINDApplication/Overview/default.htm).
7. CDER MaPPs (http://www.fda.gov/AboutFDA/CentersOffices/OfficeofMedicalProductsandTobacco/CDER/ManualofPoliciesProcedures/default.htm).
    a. MaPP 6030.1 INDs—process and review procedures
    b. MaPP 6030.2 INDs—review of informed consent documents
    c. MaPP 6030.4 INDs—screening INDs
    d. MaPP 6030.8 INDs—exception from informed consent requirements for emergency research
8. CBER Manual of Regulatory Standard Operating Procedures and Policies (SOPPs) (http://www.fda.gov/BiologicsBloodVaccines/GuidanceComplianceRegulatoryInformation/ProceduresSOPPs/default.htm).
    a. SOPP 8201 issuance of and response to clinical hold letters for INDs
9. Good clinical practice in FDA-regulated clinical trials (http://www.fda.gov/ScienceResearch/SpecialTopics/RunningClinicalTrials/default.htm).

OTHER WEBSITES

1. RegSource.com (www.regsource.com/default.html). A comprehensive site that contains a wealth of information on many topics within regulatory affairs including INDs.

REFERENCES

1. Federal Food, Drug & Cosmetic Act, Chapter II, Section 201(g)(1).
2. Code of Federal Regulations, Title 21, Section 312.2.
3. Code of Federal Regulations, Title 21, Section 312.2 (b)(5).
4. Draft Guidance for Industry: IND Exemptions for Studies of Lawfully Marketed Cancer Drug or Biologic Products. FDA, Silver Spring, MD, January 2004.
5. Center for Drug Evaluation and Research Information for Sponsor-Investigators Submitting Investigational New Drug Applications. Available at http://www.fda.gov/cder/forms/1571-1572-help.html
6. Guidance for Industry: Content and Format of Investigation New Drug (IND) Applications for Phase I Studies of Drugs, Including Well-Characterized, Therapeutic, Biotechnology Drugs. FDA, Silver Spring, MD, November 1995.

7. Guidance for Industry: ICH E6 Good Clinical Practice: Consolidated Guidance, May 1997.
8. Guidance for Industry: ICH Q7A Good Manufacturing Practice Guidance for Active Pharmaceutical Ingredients. FDA, Silver Spring, MD, August 2001.
9. Guidance for Industry: INDs for Phase 2 and Phase 3 Studies. Chemistry, Manufacturing, and Controls Information. FDA, Silver Spring, MD, May 2003.
10. Code of Federal Regulations, Title 21, Section 312.6(a).
11. Guidance for Industry: Environmental Assessment of Human Drug and Biologics Applications. FDA, Silver Spring, MD, July 1998.
12. Code of Federal Regulations, Title 21, Section 25.31(e).
13. Guidance for Industry: ICH M3(R2) Nonclinical Safety Studies for the Conduct of Human Clinical Trials for Pharmaceuticals. FDA, Silver Spring, MD, January 2010.
14. Guidance for Industry: ICH E14 Clinical Evaluation of QT/QTc Interval Prolongation and Proarrhythmic Potential for Non-Antiarrhythmic Drugs. FDA, Silver Spring, MD, October 2005.
15. Guidance for Industry: ICH S7B Safety Pharmacology Studies for Assessing the Potential for Delayed Ventricular Repolarization by Human Pharmaceuticals. FDA, Silver Spring, MD, October 2005.
16. Center for Drug Evaluation and Research Guidance Documents. Available at http://www.fda.gov/cder/guidance/index.htm
17. GLP Regulations, 21 CFR Part 58
18. Guidance for Industry Q&A: Content and Format of INDs for Phase 1 Studies of Drugs, Including Well-Characterized, Therapeutic, Biotechnology-Derived Products. FDA, Silver Spring, MD, October 2000.
19. Code of Federal Regulations, Title 21, Section 312.53(c)(4).
20. Code of Federal Regulations, Title 21, Part 54—Financial Disclosure by Clinical Investigators.
21. Draft Guidance for Industry: Financial Disclosure by Clinical Investigators. FDA, Silver Spring, MD, May 2011.
22. Center for Drug Evaluation and Research IND, NDA, ANDA, or Drug Master File Binders. Available at http://www.fda.gov/Drugs/DevelopmentApprovalProcess/FormsSubmissionRequirements/DrugMasterFilesDMFs/ucm073080.htm
23. Guidance for Industry: Providing Regulatory Submissions in Electronic Format—Human Pharmaceutical Product Applications and Related Submissions Using the eCTD Specifications. FDA, Silver Spring, MD June, 2008.
24. Code of Federal Regulations, Title 21, Section 312.32(a).
25. Guidance for Industry and Investigators: Safety Reporting Requirements for INDs and BA/BE Studies, FDA, Silver Spring, MD September, 2010.
26. Code of Federal Regulations, Title 21, Section 312.32(e).
27. Draft Guidance for Industry: Protecting the Rights, Safety, and Welfare of Study Subjects—Supervisory Responsibilities of Investigators. FDA, Silver Spring, MD, May 2007.
28. Code of Federal Regulations, Title 21, Section 312.53.
29. Code of Federal Regulations, Title 21, Section 312.33.
30. Code of Federal Regulations, Title 21, Section 312.34.
31. Guidance for Industry, Investigators, and Reviewers—Exploratory IND Studies. FDA, Silver Spring, MD, January 2006.
32. Code of Federal Regulations, Title 21, Section 312.7(a).
33. Code of Federal Regulations, Title 21, 312.7(d)(1).
34. Code of Federal Regulations, Title 21, Section 312.7(d)(2).

3 The New Drug Application

Charles Monahan and Josephine C. Babiarz

CONTENTS

Overview 78
Laws, Regulations, and Guidances 79
Development of the NDA 83
Format and Content of the NDA 86
    Module 1: Administrative and Prescribing Information 87
        Section 1.1: Forms—Application Form [21 CFR 314.50(a)].................. 88
        Section 1.1: Forms [User Fee Cover Sheet (Form FDA 3397)]............. 92
        Section 1.2: Cover Letter/Index [21 CFR 314.50(b)]............................. 92
        Section 1.3.2: Field Copy Certification [21 CFR 314.50(d)(1)(v)]......... 92
        Section 1.3.3: Debarment Certification [FD&C Act 306(k)(1)]............ 92
        Section 1.3.4: Financial Certification and Disclosure [21 CFR Part 54] .. 92
        Section 1.3.5.1: Patent Information [21 CFR 314.50(h) and 314.53] 93
        Section 1.3.5.2: Patent Certification [21 CFR 314.50(i) and 314.52] 94
        Section 1.12: Other Information [21 CFR 314.50(g)].............................. 94
        Section 1.14: Labeling [21 CFR 314.50(e)].. 94
    Module 2: Common Technical Document–Summary (21 CFR 314.50) 94
        Section 2.2: Introduction to the Summary Documents 94
        Section 2.3: Quality Overall Summary 94
        Section 2.4: Nonclinical Overview.. 95
        Section 2.5: Clinical Overview.. 95
        Section 2.6: Nonclinical Written and Tabulated Summaries............... 95
        Section 2.7: Clinical Summary.. 95
    Module 3: Quality 95
    Module 4: Nonclinical Study Reports.. 95
    Module 5: Clinical Study Reports... 95
    Mapping the Content of the NDA to the CTD Format...................... 95
Submission and Review of the NDA 97
    Step 1: Ensure Readiness for Application through Presubmission
    Activities... 97
    Step 2: Process Submission... 99
    Step 3: Plan Review of the Application 99
    Step 4: Conduct Scientific/Regulatory Review of the Application 100

>   Step 5: Take Official Action on the Application 100
>   Step 6: Provide Postaction Feedback to the Applicant 100
> Maintenance of the NDA 101
>   Postmarketing 15-Day Alert Reports 101
>   Postmarketing 15-Day Alert Reports Follow-Up 101
>   Periodic ADE Reports 101
>   NDA Annual Report 101
> Conclusion 102
> Notes 102

OVERVIEW

The submission of a new drug application (NDA) to the Food and Drug Administration (FDA) is an official request by a pharmaceutical company (applicant) to sell and market a drug in the United States. When complete, an NDA will contain thousands of pages of nonclinical, clinical, and drug chemistry information that supports the proposed labeling of the product.

Pharmaceutical companies take years developing the content of the NDA during the investigational new drug (IND) stage of the drug development process. Although the content of an NDA is defined by regulation, each NDA will be unique due to the disease or condition being treated and the characteristics of the investigational drug. Since each application will be unique, it is critical for a pharmaceutical company to obtain guidance from the FDA at each phase of drug development. This will ensure that the proper efficacy and safety data are developed to support the filing of the NDA.

The presentation of the efficacy and safety information in the NDA is critical to a successful review and subsequent approval of the application. Therefore, the applicant should ensure that the information is presented clearly and consistently throughout the NDA. The submission should be organized following the format of the Common Technical Document (CTD) and published electronically to facilitate review by the FDA. The electronic format also facilitates the submission of the NDA through the Electronic Submission Gateway at the FDA.

Once the NDA is submitted, the application will move through the review and approval process. Over the course of 8–12 months, specialized review teams at the FDA will evaluate the different technical sections of the NDA to see if the data support the proposed product label. In addition, representatives from the FDA will conduct audits of the applicant, clinical study sites, and drug manufacturing facilities to ensure the integrity of the information in the application. During this time, the applicant will work closely with the FDA review team to respond to questions, facilitate the audits, and negotiate final labeling for the drug.

When evaluating the NDA, the review team is guided by the axiom that no drug is truly safe but that the benefits to the patients outweigh the risks of using it. While the statute governing the NDA process requires that the article be "safe for

The New Drug Application

use" and "effective for use,"[1] it does not define these terms.[2] The US Supreme Court interpreted these requirements as follows[3]:

> A drug is effective if there is general recognition among experts, founded on substantial evidence, that the drug in fact produces the results claimed for it under prescribed conditions. Effectiveness does not necessarily denote capacity to cure. In the treatment of any illness, terminal or otherwise, a drug is effective if it fulfills, by objective indices, its sponsor's claims of prolonged life, improved physical condition, or reduced pain ... Few if any drugs are completely safe in the sense that they may be taken by all persons in all circumstances without risk. Thus, [the FDA] generally considers a drug safe when the expected therapeutic gain justifies the risk entailed by its use.

The FDA has not only adopted this interpretation but has also developed content requirements for an NDA that implement these principles. The required content of an NDA is outlined in the Food, Drug, and Cosmetic Act (FD&C Act) and Title 21 of the US Code of Federal Regulations (CFR). Applicants should follow these requirements to assure that their NDA provides enough information to enable the FDA reviewers to reach the following key decisions[4]:

- Whether the drug is safe and effective in its proposed use(s) and whether the benefits of the drug outweigh the risks
- Whether the drug's proposed labeling (package insert) is appropriate and what it should contain
- Whether the methods used in manufacturing the drug and the controls used to maintain the drug's quality are adequate to preserve the drug's identity, strength, quality, and purity

Based upon the outcomes of their review, the teams will make a final approval decision. If approved, the applicant may market the drug according to its approved labeling and initiate postapproval monitoring of the drug to maintain the NDA.

This chapter will provide an overview of an NDA by presenting the regulatory requirements, the development and presentation of the content, the review and approval process, and the required maintenance of the NDA.

LAWS, REGULATIONS, AND GUIDANCES

The Federal FD&C Act is a federal law (statute), enacted by Congress, granting the FDA the authority to oversee the safety and efficacy of drugs in the United States. Section 505 of the FD&C Act [21 United States Code (USC) 355] clearly establishes the requirement for and approval of an NDA prior to an applicant marketing a new drug in the United States. The law states that "No person shall introduce or deliver for introduction into interstate commerce any new drug, unless an approval of an application is effective with respect to such drug."[5]

The Act requires that the application contain[6]:

(A) full reports of investigations which have been made to show whether or not such drug is safe for use and whether such drug is effective in use; (B) a full list of the articles used as components of such drug; (C) a full statement of the composition of such drug; (D) a full description of the methods used in, and the facilities and controls used for, the manufacture, processing, and packing of such drug; (E) such samples of such drug and of the articles used as components thereof as the Secretary may require; (F) specimens of the labeling proposed to be used for such drug; and (G) any assessments required under section 505B. The applicant shall file with the application the patent number and the expiration date of any patent which claims the drug for which the applicant submitted the application or which claims a method of using such drug and with respect to which a claim of patent infringement could reasonably be asserted if a person not licensed by the owner engaged in the manufacture, use, or sale of the drug.

In order to ensure that the requirements of the FD&C Act are met, the FDA issues regulations. There are many regulations that are pertinent to this chapter; however, the regulations defining a "new drug" and the requirements regarding applications to market a new drug are critical to understanding this chapter.

The regulations in 21 CFR Part 310 titled "New Drugs" outline the scope of a new drug as any changes to a molecular entity, no matter how small, which have not been the subject of an approved NDA or "grandfathered" (those drugs sold prior to 1938). Consequently, articles that are "new" and not marketable without further testing include a new substance, even a coating, excipient, or carrier of the drug; a new combination, even of individually approved drugs or if the proportion of ingredients in the combination has changed; a new use; or a new dosage, duration, or method of administration.[7]

The regulations in 21 CFR Part 314 titled "Applications for FDA Approval to Market a New Drug" can be found using the e-CFR page of the Government Printing Office at http://www.ecfr.gov/cgi-bin/text-idx?c = ecfr&SID = a2f4bb2eab0 df8c01fa5f2ee9d55aaf0&rgn = div5&view = text&node = 21:5.0.1.1.4&idno = 21.[8] The purpose of the regulations, as outlined in Part 314.2, is to establish an efficient and thorough drug review process to (1) facilitate the approval of drugs shown to be safe and effective and (2) ensure the disapproval of drugs not shown to be safe and effective. These regulations are also intended to establish an effective system for FDA's surveillance of marketed drugs. These regulations shall be construed in light of these objectives.[9]

Of particular relevance to this chapter are the regulations in 21 CFR Part 314.50, which outline the primary content and format of the NDA. The required content will be discussed in greater detail later in this chapter and presented in the format of the CTD.

In addition to the regulations, numerous guidance documents have been established by the FDA that represents the current thinking of the Agency on the content and format of an NDA. Guidance documents can be best accessed through the FDA Website, http://www.fda.gov/RegulatoryInformation/Guidances/default.htm.

The FDA publishes these documents to provide applicants with guidance on how to be compliant with the regulations. Guidance documents are not considered law; therefore, they are not legally binding on the public or the FDA. Thus, applicants may use alternative approaches to satisfy the applicable statutes and regulations.

The FDA has published a series of guidance documents regarding the format and submission of an NDA using the electronic CTD (eCTD) format. Certain guidances, such as "Guidance for Industry Providing Regulatory Submissions in Electronic Format—Human Pharmaceutical Product Applications and Related Submissions Using the eCTD Specifications," provide general guidance on how to organize information in the application while others, such as "M4: Organization of the CTD, M4Q: The CTD—Quality; M4S—The CTD Safety; and M4E: The CTD—Efficacy," provide guidance on the information to be included in the technical sections of the application.

When preparing for the submission of an NDA, applicants should be aware of ongoing changes to laws, regulations, and guidances as they may impact their NDA submission. In 2012, The FDA Safety and Innovation Act (FDASIA) was passed. This law included the reauthorization of the Prescription Drug User Fee Act (PDUFA) which authorizes the FDA to collect fees from companies that produce certain human drug and biological products. The fees ensure that the FDA has the necessary resources to maintain a predictable and efficient review process to expedite the drug approval process. PDUFA must be reauthorized every five years. Congress originally created PDUFA in 1992 and renewed it in 1997 (PDUFA II), 2002 (PDUFA III), 2007 (PDUFA IV), and 2012 (PDUFA V).[10]

The government has established multiple performance goals with the reauthorization of PDUFA V through the year 2017 which will impact the submission of an NDA. To begin with, an applicant will continue to pay user fees to file an NDA and begin the review process. For full NDA applications requiring review of clinical data, the fee is US$1,958,800. If the application does not require review of clinical data or is a supplemental application requiring review of clinical data, the fee is US$979,400. These amounts became effective on October 1, 2012, and are updated annually.[11] There are no fees payable if the FDA refuses to file the NDA.[12]

One of the major commitments of PDUFA V is the introduction of a new review program for all new molecular entity NDAs (NME-NDAs) and original biologics license applications (BLAs). The purpose of "the Program," which will be discussed in detail later in the chapter, is to promote greater transparency and increased communication between the FDA review team and the applicant. By improving communication, the Agency hopes to increase the efficiency and effectiveness of the application's first review cycle and decrease the number of review cycles necessary for approval.[13]

Given the complexity of the applications and the need for additional time to meet with applicants, PDUFA V also increases the time allowed for the Agency to complete its review of the application. Priority applications being reviewed under the PDUFA V program will now be reviewed in eight months instead of six months, and standard applications will be reviewed in 12 months instead of 10 months.

Another goal of PDUFA V is to further improve the efficiency of the NDA review process by requiring electronic submissions and standardization of electronic drug application data. Section 745A(a) of the FD&C Act, added by Section 1136 of FDASIA (Pub. L. 112–144), requires that submissions under Section 505(b), (i), or (j) of the FD&C Act, and submissions under section 351(a) or (k) of the Public Health Service (PHS) Act, be submitted in electronic format specified by the FDA (or the Agency).[14] The FDA has issued a draft revised guidance document titled "Guidance for Industry Providing Regulatory Submissions in Electronic Format—Certain Human Pharmaceutical Product Applications and Related Submissions Using the eCTD Specifications," January 2013. Unlike other guidance documents, this guidance contains both binding and nonbinding provisions that will go into effect 24 months after the document is finalized.

As can be seen, an applicant needs to be aware of changes to laws, regulations, and guidances when preparing an NDA. Revised user fees, revised review processes, and timelines and electronic submission requirements are all examples of changes that can affect the submission of an NDA.

In addition to knowing the laws and regulations that outline the requirements of an NDA submission, applicants should be aware of the laws and regulations that provide exceptions to those listed above. The Orphan Drug Act (see Chapter 7) allows the FDA to grant special status to a drug intended to treat a rare disease or condition. To qualify for Orphan status, the drug must be intended to treat a disease or condition that affects fewer than 200,000 people in the United States each year. With so few patients, it would be difficult for the applicant to recoup the development cost. Therefore, the Act provides incentives to applicants to develop these drugs. One incentive is a waiver of the PDUFA fee.

The regulations for accelerated approval of drugs for serious or life-threatening illnesses are codified in 21 CFR 314 Subpart H. Under these regulations, the FDA may grant marketing approval for a new drug product on the basis of adequate and well-controlled clinical trials establishing that the drug product has an effect on a surrogate endpoint that is reasonably likely to predict clinical benefit.[15] Accelerated approval of an NDA allows the applicant to market its drug while conducting confirmatory studies to establish clinical benefit.

Section 112 of the FDA Modernization Act (FDAMA) of 1997 amended the FD&C Act by adding new Section 506 (21 USC 356). This section "Expediting study and approval of fast track drugs" mandates the Agency facilitate the development and expedite the review of drugs and biologics intended to treat serious or life-threatening conditions that demonstrate the potential to address unmet medical needs. *Fast track* adds to existing programs, such as accelerated approval, the possibility of a "rolling submission" for the NDA. This allows sponsors to submit sections of the NDA as they are completed, expediting the review process. An important feature of *fast track* is that it emphasizes the critical nature of close early communication between the FDA and sponsor to improve the efficiency of product development.[16]

As can be seen, the use of the different laws and regulations can have a strategic impact on the development of the NDA. Therefore, applicants should identify

The New Drug Application

which laws and regulations are applicable to their drug, review the applicable guidance documents to understand the FDAs thinking on a topic, and discuss them with the FDA early in development.

DEVELOPMENT OF THE NDA

Pharmaceutical companies take years developing the content of the NDA as part of the drug development process. This process, as shown in Figure 3.1, is not only time-consuming but is also costly. The cost of bringing a drug to market has been reported to be over US$1 billion; however, when you factor in the costs associated with drug development failures, the estimated cost in research dollars spent for every drug that is approved soars to US$4 billion.[17] Given the high costs of time and money, an applicant should not wait until the end of its pivotal studies to start thinking about the submission of an NDA. To ensure that the necessary information is available for the NDA, careful planning should begin and continue throughout all stages of development. This can be accomplished by outlining an approval pathway for the new drug in a regulatory development plan (RDP).

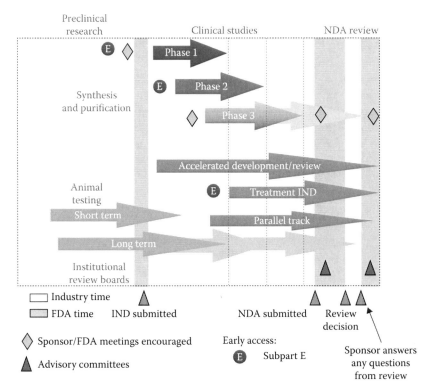

FIGURE 3.1 The drug development process.

An RDP outlines the clinical, nonclinical, and chemistry, manufacturing, and controls (CMC) activities, required at each stage of development, to support the NDA and the timeline to complete them. The RDP should also outline the time points when the applicant will meet with the FDA to discuss the ongoing development of the drug and receive feedback from the FDA regarding the plan.

The RDP is usually prepared by the regulatory affairs representative on the development team. In addition to knowing the contents of applicable laws, regulations, and guidance documents, this individual should have an understanding of the disease or condition being investigated, the affected patient population, the approved drugs available to treat patients with the disease or condition, and the basis of approval for those drugs. It is also the regulatory representative's responsibility to know the expectations of the FDA division that will be reviewing the NDA.

The development of the RDP will depend upon the type of NDA being submitted. For the purposes of this chapter, we will discuss the contents of a full or complete NDA. This type of NDA is referred to as a 505(b)(1) application. This type of NDA contains full reports of investigations of safety and effectiveness to support approval. The investigations that support this type of approval are conducted by or for the applicant. If the investigations are conducted by another party, the applicant can obtain a right of reference to use the information in support of the NDA.[18] The following are additional types of NDAs that may require different content, and thus a change to the RDP[18]:

- *505(b)(2) application.* This application is one described under Section 505(b)(2) of the FD&C Act as an application for which one or more of the investigations relied upon by the applicant for approval "were not conducted by or for the applicant and for which the applicant has not obtained a right of reference or use from the person by or for whom the investigations were conducted" [21 USC 355(b)(2)]. This provision permits the FDA to rely on a previous finding of safety and effectiveness that led to the approval of an NDA or on data not developed by the applicant such as a published literature.

    505(b)(2) applications are submitted under section 505(b) of the Act and are therefore subject to the same statutory provisions that govern 505(b)(1) applications that require, among other things, "full reports" of safety and effectiveness.
- *Abbreviated NDA (ANDA).* An ANDA is described under Section 505(j) of the Act as an application that contains information to show that the proposed product is identical in active ingredient, dosage form, strength, route of administration, labeling, quality, performance characteristics, and intended use, among other things to a previously approved application [the reference listed drug (RLD)]. ANDAs do not contain clinical studies as required in NDAs but are required to contain information establishing bioequivalence to the RLD. In general, the bioequivalence determination allows the ANDA to rely on the Agency's finding of safety and efficacy for the RLD.

The New Drug Application

- *Supplemental NDA (sNDA).* According to the FD&C Act, the term supplement means a request to the Secretary of Health and Human Services to approve a change in a human drug application which has been approved. The applicant will submit an sNDA for each new indication or claim to be added to the product label.

There are many sources of information that a regulatory person can access to develop the RDP. In addition to the laws, regulations, and guidance documents presented earlier, the FDA Website provides a vast amount of information such as approval summaries, approved product labeling, and transcripts of Advisory Committee meetings that can guide the development of the RDP. By accessing the Drugs@FDA section of the Website, the approval history and approval documents for most drugs approved by the FDA can be seen. This can provide insight into nonclinical and clinical trial designs, and endpoints to support approval of drugs with similar indications. In addition the site provides the approval letters, which may provide insight into required postmarketing studies, and approved labeling. By accessing the transcripts from Advisory Committee meetings, a regulatory professional can obtain an understanding of the FDA's concerns about medications to treat a proposed indication or the use of medical criteria to assess benefit. The reader will see the FDA's questions that are posed to a panel of experts and the responses of the experts, which the FDA usually follows.

Additional sources of information to support the development of the RDP include patient disease organizations such as The National Organization for Rare Diseases (NORDs). NORD is a unique federation of voluntary health organizations dedicated to helping people with rare "orphan" diseases and assisting the organizations that serve them. NORD is committed to the identification, treatment, and cure of rare disorders through programs of education, advocacy, research, and service.[19] This type of Website may provide insight into the concerns of patients, how the disease affects their life or health-care providers, and what are the hurdles to treating a patient. The FDA will listen to these concerns when making an approval decision about a drug.

Another valuable source of information is review articles that are written by members of the FDA review teams describing the approval of a drug. Using PubMed, a regulatory professional can search for articles summarizing the approval of a similar medication. This type of an article will provide insight into the type of studies that supported approval, the clinical benefit provided to the patients, and how the review team determined that the drug demonstrated a favorable benefit–risk profile in the proposed patient population.

Using all of the information above, a regulatory professional can develop an RDP to advise the development team about the types of nonclinical, clinical, and CMC activities required to support the NDA and the timeline to conduct those activities during development process.

Based upon the study population being treated and the results of studies, the regulatory professional may recommend that the team apply for Orphan status, Fast-Track Designation, accelerated approval, or priority review. All of these mechanisms will provide advantages to the applicant filling an NDA. For example, Orphan status would exempt the applicant from paying the PDUFA fee, Fast-Track Designation would allow for rolling review of the NDA, accelerated approval would allow marketing of the drug while confirmatory studies are being completed, and priority review if granted would require the FDA to complete the review of the NDA in eight months versus 12 months.

Of all the activities that take place during the development process, meetings between the applicant and the review division at the FDA are critical. The RDP should include timelines to meet with the FDA (see Chapter 4) at each stage of clinical development to discuss issues and ensure that the evidence necessary to support a marketing approval will be developed.

Prior to submitting an NDA, the applicant should schedule a pre-NDA meeting with the FDA. The purpose of a pre-NDA/BLA meeting is to discuss format and content of the anticipated application, including labeling and risk evaluation and mitigation strategy (REMS), if applicable, presentation of data, dataset structure, acceptability of data for submission, and the projected submission date of the application. The meeting should be held sufficiently before the planned submission of the application so that the applicant has time to incorporate the feedback received from the FDA. In general, the meeting should not occur less than two months prior to the planned submission.[20]

Given the vast number of activities that need to be completed prior to submitting an NDA, a careful planning should be an ongoing practice throughout the development process for a new drug. A detailed RDP will improve the success of the NDA filing and successful review of the application.

FORMAT AND CONTENT OF THE NDA

The presentation and organization of the NDA can be instrumental in gaining a positive approval decision from the FDA. As discussed in the previous section of this chapter, an NDA contains vast amounts of complex information and data that must be analyzed by the FDA before an approval decision can be made. By presenting the information and data in a clear and organized manner, an applicant can direct the reviewer to the required information that will support the claims contained in the proposed labeling and possibly decrease the time needed to review the application.

Until the reauthorization of PDUFA V, there is no regulation that required the submission of an NDA in any particular format. However, the FDA published numerous guidance documents regarding the format, assembly, content, and submission of the NDA using the eCTD format. Therefore, this was the expected format and media for submission of an NDA to the FDA. The format directs the reviewer where to find information, and the electronic publication facilitates review of the vast amount of information.

The CTD is an agreed-upon format for the preparation of a well-organized application that will be submitted to regulatory authorities to support the registration of pharmaceuticals for human use. This format was developed and agreed upon by the parties involved with the International Conference on Harmonization of Technical Requirements for Registration of Pharmaceuticals for Human Use (ICH). The ICH is a joint initiative involving both regulators and research-based industry representatives of the European Union, Japan, and the United States. ICH's mission is to achieve greater harmonization to ensure that safe, effective, and high-quality medicines are developed and registered in the most resource-efficient manner.[21]

The initial goal of the ICH initiative was to harmonize the technical requirements for the registration of pharmaceuticals. By establishing the CTD format, the organization of the technical requirement sections in the CTD has now been harmonized. It is important for applicants to understand that the CTD format provides guidance on the organization of the documents in the submission. The content of the CTD is determined by regulations and discussions with regional regulatory authorities. In our case, a sponsor may decide to use the CTD format, but the contents of the NDA are dictated by the regulations discussed earlier, especially 21 CFR Part 314.50.

The use of the CTD format assists applicants with the preparation of global submissions. By eliminating the need to prepare multiple region-specific submissions, applicants can save valuable resources and reduce costs. In addition, the use of the CTD format can prevent the omission of critical data or analyses that could cause the FDA to refuse to file the application. This common format also facilitates the exchange of regulatory information between regulatory authorities.

The CTD format and organization are outlined in the following four guidance documents that were issued by the FDA. "M4: Organization of the CTD"[22]; "M4E: The CTD—Efficacy"[23]; "M4Q: the CTD—Quality"[24]; and "M4S: the CTD—Safety."[25]

A diagrammatic representation of the ICH CTD is depicted in Figure 3.2.[22]

The CTD format organizes the NDA submission into five separate modules. Module 1 is not part of the CTD because it is not harmonized. The contents of this module differ because it contains region-specific information. Modules 2 through 5 are harmonized and contain the technical information required in a registration submission.

Since the reauthorization of PDUFA V will mandate the submission of an NDA as an eCTD, the regulatory content of an NDA will be presented in the format of the CTD. Readers are reminded that the content of each NDA is defined by regulation and agreements made with the reviewing division at the FDA, prior to submission.

MODULE 1: ADMINISTRATIVE AND PRESCRIBING INFORMATION

For an NDA submission in the United States, this module contains all of the administrative and labeling documents required for the submission. This includes all application forms, administrative documents, and REMS if needed. Module 1 should also contain a comprehensive table of contents (TOC) and the index for the

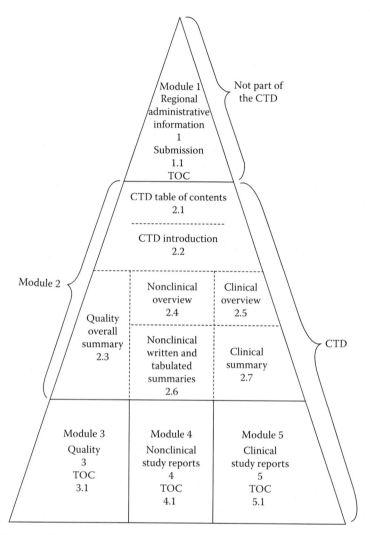

FIGURE 3.2 Diagrammatic representation of the ICH common technical document.

entire submission. The TOC should list all documents contained in the submission and their location.

Section 1.1: Forms—Application Form [21 CFR 314.50(a)]

Each applicant is required to submit a signed Form FDA 356h,[26] (see Figure 3.3). This form is published by the FDA and updated periodically. The form contains information about the sponsor, the drug, and the proposed indication, as well as a checklist of the items contained in the NDA. By signing the form, the responsible official or agent of the NDA certifies that all information in the application is true

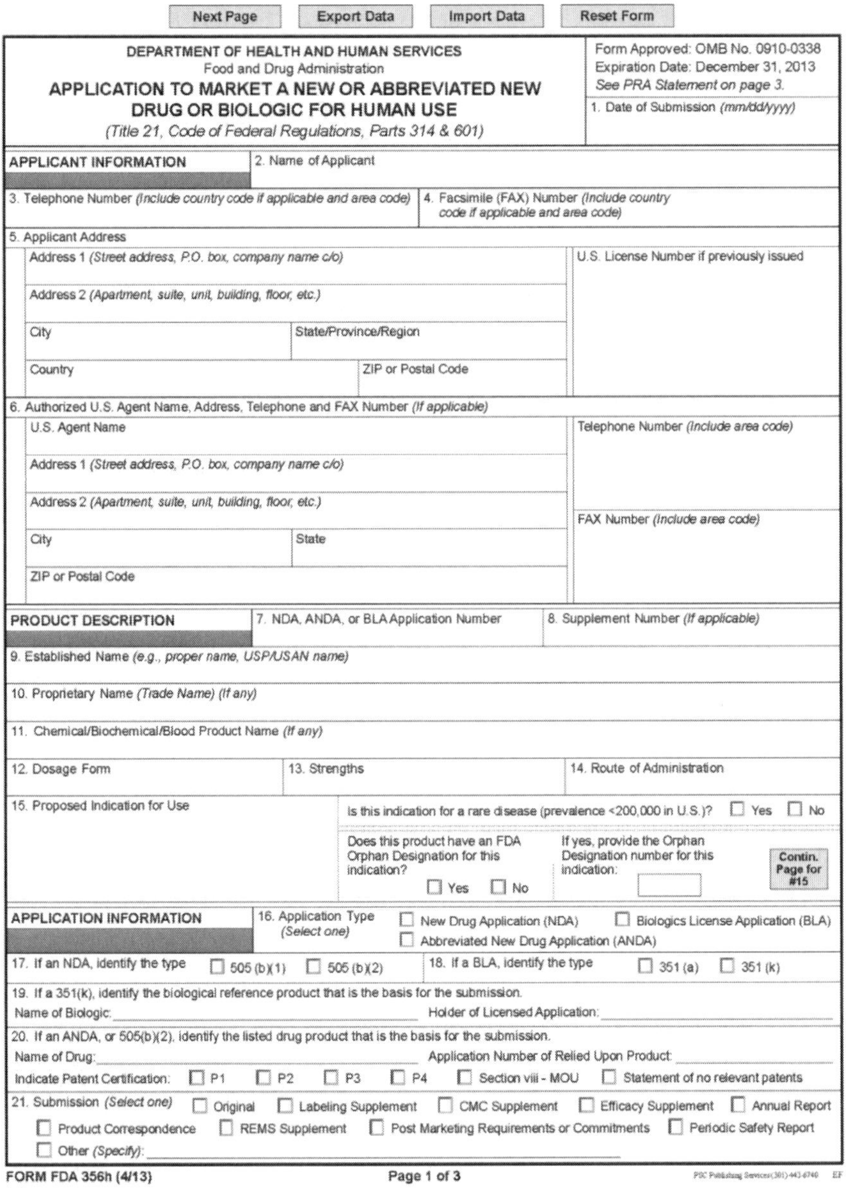

The New Drug Application

(a)

FIGURE 3.3 Form FDA 356h.

22. Submission Sub-Type	☐ Presubmission ☐ Initial Submission	☐ Amendment ☐ Resubmission	23. If a supplement, identify the appropriate category.	☐ CBE ☐ CBE-30	☐ Prior Approval (PA)
24. Does this submission contain only pediatric data?			☐ Yes ☐ No		
25. Reasons for Submission					
26. Proposed Marketing Status *(Select one)*		☐ Prescription Product (Rx)	☐ Over-The-Counter Product (OTC)		
27. This application is *(Select one)*	☐ Paper	☐ Paper and Electronic	☐ Electronic	28. Number of Volumes Submitted	

29. Establishment Information *(Full establishment information should be provided in the body of the application.)*

Provide locations of all manufacturing, packaging and control sites for drug substance and drug product (continuation sheets may be used if necessary). Include name, address, registration number (FEI), MF number, Establishment DUNS number, and manufacturing steps and/or type of testing (e.g., final dosage form, stability testing) conducted at the site. Please indicate whether the site is ready for inspection or, if not, when it will be ready.

Establishment Name		
Address 1 *(Street address, P.O. box, company name c/o)*		Registration (FEI) Number
Address 2 *(Apartment, suite, unit, building, floor, etc.)*		MF Number
City	State/Province/Region	
Country	ZIP or Postal Code	Establishment DUNS Number
Manufacturing Steps, Type of Testing, and Site Contact Information		Is the site ready for inspection? ☐ Yes ☐ No If No, when will site be ready? *(mm/dd/yyyy)*
		Continuation Page for #29

30. Cross References *(List related BLAs, INDs, NDAs, PMAs, 510(k)s, IDEs, BMFs, MAFs, and DMFs referenced in the current application.)*

Contin. Page for #30

31. This application contains the following items *(Select all that apply)*

☐ 1. Index   ☐ 2. Labeling *(Select one)*: ☐ Draft Labeling ☐ Final Printed Labeling   ☐ 3. Summary *(21 CFR 314.50 (c))*

☐ 4. Chemistry Section
  ☐ A. Chemistry, manufacturing, and controls information *(e.g., 21 CFR 314.50(d)(1); 21 CFR 601.2)*
  ☐ B. Samples *(21 CFR 314.50 (e)(1); 21 CFR 601.2 (a)) (Submit only upon FDA's request)*
  ☐ C. Methods validation package *(e.g., 21 CFR 314.50(e)(2)(i); 21 CFR 601.2)*

☐ 5. Nonclinical pharmacology and toxicology section *(e.g., 21 CFR 314.50(d)(2); 21 CFR 601.2)*   ☐ 6. Human pharmacokinetics and bioavailability section *(e.g., 21 CFR 314.50(d)(3); 21 CFR 601.2)*

☐ 7. Clinical microbiology section *(e.g., 21 CFR 314.50(d)(4))*   ☐ 8. Clinical data section *(e.g., 21 CFR 314.50(d)(5); 21 CFR 601.2)*

☐ 9. Safety update report *(e.g., 21 CFR 314.50(d)(5)(vi)(b); 21 CFR 601.2)*   ☐ 10. Statistical section *(e.g., 21 CFR 314.50(d)(6); 21 CFR 601.2)*

☐ 11. Case report tabulations *(e.g., 21 CFR 314.50(f)(1); 21 CFR 601.2)*   ☐ 12. Case report forms *(e.g., 21 CFR 314.50 (f)(2); 21 CFR 601.2)*

☐ 13. Patent information on any patent that claims the drug/biologic *(21 U.S.C. 355(b) or (c))*   ☐ 14. A patent certification with respect to any patent that claims the drug/biologic *(21 U.S.C. 355 (b)(2) or (j)(2)(A))*

☐ 15. Establishment description *(21 CFR Part 600, if applicable)*   ☐ 16. Debarment certification *(FD&C Act 306 (k)(1))*

☐ 17. Field copy certification *(21 CFR 314.50 (l)(3))*   ☐ 18. User Fee Cover Sheet *(PDUFA Form FDA 3397, GDUFA Form FDA 3794, BsUFA Form FDA 3792, or MDUFMA Form FDA 3601)*

☐ 19. Financial Disclosure Information *(21 CFR Part 54)*

☐ 20. Other *(Specify)*: _____

FORM FDA 356h (4/13)

(b)

FIGURE 3.3 (Continued) Form FDA 356h.

The New Drug Application

> [Previous Page] [Next Page]
>
> **CERTIFICATION**
> I agree to update this application with new safety information about the product that may reasonably affect the statement of contraindications, warnings, precautions, or adverse reactions in the draft labeling. I agree to submit safety update reports as provided for by regulation or as requested by FDA. If this application is approved, I agree to comply with all applicable laws and regulations that apply to approved applications, including, but not limited to, the following:
> 1. Good manufacturing practice regulations in 21 CFR Parts 210, 211 or applicable regulations, Parts 606, and/or 820.
> 2. Biological establishment standards in 21 CFR Part 600.
> 3. Labeling regulations in 21 CFR Parts 201, 606, 610, 660, and/or 809.
> 4. In the case of a prescription drug or biological product, prescription drug advertising regulations in 21 CFR Part 202.
> 5. Regulations on making changes in application in FD&C Act section 506A, 21 CFR 314.71, 314.72, 314.97, 314.99, and 601.12.
> 6. Regulations on Reports in 21 CFR 314.80, 314.81, 600.80, and 600.81.
> 7. Local, state, and Federal environmental impact laws.
>
> If this application applies to a drug product that FDA has proposed for scheduling under the Controlled Substances Act, I agree not to market the product until the Drug Enforcement Administration makes a final scheduling decision.
>
> The data and information in this submission have been reviewed and, to the best of my knowledge, are certified to be true and accurate.
>
> **Warning:** A willfully false statement is a criminal offense, U.S. Code, title 18, section 1001.
>
> | 32. Typed Name and Title of Responsible Official or Agent signing this form | | 33. Date *(mm/dd/yyyy)* |
> |---|---|---|
> | 34. Telephone Number *(Include country code if applicable and area code)* | 35. FAX Number *(Include country code if applicable and area code)* | 36. Email Address |
>
> **37. Address**
> | Address 1 *(Street address, P.O. box, company name c/o)* | |
> |---|---|
> | Address 2 *(Apartment, suite, unit, building, floor, etc.)* | |
> | City | State/Province/Region |
> | Country | ZIP or Postal Code |
>
> | 38. Signature of Applicant's Responsible Official | [Sign] | 39. Signature of Authorized U.S. Agent | [Sign] |
>
> **The information below applies only to requirements of the Paperwork Reduction Act of 1995.**
>
> The burden time for this collection of information is estimated to average 24 hours per response, including the time to review instructions, search existing data sources, gather and maintain the data needed and complete and review the collection of information. Send comments regarding this burden estimate or any other aspect of this information collection, including suggestions for reducing this burden to the address to the right:
>
> "An agency may not conduct or sponsor, and a person is not required to respond to, a collection of information unless it displays a currently valid OMB number."
>
> Department of Health and Human Services
> Food and Drug Administration
> Office of Chief Information Officer
> Paperwork Reduction Act (PRA) Staff
> PRAStaff@fda.hhs.gov
>
> **DO NOT SEND YOUR COMPLETED FORM TO THIS PRA STAFF ADDRESS.**
>
> FORM FDA 356h (4/13) Page 3 of 3
> (c)

FIGURE 3.3 (Continued) Form FDA 356h.

and accurate and, in addition, that the applicant will comply with a range of legal and regulatory requirements. If the applicant is not located in the United States, the form must name an agent with a US address.

Section 1.1: Forms [User Fee Cover Sheet (Form FDA 3397)]

A User Fee Cover Sheet is to be completed and submitted with each new drug or biologic product NDA. The form provides a cross reference to the user fee paid by the applicant.

Section 1.2: Cover Letter/Index [21 CFR 314.50(b)]

The NDA index is a comprehensive TOC that enables the reviewers to quickly find specific information in this massive document. It must show the location of every section in the archival NDA by volume and page number. It should guide reviewers to data in the technical sections, the summary, and the supporting documents.

Section 1.3.2: Field Copy Certification [21 CFR 314.50(d)(1)(v)]

The NDA must include a certification statement noting that the field copy, submitted to the local FDA office, is a true copy of the CMCs section that was submitted in the archival and review copies of the application. However, FDA district offices have access to documents submitted in electronic format. Therefore, when sending submissions in electronic format, any duplicate documentation to the FDA Office of Regulatory Affairs District Office must be provided.[14] To meet the requirements of the regulation, a letter certifying that the electronic CMC section has been submitted should be provided to the Office of Regulatory Affairs District Office and a letter certifying that the letters were submitted should be included in the NDA.

Section 1.3.3: Debarment Certification [FD&C Act 306(k)(1)]

Section 306(k)(1) of the FD&C Act requires an NDA to contain a statement certifying that the applicant did not and will not use in *any* capacity the services of any person debarred by the FDA. The certification statement should not use conditional or qualifying language, such as "to the best of my knowledge." The following wording is considered the most acceptable form of certification by the FDA[27]:

> [Name of the applicant] hereby certifies that it did not and will not use in any capacity the services of any person debarred under section 306 of the Federal Food, Drug, and Cosmetic Act in connection with this application.

Section 1.3.4: Financial Certification and Disclosure [21 CFR Part 54]

The NDA is required to contain information regarding all financial interests or arrangements between clinical investigators, their spouses and immediate family members, and the sponsor of the clinical trials that support the NDA. The applicant

The New Drug Application

should submit a Form FDA 3454 to certify which investigators had no financial interests or arrangements. A Form FDA 3455 is submitted to disclose any financial interests or arrangements with an investigator that could affect the outcome of the study and a description of steps taken to minimize the potential bias of the study results.

An investigator who had financial interests to disclose is not disqualified from the application per se; a financially incented investigator should not enroll a majority of subjects nor be the principal investigator for the larger testing sites.

Section 1.3.5.1: Patent Information [21 CFR 314.50(h) and 314.53]

The law requires patent information to be submitted with the NDA. An applicant is required to disclose all patent information that is related to the drug for which the NDA is being filed and to verify that the sponsor has all rights necessary to legally manufacture, use, and sell the drug, if the NDA is approved. The patent inquiry is a broad one and covers drug substance (active ingredient) patents, drug product (formulation and composition) patents, and method-of-use patents. In all likelihood, it should be signed only after review by a qualified attorney or patent agent who can provide an opinion as to the truth and accuracy of the completed form. The signature on the form called a "verification" reads thus:

> The undersigned declares that this is an accurate and complete submission of patent information for the NDA, amendment or supplement pending under section 505 of the Federal FD&C Act. This time-sensitive patent information is submitted pursuant to 21 CFR 314.53. I attest that I am familiar with 21 CFR 314.53 and this submission complies with the requirements of the regulation. I verify under penalty of perjury that the foregoing is true and correct.

In addition, applicants must maintain these patent statements and are required to submit updates before and after approval using Form FDA 3542(a) for each patent.[28]

We also note that there are "safe harbors" protecting a person from claims of patent infringement which apply expressly to drugs; this exemption essentially allows generic manufacturers or name-brand competitors to "jump start" the approval process by conducting required testing even though the original patent has not expired. The law reads thus:

> It shall not be an act of [patent] infringement to make, use, offer to sell or sell within the United States or import into the United States a patented invention ... solely for uses reasonably related to the development and submission of information under a federal law which regulates the manufacture, use or sale of drugs or veterinary biological products.[29]

Again, because patent infringement carries criminal penalties, consulting with a qualified patent attorney or an agent is essential.

Section 1.3.5.2: Patent Certification [21 CFR 314.50(i) and 314.52]

If the new drug is covered by a patent or patents, which the applicant(s) believe(s) to be invalid, a different procedure and format are used. Under this regulation, there is no specific form to file; there is a requirement to certify specific items, all as stated in the regulation. In this case, the patents must still be disclosed, but also, the applicant must certify under 21 CFR 314.50(i)(1)(i)(A)(4) that the patent is invalid, unenforceable, or will not be infringed, and further, the applicant is required to send a specific notice by registered or certified mail, and a return receipt requested to specified interested parties.[20] Again the purpose is to prevent the Agency from essentially wasting its time to review an application for a drug that cannot be legally manufactured.

Section 1.12: Other Information [21 CFR 314.50(g)]

The applicant can use this item to provide additional information, requested by the FDA, as needed for the NDA.

Section 1.14: Labeling [21 CFR 314.50(e)]

The labeling section must include all draft labeling that is intended for use on the product container, cartons or packages, including the proposed package insert.

The labeling requirements are very specific and detailed. Applicants must be familiar with all regulatory requirements, especially those under Sections 21 CFR 201.56(d)(1) and 201.57. The pertinent regulation at 201.56(d)(1) mandates that the labeling must contain the specific information required under section 201.57(a), (b), and (c).

Each section of the labeling must include annotations referencing the information in the summary and technical sections of the application that support the inclusion of each statement in the labeling with respect to animal pharmacology and/or animal toxicology, clinical studies, and integrated summary of safety (ISS) and integrated summary of effectiveness (ISE).

MODULE 2: COMMON TECHNICAL DOCUMENT–SUMMARY (21 CFR 314.50)

Module 2 contains a comprehensive TOC of modules 2 through 5 as well as the following overviews and summaries of the technical data in modules 3 through 5.

Section 2.2: Introduction to the Summary Documents

The introduction to the summary documents should be a one-page general introduction about the pharmaceutical product in the application. Applicants should provide information regarding the pharmacologic class, mode of action, and proposed clinical use of the drug.

Section 2.3: Quality Overall Summary

The quality overall summary should provide the reviewer with an overview of the CMC information contained in module 3. The summary should not restate the detailed CMC information contained in module 3. Instead, the summary should address key parameters of the product and discuss how the CMC information in module 3 relates to the other modules in the submission.[24]

Section 2.4: Nonclinical Overview

The nonclinical overview should provide an interpretation of the data, the clinical relevance of the findings cross-linked to the quality aspects of the pharmaceutical, and the implications of the nonclinical findings for the safe use of the pharmaceutical.[25]

Section 2.5: Clinical Overview

The clinical overview should provide a succinct discussion and interpretation of the clinical findings that support the application together with any other relevant information such as pertinent animal data or product quality issues that may have clinical implications.[30]

Section 2.6: Nonclinical Written and Tabulated Summaries

The nonclinical written and tabulated summaries should provide a comprehensive, factual synopsis of the nonclinical data.[25]

Section 2.7: Clinical Summary

The clinical summary should provide a detailed factual summarization of the clinical information in the application.[30]

MODULE 3: QUALITY

Module 3 contains a TOC for module 3 only and detailed data on CMC, including references.

MODULE 4: NONCLINICAL STUDY REPORTS

Module 4 contains a TOC for module 4 only, nonclinical study reports contained in the application, and literature references.

MODULE 5: CLINICAL STUDY REPORTS

Module 5 contains a TOC for module 5 only and a tabular listing of all clinical studies, clinical study reports, and literature references.

MAPPING THE CONTENT OF THE NDA TO THE CTD FORMAT

The required content of an NDA, as determined by US regulations, can be easily mapped to the CTD format. Table 3.1 lists the location of regulatory requirements for an NDA in relation to the CTD modules.[31]

TABLE 3.1
Regulatory Requirements and the CTD

CTD	NDA: 314.50	ANDA: 314.94 (unless otherwise indicated)	BLA: 601.2 (unless otherwise indicated)
Module 1	(a) Application form	(a)(1) Application form	(a) Application form
	(c)(2)(i) Annotated text of proposed labeling	(a)(2) Table of contents	
	(d)(1)(v) Statement of field copy	(a)(3) Basis for ANDA submission	
	(e) Samples and labeling	(a)(4) Conditions of use	(a) Labels, enclosures, and medication guides
	(h) Patent information	(a)(5) Active ingredients	306(k)(1) and (2) Debarment certification/list of convictions (FD&C Act)
	(i) Patent certification	Section 306(k)(1) and (2) debarment certification/list of convictions (FD&C Act)	
	(j) Claimed exclusivity	(a)(6) Route of administration, dosage form, and strength	
	(k) Financial certification or disclosure	(a)(8) Labeling requirements	
		(a)(12) Patent certification	
		(a)(13) Financial certifications or disclosure statement	(a) Financial certification or disclosure statement
		(d)(5) certification of field copy	(a) Claim of categorical exclusion or environmental assessment
Module 2	(b) Comprehensive table of contents	N/A	
	(c) Summaries		(a) Summaries
	(d)(5)(vii) Abuse potential		
Module 3	(d)(1) CMC	(a)(9) CMC	(a) Full description of manufacturing methods
		(a)(10) Samples	(a) Samples
Module 4	(d)(2) Nonclinical pharmtox	N/A	(a) Data from non-clinical studies
Module 5	(d)(3) Human pharmacokinetics	(a)(7) Bioequivalence/bioavailability information	

(Continued)

The New Drug Application

TABLE 3.1
(Continued) Regulatory Requirements and the CTD

CTD	NDA: 314.50	ANDA: 314.94 (unless otherwise indicated)	BLA: 601.2 (unless otherwise indicated)
	(d)(4) Microbiology	320.22(d)(2)(i) Waiver of in vivo BA/BE	
	(d)(5) Clinical data		(a) Data from clinical studies
	(d)(6) Statistical section		
	(d)(7) Pediatric use		
	(f) CRF and CRT		

Source: Reproduced from Guidance for Industry, http://www.fda.gov/downloads/drugs/guidance complianceregulatoryinformation/guidances/ucm073308.pdf
ANDA, abbreviated new drug application; BA, bioavailability; BE, bioequivalence; CMC, chemistry, manufacturing, and controls; CRF, case report form; CRT, case report tabulation; FD&C Act, Food, Drug, and Cosmetic Act.

SUBMISSION AND REVIEW OF THE NDA

As discussed earlier in the chapter, one of the commitments of PDUFA V, was the implementation of a new review program (the Program) for NME-NDAs and original BLAs to promote greater transparency and increased communication between the FDA review team and the applicant. The goals of "the Program" are to increase the efficiency and effectiveness of the first review cycle and decrease the number of review cycles necessary for approval so that patients have timely access to safe, effective, and high-quality new drugs and biologics.

The revised program is a six-step review process depicted in Figure 3.4 (titled Overview of the NDA/BLA Review Process—Major Milestones and Timelines). Two of the six steps, presubmission activities and postaction feedback to the applicant, occur outside of the review time frame that has been extended by two months. The timelines for NMEs and BLAs that fall under PDUFA V's "program" review model are 10 months for standard applications and six months for priority reviews *from the 60-day filing date* or 12 months and eight months, respectively, from the date of submission of the application.[20]

Step 1: Ensure Readiness for Application through Presubmission Activities

The first step in the new process involves multiple activities that applicants can do to improve the quality and content of their NDA/BLA application prior to its submission to the FDA.

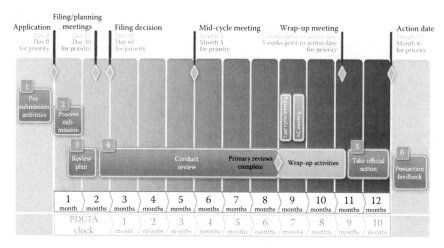

FIGURE 3.4 Overview of the NDA/BLA review process—major milestones and timelines.

During this time, the applicant should be requesting a pre-NDA meeting. As discussed earlier, the purpose of this meeting is to discuss format and content of the anticipated application, including labeling, REMS, if applicable, presentation of data, dataset structure, acceptability of data for submission, and the projected submission date of the application.

In addition to the pre-NDA meeting, applicants are also encouraged to schedule an electronic pre-submission meeting with the reviewing division to address the technical aspects of the submission. The focus of the meeting is on navigation, formatting of electronic files, and layout of the application.[20]

To prepare for the meetings, the applicant will need to submit a briefing document containing questions and supportive information for the division to review. The applicant may submit technical information such as the results of pivotal studies and proposed datasets, highlights of potential problems such as quality issues or safety signals, or a draft index of the NDA submission and ask the Agency how to address the issues in the NDA submission.

One of the goals of PDUFA V is to improve the first cycle review of applications. As such the applications are expected to be complete at the time of filing. However, during the presubmission meeting, the FDA and the applicant may reach agreement on the submission of limited application components no later than 30 days after the submission of the original application. Examples of application components that may be appropriate for delayed submission include updated stability data (e.g., 15-month data to update 12-month data submitted with the original submission) or the final audited report of a preclinical study (e.g., carcinogenicity) where the final draft report is submitted with the original application.[20] All agreements should be summarized in the minutes of meeting from the FDA.

Presubmission meetings between the applicant and the FDA, in conjunction with good FDA–industry IND interactions will help ensure that the NDA application will be complete and fileable.

The New Drug Application

STEP 2: PROCESS SUBMISSION

The review process for an NDA application begins when the NDA is submitted to the FDA. However, the PDUFA time clock starts 60 days following the submission of an NDA.

The NDA should be submitted to the FDA on physical media, or preferably through the Electronic Submission Gateway at the FDA.

The regulatory project manager (RPM) at the FDA will begin by ensuring that the PDUFA fee was paid or exempted. The PDUFA fee must be paid within 5 days of submitting the NDA, or the RPM will send the applicant an "unacceptable for filing" letter. If the fee was paid, the RPM is responsible to ensure that the NDA is administratively complete and compliant with regulatory requirements. Remember that the application must be complete unless the applicant had a previous agreement with the FDA.

If complete, the application is distributed to the discipline team leaders (DTLs) that determine if a reviewer assignment is needed. If so, a reviewer from each discipline is assigned to review the NDA and receives a copy of the NDA by day 14. The RPM is responsible to send the applicant a letter acknowledging the receipt of the NDA by day 14.

STEP 3: PLAN REVIEW OF THE APPLICATION

During the first 60 days following the submission of the NDA, the FDA has to (1) determine the fileability of the application and (2) plan the review.

The review team conducts an initial assessment of the NDA/BLA and associated labeling to identify and address any potential filing issues. The RPM will convey any potential filing issues to the applicant as soon as possible to promote resolution. Within 14 days, a tentative decision should be made if the NDA will receive priority designation. Within 45 days of receiving the application, the FDA may request an applicant orientation presentation meeting. The purpose of this meeting is to have the applicant orient the review team to the application.

During the filing meeting, day 45 of the review or day 30 for priority reviews, the review team will decide if the NDA is fileable, identify significant review issues, and determine if the NDA review will be classified as priority or standard. There are three potential filing decisions: (1) file the application, (2) potentially refuse to file the application, and (3) refuse to file the application. The decision is sent to the applicant by day 60 in a filing notification letter.

The RPM will then prepare a filing communication letter (74-day letter) that informs the applicant of deficiencies, filing review issues, and the planned timeline, including the internal mid-cycle meeting, for review activities. The letter must include the target dates for transmitting initial labeling and postmarketing requirements (PMRs) and PM commitments (PMCs) comments and final review designation (within 60 days for priority review). This letter will also include preliminary plans for an Advisory Committee meeting to discuss the application.

If the NDA is deemed fileable, the RPM will initiate a planning meeting. The purpose of the planning meeting is to organize review tasks, minimize review overlap across review disciplines, and establish an agreed-upon internal review timeline, including a schedule of team meetings and deliverables.[20] The review team will determine if any consultant reviewers are needed, establish a plan for labeling review, and finalize the need for an Advisory Committee.

The review team will also identify sites to conduct inspections on good laboratory practices (GLPs), good clinical practices (GCPs), and good manufacturing practices (GMPs). The FDA's program to inspect sites for GLPs and GCPs is called the bioresearch monitoring program (BIMO). In preparation for an inspection, the applicant can review the BIMO procedures manual and have the information ready when the FDA inspector arrives.

Step 4: Conduct Scientific/Regulatory Review of the Application

During the review phase, the reviewers from each discipline conduct their in-depth reviews. Reviewer's requests for additional information or analysis will be communicated to the applicant through the RPM.

A mid-cycle meeting is held by month 5 for standard reviews and month 3 for priority reviews. The objectives of this meeting are to present the status of the review and any key findings, confirm the decision regarding the need for an Advisory Committee, identify approvability issues, and discuss labeling and the need for an REMS. The mid-cycle meeting also gives the review team a chance to get feedback from signatory authorities and other discipline directors. The RPM will provide the applicant with an update on the status of the review within two weeks of the mid-cycle meeting.

For PDUFA V "program" reviews, a late-cycle meeting is held between the review team and the applicant. An additional two months is available for PDUFA V "program" applications to address complex review issues and attempt to remedy minor problems with the application.

Step 5: Take Official Action on the Application

Based on the signatory authority's review of the action package and on discussions with the review team, the signatory authority determines the action to be taken on the application. The final action decision is conveyed to all team members.

Step 6: Provide Postaction Feedback to the Applicant

The focus of this activity is to learn from the review experience. This optional meeting can take place as either an end-of-review conference, typically held following an action other than an approval, and/or a postaction feedback/lessons learned meeting. These two meetings can be combined into a single meeting if appropriate.

MAINTENANCE OF THE NDA

After an NDA receives approval, the applicant must conduct extensive postmarketing surveillance of the drug to monitor safety. Applicants are required to review all of the adverse drug experience (ADE) information that they receive about their drug from any source such as postmarketing clinical investigations, commercial marketing experience, postmarketing epidemiological/surveillance studies, or reports in scientific papers.

POSTMARKETING 15-DAY ALERT REPORTS

For reports of ADEs that are both serious and unexpected, the applicant must report the ADE to the FDA within 15 days of receipt of the information. The report is sent to the FDA electronically or on paper using a Form FDA 3500A, also known as a MedWatch form.

An ADE is serious if it results in any of the following outcomes: death, a life-threatening ADE, inpatient hospitalization or prolongation of existing hospitalization, a persistent or significant disability/incapacity, or a congenital anomaly/birth defect. These events can occur at any dose. An ADE is unexpected if it is not listed in the current labeling of the product.

POSTMARKETING 15-DAY ALERT REPORTS FOLLOW-UP

Applicants are required investigate the ADEs that are the subject of a 15-day report and provide follow-up information to the FDA within 15 days of receiving the new information.

PERIODIC ADE REPORTS

Following the approval of an NDA application, an applicant is required to submit a periodic ADE report. This report is submitted quarterly for the first three years after approval and then at annual intervals. This report contains all ADEs not reported as a 15-day alert report during the reporting period. Periodic reporting does not apply to reports from postmarketing studies, reports in scientific literature, or reports from foreign marketing experience.

NDA ANNUAL REPORT

Within 60 days of the anniversary of the NDA approval, the applicant is required to submit an NDA annual report. The content of this report is defined in 21 CFR 314.81. The purpose of the report is to provide the FDA with an update on any new information about the drug, the status of clinical and nonclinical development, and product labeling changes, thereby preserving the marketing approval.

CONCLUSION

The NDA is the capstone of drug development. It is a process that deserves intense scrutiny; it balances the need for drugs whose benefits outweigh the risks of their side effects. The public hopes that the Agency "gets it right"; but ultimately, the decisions are reflective of the science, the data, and the uncertainties presented by the human condition.

As new discoveries and technologies permit new therapies and approaches to curing, mitigating, and diagnosing disease and the public demand for even more rapid access to safe medications, increases, we can anticipate that the rules and regulations governing the Agency will continue to evolve.

NOTES

1. 21 USC Section 355. New Drugs.
2. FDA guidance notes that "safe" is proven through adequate scientific evidence, while "effective" is determined by substantial objective evidence. CDER's main guidance page is at http://www.fda.gov/Drugs/GuidanceComplianceRegulatoryInformation/default.htm
3. *United States v. Rutherford*, 442 US 544 (1979).
4. http://www.fda.gov/Drugs/DevelopmentApprovalProcess/HowDrugsareDevelopedandApproved/ApprovalApplications/NewDrugApplicationNDA/ucm2007029.htm
5. 21 USC Section 355(a), http://www.gpo.gov/fdsys/pkg/USCODE-2010-title21/html/USCODE-2010-title21-chap9-subchapV-partA-sec355.htm
6. 21 USC Section 355(b), http://www.gpo.gov/fdsys/pkg/USCODE-2010-title21/html/USCODE-2010-title21-chap9-subchapV-partA-sec355.htm
7. 21 CFR 310.100, http://www.ecfr.gov/cgi-bin/text-idx?c=ecfr&SID=2c9da9210aee4e2ffdf036ef06b2c940&rgn=div5&view=text&node=21:5.0.1.1.2&idno=21#21:5.0.1.1.2.2.1.1
8. 21 CFR 314, http://www.ecfr.gov/cgi-bin/text-idx?c=ecfr&SID=a2f4bb2eab0df8c01fa5f2ee9d55aaf0&rgn=div5&view=text&node=21:5.0.1.1.4&idno=21
9. 21 CFR 314.2, http://www.ecfr.gov/cgi-bin/text-idx?c=ecfr&SID=a2f4bb2eab0df8c01fa5f2ee9d55aaf0&rgn=div5&view=text&node=21:5.0.1.1.4&idno=21#21:5.0.1.1.4.1.1.2
10. PDUFA Legislation and Background; http://www.fda.gov/ForIndustry/UserFees/PrescriptionDrugUserFee/ucm144411.htm
11. *Federal Register*/Vol. 77, No. 148/Wednesday, August 1, 2012/Notices; http://www.gpo.gov/fdsys/pkg/FR-2012-08-01/pdf/2012-18711.pdf
12. Public Law No. 110-85; see chapter 1 for an overview of this legislation.
13. FDA Website, PDUFA V Fiscal years 2013–2017; http://www.fda.gov/ForIndustry/UserFees/PrescriptionDrugUserFee/ucm272170.htm
14. Guidance for industry providing regulatory submissions in electronic format—Certain human pharmaceutical product applications and related submissions using the eCTD specifications, Draft Guidance, January 2013, http://www.fda.gov/downloads/Drugs/GuidanceComplianceRegulatoryInformation/Guidances/UCM333969.pdf
15. 21 Code of federal regulations 314.510, http://www.accessdata.fda.gov/scripts/cdrh/cfdocs/cfcfr/CFRSearch.cfm?CFRPart=314&showFR=1&subpartNode=21:5.0.1.1.4.8

16. FDA Website, Fast track designation request performance, http://www.fda.gov/AboutFDA/CentersOffices/OfficeofMedicalProductsandTobacco/CBER/ucm122932.htm
17. Herper M., The truly staggering cost of inventing new drugs, Forbes Website, February 12, 2012, http://www.forbes.com/sites/matthewherper/2012/02/10/the-truly-staggering-cost-of-inventing-new-drugs/.
18. Small business assistance: Frequently asked questions on the regulatory process of over-the-counter (OTC) drugs, drug product applications, http://www.fda.gov/Drugs/DevelopmentApprovalProcess/SmallBusinessAssistance/ucm069917.htm
19. National Organization for Rare Diseases, http://rarediseases.org/about
20. CDER 21st-century review process desk reference guide, V. September 2012, p. 9, http://www.fda.gov/downloads/AboutFDA/CentersOffices/CDER/ManualofPoliciesProcedures/UCM218757.pdf
21. ICH official Website, http://www.ich.org/.
22. See M4: Organization of the CTD, http://www.fda.gov/downloads/Drugs/GuidanceComplianceRegulatoryInformation/Guidances/ucm073257.pdf
23. See M4E: The CTD—Efficacy, http://www.fda.gov/downloads/RegulatoryInformation/Guidances/UCM129865.pdf
24. See M4Q: the CTD—Quality, http://www.fda.gov/downloads/RegulatoryInformation/Guidances/UCM129904.pdf
25. See M4S: the CTD—Safety, http://www.fda.gov/RegulatoryInformation/Guidances/ucm129905.htm
26. Form FDA 356h, http://www.fda.gov/downloads/AboutFDA/ReportsManualsForms/Forms/UCM082348.pdf
27. Guidance for industry providing regulatory submissions in electronic format—Certain human pharmaceutical product applications and related submissions using the eCTD specifications, January 13, 2013.
28. Guidance for industry submitting debarment certification statements, Draft Guidance, September 1998, p. 2, http://www.fda.gov/downloads/Drugs/GuidanceComplianceRegulatoryInformation/Guidances/UCM080584.pdf
29. Form FDA 3542a, http://www.fda.gov/downloads/AboutFDA/ReportsManualsForms/Forms/UCM048352.pdf
30. 35 USC Section 271(e)(1), Infringement of Patent.
31. Guidance for Industry, submitting marketing applications, according to the ICH-CTD format—General considerations, Draft Guidance, August 2001, http://www.fda.gov/downloads/drugs/guidancecomplianceregulatoryinformation/guidances/ucm073308.pdf

4 Meetings with the FDA

Alberto Grignolo and Sally Choe

CONTENTS

Introduction	106
Types of FDA Meetings	106
Pre-IND Meetings	106
End-of-Phase II Meetings	107
Special Protocol Assessment and *Ad Hoc* Technical Meetings	107
Pre-NDA/BLA Meetings	107
Advisory Committee Meetings	107
Late Cycle Meetings	108
Labeling Meetings	108
Postmarketing Requirement/Postmarketing Commitment, and Risk Evaluation and Mitigation Strategy Meetings	108
Postaction Meetings	108
Biosimilar Biological Product Development Meetings	108
FDA Expectations	109
Preparing for FDA Meetings	111
Conduct at FDA Meetings	114
Avoiding the Pitfalls	115
Open-Ended Questions	115
Reasoned Proposals	116
Specific Meeting Objectives	116
Pre-IND Meetings	117
End-of-Phase II Meetings	117
Special Protocol Assessment Meetings	118
Pre-NDA/BLA Meetings	119
Advisory Committee Meetings	119
Late-Cycle Meetings	120
Labeling Meetings	121
PMR/PMC and REMS Meetings	121
Postaction Meetings	121
BPD Meetings	122
Conclusion	122
Reference	123

INTRODUCTION

Meetings with the Food and Drug Administration (FDA) are a critical component of the regulatory review and approval process for new prescription drugs, biologics, and medical devices. These direct exchanges between Agency personnel and company scientists provide a forum for the sharing of information that is essential to demonstrating the safety, efficacy, and quality of a product to the "FDA's satisfaction." The purpose of this chapter is to illustrate the types and objectives of various meetings with the FDA and to highlight some of the critical success factors and pitfalls associated with Agency interactions. While the main focus of the chapter is on drugs and biologics, the principles apply broadly to all meetings with the FDA.

Successful meetings with the FDA depend on three key factors: (1) good science and good medicine, (2) regulatory knowledge, and (3) sound management of the meeting process. While a pharmaceutical product's approval is ultimately determined by the strength and adequacy of its scientific data, the way a sponsor interacts with the FDA throughout the lengthy drug development and regulatory review process can spell the difference between a relatively smooth, timely approval and a costly delay or rejection of an application. A product's chances for approval can be substantially increased if the sponsor manages the meeting process in a way that presents the scientific data effectively and facilitates reaching consensus on key issues.

If handled properly, these meetings can actually reduce the approval time for a new product. A study by the Tufts Center for the Study of Drug Development indicated that companies that hold effective pre-IND and end-of-phase II meetings with the FDA achieve shorter clinical development times [1]. This is a significant finding for the highly competitive pharmaceutical industry, where time to market is a crucial success factor. By employing the right resources and the right approach—and by avoiding some common pitfalls—sponsors can take full advantage of the opportunities presented by the FDA meetings to expedite the review process and help their products reach the market more quickly.

TYPES OF FDA MEETINGS

The purpose of meeting with the FDA and its Review Divisions is to present proposals, provide answers, and resolve scientific and technical issues that arise concerning the development of a pharmaceutical product at various stages of the regulatory review process. These meetings also mark major development milestones, helping to determine if a product will be able to move forward to the next stage. The most important types of FDA meetings are described in the following sections.

Pre-IND Meetings

In these meetings, a sponsor presents characterization, manufacturing and non-clinical test data, and other information and discusses the initial plan and protocols for human clinical trials. The goal of these meetings is to receive the FDA feedback on the proposed studies and to reach agreement on what information the sponsor needs to submit in the investigational new drug (IND) application so that it is likely to be placed on active status by the FDA (rather than being placed "on hold" because of safety concerns on the part of the Agency). While the Agency (in particular, the Review Division) will not commit to not placing the IND on hold until the IND is submitted by the sponsor, it may be willing to state that at the time of the pre-IND meeting there appear to be no issues that would require a hold. With the current emphasis on global drug development in the pharmaceutical industry, many sponsors may present clinical data they have accumulated from outside of the United States at this pre-IND meeting. In such cases, notwithstanding the meeting being classified as a pre-IND meeting, the objective and content of the meeting may be similar to those of an end-of-phase II or a pre-new drug application (NDA)/biologics license application (BLA) meeting depending on the stage of the drug/biological product development.

END-OF-PHASE II MEETINGS

These are, perhaps, the most critical regulatory meetings during the development process. The sponsor is expected to provide "proof of concept" for the product through early efficacy data and other information demonstrating that the drug is performing a desired function. Equally important, phase III trial designs are discussed during these meetings, including the types of information on indications, dosing, safety, and manufacturing that the FDA would expect to see in a "strong" NDA or BLA.

SPECIAL PROTOCOL ASSESSMENT AND *AD HOC* TECHNICAL MEETINGS

These are held to discuss and resolve specific technical issues that arise during drug development, which may include detailed review of key clinical protocols, discussion of challenging manufacturing issues, or review of carcinogenicity study protocols.

PRE-NDA/BLA MEETINGS

In these meetings, a sponsor and the FDA typically discuss process-oriented issues concerning an upcoming application—how the data will be presented and how the application will be organized.

ADVISORY COMMITTEE MEETINGS

These meetings, which take place as a public forum after an NDA/BLA submission, are conducted for certain products when the FDA wants to obtain the advice of academic, medical, and other external experts about the approvability of an application. Essentially, the FDA convenes a panel of experts to hold public meetings and receives the recommendations of key opinion leaders on whether and under what conditions a marketing application might be approved by the Agency. Advisory Committees (ACs) are not empowered to approve or reject an application.

Late Cycle Meetings

These meetings are applicable only to new molecular entity (NME) NDAs and original BLAs. This meeting takes place in the late stage of the review cycle after an NDA/BLA is submitted where the FDA reviewers present to the sponsor their assessment of the application to date.

Labeling Meetings

In these meetings, negotiations take place between the FDA and the sponsor on the specific language of the product labeling, or prescribing information. These meetings are held after an NDA/BLA is submitted and constitute the final and critical stages in drug development prior to the FDA approval of a drug.

Postmarketing Requirement/Postmarketing Commitment, and Risk Evaluation and Mitigation Strategy Meetings

In postmarketing requirement/postmarketing commitment (PMR/PMC) meetings, the FDA informs the sponsor of PMRs or PMCs that it has identified during its review cycle. If the application is in need of REMS, a meeting can be arranged to discuss the risk evaluation and mitigation strategy (REMS) between the FDA and the sponsor. These meetings typically occur at a similar time as the labeling meetings.

Postaction Meetings

There are two optional meetings that can be held after the FDA takes a final action on the approvability of the NDA/BLA application: postaction feedback meeting and end-of-review conference. These meetings will be described later in this chapter.

Biosimilar Biological Product Development Meetings

These meetings are for sponsors developing biosimilar biological products. The series of meetings starts with a biosimilar initial advisory meeting during which an initial assessment is conducted to determine whether the product qualifies as a biosimilar biological product under Section 351(k) of the Public Health Service Act. After this initial assessment, there are four types of meetings that the sponsor can request throughout the different stages of biological product development (BPD) depending on the stages of the development and issues: Types 1, 2, 3, and 4.

There are some variations among the three FDA centers focused on drugs, biologics, and medical devices for human use—Center for Drug Evaluation and Research (CDER), Center for Biologics Evaluation and Research (CBER), and Center for Devices and Radiological Health (CDRH)—concerning the different types of meetings as well as differences among the divisions within each center. A guidance document "Guidance for Industry: Formal Meetings with Sponsors and Applicants for PDUFA Products" is available from the FDA that details the regulations covering these meetings. Meeting guidelines and relevant documents are also published by each of the centers (see Table 4.1).

In addition, meetings with the FDA are classified, with exception of BPD meetings, as one of three different types—Type A, Type B, or Type C—for the purpose of setting priorities and timelines for action based on their urgency. A Type A meeting is one that is immediately necessary to resolve an issue that prevents a drug development program from moving forward—a high-priority or "critical path" meeting. An example is a phase III study in which the dosage specified in the trial protocol is not effective, requiring a new study design or protocol. It is vital for the sponsor to discuss and agree with the FDA on the new proposed design. Further, if an end-of-review conference is requested within three months of complete response (CR) action, then this will also be considered a Type A meeting. Type B meetings, often called milestone meetings, are those with normal priorities and include pre-IND, end-of-phase II, and pre-NDA/BLA meetings. Type C meetings, with the lowest priority, encompass any other types of meetings. Meetings involving issues with a submitted NDA/BLA take priority over other meetings because of the performance targets established by the Prescription Drug User Fee Act (PDUFA) for the FDA to process submissions. A meeting's classification determines its scheduling: Type A meetings should occur within 30 calendar days of the FDA receipt of the meeting request; Type B, within 60 days; and Type C, within 75 days. While the sponsor makes the request for a certain meeting classification, it is the FDA that makes the final classification and determination of a meeting's priority.

FDA EXPECTATIONS

In addition to the FDA's formal regulations covering these different types of meetings, an informal "FDA meetings way" has evolved over time with common criteria and characteristics as to how the Agency generally expects its interactions with the pharmaceutical industry to be conducted in any type of meeting. Understanding and abiding by these expectations are just as important as following the formal regulations.

The most important characteristic to remember is that all the FDA meetings are *serious* and *formal*. The main order of business in every meeting is that of a discussion on science and medicine, and the orientation of the discussion is from

TABLE 4.1
How to Obtain Meeting Guidance Information from the FDA

Guidance and Relevant Documents	Website Address
Guidance for Industry: Formal Meetings between the FDA and Sponsors or Applicants	http://www.fda.gov/downloads/Drugs/GuidanceComplianceRegulatoryInformation/Guidances/UCM153222.pdf
Guidance for Industry: IND Meetings for Human Drugs and Biologics: Chemistry, Manufacturing, and Controls Information	http://www.fda.gov/downloads/Drugs/GuidanceComplianceRegulatoryInformation/Guidances/UCM070568.pdf
Disclosure of Materials Provided to Advisory Committees in Connection with Open Advisory Committee Meetings Convened by the Center for Drug Evaluation and Research Beginning on January 1, 2000	http://www.fda.gov/downloads/Drugs/GuidanceComplianceRegulatoryInformation/Guidances/UCM079679.pdf
Preparation and Public Availability of Information Given to Advisory Committee Members—Final Guidance—August 1, 2008	http://www.fda.gov/downloads/RegulatoryInformation/Guidances/UCM125650.pdf
Early Collaboration Meetings under the FDA Modernization Act (FDAMA) Final Guidance for Industry and CDRH Staff	http://www.fda.gov/MedicalDevices/DeviceRegulationandGuidance/GuidanceDocuments/ucm073604.htm
Guidance for Industry: Special Protocol Assessment	http://www.fda.gov/downloads/Drugs/GuidanceComplianceRegulatoryInformation/Guidances/UCM080571.pdf
PDUFA Reauthorization Performance Goals and Procedures Fiscal Years 2013 through 2017	http://www.fda.gov/downloads/ForIndustry/UserFees/PrescriptionDrugUserFee/UCM270412.pdf
Biosimilar Authorization Performance Goals and Procedures Fiscal Years 2013 through 2017	http://www.fda.gov/downloads/Drugs/DevelopmentApprovalProcess/HowDrugsareDevelopedandApproved/ApprovalApplications/TherapeuticBiologicApplications/Biosimilars/UCM281991.pdf

scientist to scientist. A typical FDA meeting might be compared with a scientific "summit," with chief negotiators, numerous people in attendance, a limited time frame, a very specific agenda, and minute-takers. Consistent with their scientific orientation, the FDA meetings emphasize building consensus on the basis of sound scientific data. This also means that the attendees representing the sponsor should mostly be scientists who are prepared to discuss the relevant data. Financial and product promotional discussions are seldom, if ever, appropriate at the FDA meetings.

What does the FDA expect from a sponsor during these meetings? First and foremost, the Agency expects a data-driven discussion of a product with the strong support of good science and good medicine. All meetings should be focused on scientific or medical issues that directly relate to the product and the relevant FDA regulations. Every meeting should also have a clear purpose. Sponsors must know what they want to accomplish, develop a meeting agenda that helps answer the key questions, and then adhere to that agenda. In addition, the sponsor is expected to be well prepared—to bring the right people who understand the issues involved. Sponsors must be knowledgeable about the applicable regulations and guidelines for their products as well, so that they speak the same language as the FDA. A sponsor should also be careful to schedule meetings with the FDA at appropriate times, when useful discussions are possible and when the company is genuinely seeking Agency input.

Another important characteristic of FDA meetings is that sponsors are expected to present positions for discussion, rather than ask the Agency open-ended questions about what the sponsors should do. The FDA is not in the business of developing drugs or designing sponsors' drug development plans. What the Agency will do is comment on a sponsor's plans, provide input, voice objections, and give advice on the basis of its broad experience with other sponsors and drugs (within the bounds of maintaining confidentiality on sponsor's proprietary information, of course). Instead of asking the FDA personnel to suggest a course of action, a sponsor should present the company's plans, provide full data-driven justification for those plans, and then seek the Agency's scientific input and concurrence.

PREPARING FOR FDA MEETINGS

Because preparation is essential for a successful FDA meeting, sponsors should allow plenty of time prior to any meeting to strategize, organize materials, select attendees, and rehearse key discussions. This preparation begins with scheduling the meetings. As discussed above, every meeting is classified as Type A, Type B, or Type C, and each classification carries its own timeline for scheduling and submission of documentation prior to the meeting. If a Type A meeting is requested, the FDA will expect the sponsor to provide justification for the claimed high priority and will make the final decision about the classification. It is also important to request the meeting through the proper person in the Review Division (usually the regulatory project manager or RPM) assigned to the product or sometimes a meeting coordinator) to avoid confusion or delay.

The sponsor requests the meeting by sending to the applicable FDA Review Division a "meeting request" letter in accordance with a specific FDA-recommended format:

1. Product name and application number (if applicable)
2. Chemical name and structure
3. Proposed indication(s)
4. The type of meeting being requested (i.e., Type A, Type B, or Type C)
5. A brief statement of the purpose of the meeting

6. A list of the specific objectives/outcomes expected from the meeting
7. A preliminary proposed agenda, including estimated time needed for each agenda item and designated speaker(s)
8. A draft list of specific questions, grouped by discipline
9. A list of all individuals (including titles) who will attend the proposed meeting from the sponsor's or applicant's organization and consultants
10. A list of Agency staff requested by the sponsor or applicant to participate in the proposed meeting
11. The approximate date on which supporting documentation (i.e., the information package) will be sent to the Review Division
12. Suggested dates and time (i.e., morning or afternoon) for the meeting

It is strongly recommended that the sponsor give considerable thought to all elements of the meeting request letter, and especially to the questions: item (8). The FDA will likely grant or decline the meeting primarily on the basis of the detailed nature and specificity of the questions.

Once a meeting has been granted and scheduled by the Review Division, the sponsor must submit supporting documentation at least two weeks prior to Type A meetings, and at least one month prior to Type B and Type C meetings. However, it is a good idea to contact the FDA RPM as soon as the sponsor decides that a meeting with the FDA is needed. For example, the Division of Antiviral Products (DAVPs) prefers that the sponsor submit the complete Briefing Document at the time of the meeting request for pre-IND meetings (see http://www.fda.gov/Drugs/DevelopmentApprovalProcess/HowDrugsareDevelopedandApproved/ApprovalApplications/InvestigationalNewDrugINDApplication/Overview/ucm077776.htm). In addition, with respect to DAVP, pre-IND meetings are typically scheduled as a teleconference meeting rather than a face-to-face meeting. The supporting documentation, variously called a briefing package, an information package, or a Briefing Document, is the most critical part of the pre-meeting preparations, because it sets the agenda for the meeting and defines the issues to be discussed. For a successful meeting, it is essential for the sponsor to provide a strong, focused Briefing Document that clearly states the purpose of the meeting and the issues upon which the sponsor seeks consensus. The documents must also provide sufficient background information on the drug (including chemistry, manufacturing, nonclinical and clinical summaries, and data tables) to orient the FDA attendees to such issues.

Once again, the FDA recommends a specific structure for the Briefing Document:

1. Product name and application number (if applicable)
2. Chemical name and structure
3. Proposed indication(s)
4. Dosage form, route of administration, and dosing regimen (frequency and duration)
5. A brief statement of the purpose of the meeting

6. A list of the specific objectives/outcomes expected from the meeting
7. A proposed agenda, including estimated time needed for each agenda item and designated speaker(s)
8. A list of specific questions grouped by discipline
9. Clinical data summary (as appropriate)
10. Preclinical data summary (as appropriate)
11. Chemistry, manufacturing, and controls (CMC) information (as appropriate)

In recent years, the Briefing Document has completely replaced the sponsor's introductory presentations at meetings with the FDA. Meetings now begin with an immediate discussion of the issues raised in the Briefing Document, which the FDA personnel have reviewed prior to the meeting. In that context, a sponsor presentation of the same information is superfluous, and a poor use of the limited time is made available by the FDA for the meeting (usually one hour).

Many review divisions provide the sponsor with detailed written preliminary responses to all the questions in the Briefing Document within 24–48 hours (and sometimes even as early as seven days) prior to the meeting. These responses are often extremely thorough and informative. The clear benefits to both the sponsor and the FDA are that these "preliminary responses" remove the guesswork surrounding the Agency's opinion of the questions and issues, allow the sponsor to prepare counterresponses, and enable the face-to-face meetings with the FDA to be more relaxed, productive, and mutually satisfying. In effect, the availability of the "preliminary responses" has dramatically increased the efficiency of the interaction and provided for far better substance and style of sponsor-FDA communication.

When planning a meeting with the FDA, the sponsor will be faced with the important decision of selecting the right people to attend the meeting. This decision can present significant internal challenges for the sponsor when dealing with corporate politics, organizational issues, and individual egos. However, the selection criteria should always be focused on choosing those who can contribute to the *scientific and technical* discussions, because that is what matters in the end. Depending on the stage of product development, a sponsor might draw on internal (or external consultant) expertise in areas such as pharmacology, toxicology, pharmacokinetics, chemistry, manufacturing, clinical development, and biostatistics, as well as regulatory affairs.

While marketing personnel are always interested in the timelines for drug approval, they should be "silent partners" at most FDA meetings (if they attend at all) except during the negotiation of the final product labeling. Because the sponsor's marketing and promotional activities will be directly affected by the FDA-approved language of the product labeling, it is appropriate for marketing personnel to participate in the labeling negotiation process.

In general, company lawyers and chief executive officers (CEOs) should not attend typical FDA meetings unless there are legal issues to be discussed (which would be unusual at scientific meetings with the Agency) or unless the CEO is also the sponsor's chief scientist, with intimate knowledge of the science behind the drug. Expert consultants can play a role if they can help a sponsor articulate particular scientific or regulatory positions.

In preparing for an FDA meeting, it is also important to recognize the decision-making authority of the people who will be attending for the Agency, so that the issues being debated are commensurate with the authority of the attendees. For example, technical commitments can only be made by a therapeutic area division director or higher, not by the RPM, who is the sponsor's usual day-to-day contact. Drug approval decisions can only be made by division directors and office directors. Policy decisions can only come from a Center Director or the FDA Commissioner's office. It is not appropriate for the sponsor to discuss high-level FDA policy (e.g., Why do INDs exist?) with a Division Director.

Rehearsals make up the final ingredient in good meeting preparation. A team leader should be appointed to coordinate the company's responses during the meeting with the FDA. The role of each team member at the meeting should be discussed and decided in advance, and all attendees should practice what they are going to say—although formal presentations are not typically made. Emphasis should be placed on keeping all attendees focused on the crucial issues to be discussed at the meeting and the outcomes desired by the sponsor. It is often useful to ask the regulatory affairs professional on the team to "role play" the FDA during rehearsals—asking tough questions and challenging the sponsor's positions to help the team members think through their answers carefully and thoroughly.

Now that many Review Divisions provide "preliminary responses" to the Briefing Document, sponsor rehearsals do not focus on anticipating possible FDA comments, but are instead devoted to preparing counterproposals to the known FDA objections, if any. Also, it is quite customary to focus only on the issues where the FDA and the sponsor disagree, leaving the agreed issues off the table. Finally, it is not uncommon for sponsors to actually cancel the meeting with the FDA if the "preliminary responses" received as late as the day before the meeting reflect FDA concurrence with the sponsor positions. Why waste precious time for a meeting?

CONDUCT AT FDA MEETINGS

How should attendees conduct themselves during an FDA meeting? The most important thing to remember is to *listen*. Introductory remarks should be brief and confined to introducing the sponsor team and stating briefly the purpose of the meeting from the sponsor's point of view. In addition, the sponsor's team should not plan to make a formal presentation to convey the company's case—although it is always a good idea to have backup material (e.g., in the form of hard copies of a relevant presentation) ready to present in case questions arise. A properly prepared Briefing Document will present the company's case in advance and spell out the issues to be discussed during the meeting.

In fact, most FDA meetings now begin with the Agency reminding the sponsor that "preliminary responses" were provided and that it is up to the sponsor to "run the meeting" and drive the agenda. Sponsors then typically discuss the issues where there is disagreement with the FDA, and the meeting focuses on those issues primarily or exclusively. Attendees should listen carefully to what the FDA reviewers say, take extensive notes, and, most importantly, *should not interrupt*. Once the discussions begin, let the sponsor team leader orchestrate the team's responses to FDA questions and statements and stay focused on the agenda and objectives of the meeting. It is essential that the sponsor's team avoid being aggressive, arrogant, condescending, or confrontational. Keep in mind that the goal of every FDA meeting—both for the sponsor and for the Agency—is to seek consensus and resolve all issues professionally and scientifically so that the drug development effort can proceed. At the end of the meeting, the sponsor should have a clear understanding about any decisions that have been made, as well as any actions that need to be taken—and by whom. If there are action items to be addressed after the meeting, the sponsor should follow up promptly with the FDA.

According to its own guidelines, the Agency is expected to provide the official minutes of the meeting within four weeks. Delays are common, but the Agency is trying to improve its performance in this regard. A sponsor can request changes to the minutes but should not expect to make wholesale alterations. The sponsor can also provide the company's own minutes of the meeting if the FDA Review Division is willing to accept them, and they should be delivered to the FDA within two to three days after the meeting to maximize the possibility that the sponsor's input will be considered in the FDA's minutes. It must be remembered that the FDA will consider its own minutes to be the only official record of the meeting.

AVOIDING THE PITFALLS

By understanding the FDA's expectations and following the above guidelines for a successful meeting, most sponsors should be able to avoid the common pitfalls that can slow the regulatory approval process and delay a product's progress toward the market. But because these mistakes continue to occur regularly, it is worthwhile to reiterate some of the more frequent slips that sponsors make during their encounters with the Agency.

One of the most common errors is to present the Agency with open-ended questions rather than reasoned proposals based on science. Following are some examples that illustrate the difference between the two.

OPEN-ENDED QUESTIONS

1. The phase II trials showed that several different dosages were effective for this condition. What would you recommend as the dosage for the phase III trials?
2. How many patients should be included in the phase III trials?
3. This drug has shown efficacy against several diseases. Which one should be selected for development first?

REASONED PROPOSALS

1. Several dosages were tried, and the 5- and 10-mg doses seem to be the most promising for the phase III trials (as shown in the Briefing Document). Do you agree?
2. A statistical power calculation shows that a phase III study with 1000 patients will provide valid results. Do you agree that 1000 will be sufficient?
3. This drug has shown efficacy against several diseases. Condition X has been chosen for the first phase III studies because there is no therapeutic alternative and enrollment can be completed rapidly. Do you concur?

Remember that it is not the FDA's role to make scientific, marketing, or drug development decisions for sponsors, but to provide insight and guidance on the basis of the regulations and the Agency's expertise.

Here are some other important "Dos" and "Don'ts" for FDA meetings.

Do	Don't
Be prepared	Waste time
Be polite	Be aggressive or rude
Reach consensus	Argue or be confrontational
Meet at the appropriate time	Meet when discussion is not useful
Discuss key product issues	Socialize or make a sales pitch
Focus on the agenda	Bring up side issues or complaints
Bring scientists and technical experts	Bring lawyers and CEOs
Present strong data	Try to rely on charm or hype
Be open and truthful	Lie or stonewall
Be clear	Obfuscate
Know key contacts	Go "blind" into the meeting
Rely on the data	Rely on political clout
Be reliable	Fail to follow through on commitments

Avoiding these meeting pitfalls can spell the difference between a successful, productive relationship with the FDA and a contentious relationship that slows the regulatory process for everyone.

SPECIFIC MEETING OBJECTIVES

In addition to understanding the characteristics and approaches that are common to all the FDA meetings, it is worthwhile to note the specific purposes and objectives of the major FDA meeting categories mentioned earlier in this chapter. It is also important to keep in mind that, while most are not mandatory, these meetings play a significant role in the successful development of any new drug.

Pre-IND Meetings

The pre-IND meeting has several important purposes—all of which are designed to prepare the FDA for the submission of the IND application for a new drug/biologic. The most important objective of these meetings is to introduce the new drug/biologic to the FDA, including the presentation and discussion of the entity's characterization, manufacturing process, and other nonclinical data collected in the laboratory. If the sponsor is a small company or does not have much prior experience working with the FDA, the pre-IND meeting presents an opportunity for the sponsor to discuss its background and qualifications and learn the FDA's way of thinking.

At this meeting, the sponsor will typically present the overall clinical investigational plan for the drug and relate that plan to the targeted labeling or prescribing information or even the Target Product Profile (TPP; http://www.fda.gov/downloads/Drugs/GuidanceComplianceRegulatoryInformation/Guidances/UCM080593.pdf).

The initial IND-opening clinical protocol might also be discussed, and there could be an agreement on some of the details of the protocol. If the sponsor is aware of any critical scientific or technical issues concerning the test product (e.g., nonclinical safety data showing slight liver enzyme elevations in an animal species), they would be introduced—and sometimes even resolved—in this meeting. The ultimate goal of the pre-IND meeting is for the sponsor to be aware of FDA's expectations and initial preliminary review issues regarding the test product so that an IND can be submitted successfully. It should be noted, however, that a successful pre-IND meeting does not guarantee that the FDA will activate the IND application after it is reviewed in detail.

With the current emphasis on global development, many sponsors initiate clinical development outside the United States, which would not require an active IND with the FDA. Therefore, in those cases, sponsors already have clinical data from the first-in-human trial, completed phase I trials, or even part of phase II trials at the time of pre-IND meeting. In this case, the pre-IND meeting may play the role of an end-of-phase II or pre-NDA/BLA meeting depending on the stage of the test product development.

End-of-Phase II Meetings

Sponsors should *always* have an end-of-phase II meeting before beginning phase III clinical trials. The end-of-phase II meeting is an indispensable step in the drug development process. With the pivotal importance—and significant cost—of phase III trials for new drugs, the end-of-phase II meeting is a vital opportunity to obtain the FDA's comments on phase III study designs and key trial end points. This meeting also gives the sponsor a chance to solicit input from the FDA on the final development plan, which can help "fine-tune" the approaches for CMC, toxicology, and other key data, as well as help shape the anticipated labeling language and claims.

When should an end-of-phase II meeting be held? It should be scheduled once the phase II trials have produced the key data needed to support expanded trials. This means that an effective dose and dose–response relationship has been established, and the understanding of the pharmacokinetic/pharmacodynamic relationship of the test product is well advanced. In particular, having a good understanding of the potential effective dose(s) to be tested in phase III trials is important because the FDA Review Division can deny the request for an end-of-phase II meeting if this is not properly presented in the meeting request or Briefing Document. It also means that the earlier trials have produced the information needed to solidify the proposed labeling and that the design for the phase III trials is essentially complete. As the name implies, an end-of-phase II meeting should be held *before* the sponsor has made a commitment to the significant financial investment required for phase III trials.

The Briefing Document for these meetings must be thorough and informative in order to solicit the most helpful feedback from the FDA, with detailed discussions of pertinent clinical and nonclinical data. Sponsors can request two separate end-of-phase II meetings, one specific for CMC information, and the other for rest of the information on the test product. With the given meeting time of one to two hours, often the amount of data presented at this stage can be too great. So the FDA recommends having two separate meetings to give the sponsor opportunities to discuss their issues more thoroughly. The best way for a sponsor to ensure a successful end-of-phase II meeting—in addition to having strong scientific data—is to present all of the relevant information about the drug openly and completely. Sponsors should state their positions about the compound and the trials clearly and present a strong, well-designed phase III development plan. There should be no attempt to hide any shortcomings of the early clinical data or to postpone difficult decisions. Any issues or problems will be even more difficult—and costly—if they are brought to the surface later in the development process. The sponsor's credibility can also be significantly damaged. Being forthright and working together with the FDA in a spirit of teamwork to resolve any issues will greatly increase the likelihood that this vital part of the regulatory process will reach a satisfactory conclusion. A fundamental requirement of end-of-phase II meeting is to lay a firm foundation for phase III. It is not uncommon for the FDA to object that a sponsor has not provided an adequate justification for the phase III study dose(s) and that additional phase II studies are therefore necessary. Sponsors are forewarned to expect this type of objection and to be prepared to address it with relevant and persuasive data.

Special Protocol Assessment Meetings

Sponsors can submit their study protocols under the Special Protocol Assessment (SPA; http://www.fda.gov/downloads/Drugs/GuidanceComplianceRegulatoryInformation/Guidances/UCM080571.pdf) process in connection with three specific aspects of the drug development process: carcinogenicity studies, stability studies, and phase III trials that will support an efficacy claim. The FDA provides this additional review opportunity because regulators understand that these types of studies are costly and time consuming. FDA communicates its review findings on protocols along with agreement or disagreement via a letter to the sponsor. If either the FDA or the sponsor feels that an SPA meeting can facilitate the FDA's protocol assessment or reaching an agreement on the content of the protocols, an SPA meeting is recommended. Once granted, this meeting will be classified as a Type A meeting.

Sponsors do not always submit their phase III studies to the SPA, which takes 45 days or more. Some are not willing to delay the start of phase III; others are skeptical that the SPA agreement is truly binding. However, careful consideration should be given to the advantage of the FDA's in-depth and documented review of a study protocol, which is a benefit to the sponsor and could pave the way for a readier acceptance of the resulting data in the NDA.

Pre-NDA/BLA Meetings

Before submitting an NDA or a BLA, sponsors should *always* schedule a pre-NDA/BLA meeting with the FDA. These meetings will uncover any unresolved issues that might delay the review of the submission, orient the reviewers to the content and format of the NDA, and help sponsors understand key FDA expectations about the NDA contents—such as identifying critical studies and discussing proposed analyses.

From the FDA's point of view, the pre-NDA/BLA meeting provides an important opportunity to review the NDA/BLA plan and understand its content, which will facilitate the Agency's processing of the document. The FDA will want to review any issues that were raised at the end-of-phase II meeting to ensure that they have been addressed. The actual submission process will also be discussed, including its timing, format (electronic vs. paper, the organization of tables, etc.), and, increasingly, agreement on the Common Technical Document (CTD) or e-CTD format of the NDA/BLA. A successful pre-NDA/BLA meeting will produce a consensus that makes it likely that the FDA will accept the NDA/BLA for review if the agreements reached at the meeting have been satisfied.

Advisory Committee Meetings

In some cases, the FDA may want to obtain outside expert opinions about an NDA/BLA and the approvability of a new drug. In such circumstances, the Agency has the authority to convene an official AC to review the NDA and hold public meetings about the approval of the product for sale. For NME-NDAs and original BLAs, the AC meeting should generally take place no later than three months for standard reviews or no later than two months for priority reviews prior to the PDUFA goal date. In addition, for these applications, the 74-day filing letter after submission of NDA/BLA should indicate preliminary plans on whether an AC meeting will be held for that NDA. The FDA maintains a number of standing ACs, each with a specific therapeutic focus (for the list of standing ACs, visit www.fda.gov/oc/advisory/default.htm). These AC meetings are unique to the FDA (compared with its counterpart agencies in other countries) and also uniquely stressful for the sponsor—primarily because they are open to the public, including competitors, financial analysts, the media, patients, patient advocates, and other consumers. Regulations require that the sponsor's presentation materials to the AC be made available to the public no later than one day before the meeting. At these meetings, the sponsor and the FDA have the opportunity to present key findings about the safety and efficacy of the product to the AC. The AC members offer their own views; discuss the benefits and risk of the drug; and, at the end of the meeting, take a vote on whether to recommend it for the FDA approval. The FDA is not obligated to follow the recommendations of its ACs, but it usually does.

AC meetings are recorded on audio and videotape, transcribed, and broadcast on the Web. This unusual public forum is particularly risky for the sponsor because years of development and investment are at stake. Extensive preparation by the sponsor is essential to ensure that the company's position is presented thoroughly, concisely, and professionally. Many sponsors utilize both in-house and external consultant resources and prepare hundreds or even thousands of backup slides that can be used to respond to detailed questions by AC members. It is not uncommon for sponsors to hold 6–10 rehearsals in the weeks leading up to an AC presentation. The main goal of the sponsor is to present the "case for approval" by demonstrating a favorable benefit–risk profile of the drug on the basis of clinical and nonclinical data. AC meetings have been convened by the FDA for other purposes as well—such as discussion of draft therapeutic drug guidelines, Rx-to-OTC switches, or assessment of drug safety in an era where this has become a "hot topic" in the public arena. AC deliberations and votes receive coverage in the business and lay media and are commonly regarded as directly relevant to the public health.

Late-Cycle Meetings

These meetings are unique to NME-NDAs and original BLAs. They provide the FDA an opportunity to proactively communicate its interim assessment of the application to the sponsors in the hope of enhancing the approvability rate during the first cycle of the review. The meeting does not need a meeting request and the RPM should notify the sponsor on the general timeline of the meeting date at the time of 74-day filing communication. For applications that will be discussed at an AC meeting, the late-cycle meeting will occur not less than 12 calendar days before the date of the AC meeting; for those applications that will not be discussed with an AC meeting, the late-cycle meeting will generally occur not later than three months (standard review) or two months (priority review) prior to the PDUFA goal date. The Briefing Document will be prepared by the FDA Review Division, and it should be sent to the sponsor not less than eight days before the meeting if an AC is to be held and 12 days before the meeting if no AC meeting is planned. The content of this late-cycle meeting can include any aspects of review status of the application, especially focusing on major deficiencies identified to date. This is a great opportunity for the sponsor to obtain preliminary understanding of the status of its application review process and also offer additional information to clarify some of the reviewers' concerns about the application. It is likely that this can be arranged via teleconference; the meeting minutes will be generated by the RPM at the FDA.

Labeling Meetings

Labeling meetings are one of the final links in the long chain of drug development. They occur at the end of the NDA review process, when the FDA and the sponsor meet to negotiate the formal language that describes to physicians what specific indications a product has been approved for, the recommended dosages, the side effects, and other specific information that physicians and patients need to know about a new prescription drug. This prescribing information is known as the product labeling.

All the effort that goes into developing a new product begins with the goal to achieve a certain target labeling, because it is this prescribing information that determines how the product will be used and, ultimately, how successful it will be on the market. This approach is commonly known as "beginning with the end in mind," and it helps sponsors focus on specific, achievable objectives for a drug/biologics at an early stage of the development process.

With so much riding on the outcome, labeling meetings can sometimes involve very difficult negotiations to reach agreement on the final language. Several rounds of meetings may be required, and extensive internal consultations within the sponsor organization (e.g., with the marketing department) occur. It is increasingly common to hold labeling meetings via teleconference; this enables both the Agency and the sponsor to put the conversation on "mute" and work out their respective positions in private before resuming negotiations. While this removes the advantage of observing each other's body language, it usually accelerates the negotiation process. The importance of the outcome makes it even more vital to maintain a spirit of cooperation and consensus during this process. The fundamental goal of both the FDA and the sponsor is to bring a useful new medicine to the market; finding labeling language that satisfies both parties benefits everyone. Once the final language has been approved, the product can be launched.

PMR/PMC and REMS Meetings

These meetings take place in a similar time frame as the labeling meeting/negotiation. Preliminary discussions on PMR/PMC and REMS often occur between the FDA and the sponsor prior to NDA/BLA submissions. During the review cycle, the FDA will communicate to sponsors their further assessment on PMR/PMC and REMS if they are needed in the context of NDA/BLA approval. These meetings might not be as iterative as labeling meetings; but if these are needed prior to approval action, the meetings will be scheduled as necessary.

Postaction Meetings

Postaction feedback meeting is where the FDA and the sponsor can discuss and review the quality of the application and the procedures during the review cycle so that future applications by the sponsor can be enhanced from the lessons learned during the FDA's review of the current application. These meetings are offered by the FDA. Alternatively, an end-of-review conference can be requested by the sponsor if the sponsor receives a CR (Complete Response) action on its application. The purpose of an end-of-review conference is to discuss the deficiencies that the reviewers at the FDA have found and to help the sponsor address those deficiencies for the resubmission of the product for approval. In many cases, the objectives of both postaction feedback meeting and end-of-review conference can be concurrently achieved during one meeting.

BPD Meetings

In developing biosimilar biological products, sponsors have five different types of meetings that can be requested of the FDA. The Biosimilar Initial Advisory Meeting, which can be scheduled within 90 days of request, is a meeting where an initial assessment is conducted to determine whether the product qualifies as a biosimilar product under Section 351(k) of the Public Health Service Act. In this meeting, the general development program can be discussed. Subsequent meetings can be categorized as Type 1, 2, 3, and 4 depending on the stages of the development and issues involved with the product. A Type 1 meeting, scheduled within 30 days of the meeting request, is an immediately needed meeting to resolve those issues necessary to move forward an otherwise stalled development program to its next development stage. Examples include clinical hold meetings, dispute resolution meetings, and SPA meetings. A Type 2 meeting, scheduled within 75 days of the meeting request, is a meeting to discuss a specific set of issues or questions with the FDA on a biological product development program where summary data, and not the full study reports, are reviewed. A Type 3 meeting, scheduled within 120 days of the meeting request, is an in-depth data review and advice meeting where full study reports are submitted and reviewed. Lastly, a Type 4 meeting, which is scheduled within 60 days of a meeting request, is a meeting to discuss the format and content of a biosimilar biological product application or supplement submitted under Section 351(k) of the Public Health Service Act. The sponsor should note that at the time of a BPD meeting request, the sponsor should simultaneously submit the completed Briefing Document, which is a departure from the general approach to other FDA meetings.

CONCLUSION

While the information in this chapter provides some guidance regarding the best way to approach meetings with the FDA, it should also illustrate how complex and demanding the regulatory review process can be. How the sponsor works with the FDA throughout the approval process can have a substantial impact on the approval time for a new product. The best way to approach this process is to assemble the right resources with the knowledge and experience to manage the meeting strategy efficiently—allowing the scientific data to be presented effectively and promoting consensus on key product issues. By applying sufficient resources with the proper background to manage the FDA meetings, a sponsor can substantially increase a product's chances for approval and significantly reduce time to market.

Fundamentally, meetings with the FDA are opportunities for the sponsor to build FDA trust in the development plan, the regulatory strategy, the data, and the sponsor itself. With trust, communication is more effective and efficient, product approval is more likely, and crisis management becomes a joint sponsor–FDA team effort. Building trust depends in good measure on not "surprising" the Agency with unwelcome news, premature claims, and unreasonable demands, and it is primarily the sponsor's responsibility.

REFERENCE

1. DiMasi, J.A. and Manocchia, M. (1997). Initiatives to speed new drug development and regulatory review: The impact of FDA-sponsor conferences. *Drug Inf J*, 31:771–788.

5 FDA Medical Device Regulation

Barry Sall

CONTENTS

Introduction 126
Is It a Device? 128
    Product Jurisdiction 128
    Types of Medical Devices 129
Medical Device Classification 129
    Determining Device Classification 130
    Reclassification 132
An Introduction to the Medical Device Approval Process 135
    Strategic Choices 135
    Modification of Marketed Devices 136
Design Controls 137
    The Difference between Research and Development 137
    Design Control Components 138
Medical Device Clinical Research 141
    Exempted Studies 142
    Nonsignificant Risk Studies 142
    Significant Risk Studies 144
    The Investigational Device Exemption 144
    Unique Aspects of Medical Device Studies 144
The 510(k) Premarket Notification 146
    Substantial Equivalence 147
    Types of 510(k)s 148
        Traditional 510(k) 148
        Abbreviated 510(k) 148
        Special 510(k) 150
        De Novo 510(k) 150
    510(k) Components 150
        The Cover Sheet (Form FDA 3514) 150
        The Cover Letter 150
        The Table of Contents 151
        User Fee Information 151
        Form FDA 3654 151
        Statement of Substantial Equivalence 151

    Labeling 151
    Advertising and Promotional Material 152
    Comparative Information 152
    Biocompatability Assessment (If Necessary) 152
    Truthful and Accurate Statement 153
    Clinical Data 153
    Shelf Life (If Necessary) 153
    Indication for Use Form 153
  510(k) Summary 154
  Practical Aspects for 510(k)s 154
  Postsubmission Considerations for 510(k)s 155
The PMA Application 155
  Introduction to the PMA 155
  The PMA Process 155
  User Fee 156
  Advisory Panels 156
  Clinical Data 157
  Use of International Data 157
  Components of the PMA 158
The Quality System Regulation 159
  Design Controls 159
  Management Controls 160
  Corrective and Preventive Action 161
  Production and Process Controls 163
Postmarketing Issues 163
  Registration and Listing 163
  Medical Device Modifications 163
  Modifications to 510(k) Devices 164
  Modifications to PMA Devices 164
  Medical Device Reporting 165
  Unique Device Identification 166
  Advertising and Promotion 167
References 167

INTRODUCTION

Since the technological advances of the 1950s and 1960s, the rate of innovation in the medical device industry has greatly accelerated. These innovations have led to very substantial therapeutic, monitoring, and diagnostic benefits in all areas of medicine. Often, these innovative devices were selected and used by health-care professionals who received their basic scientific training before these technologies were developed. By the early 1970s, many medical devices were becoming so complex that medical professionals were no longer able to fully assess their

FDA Medical Device Regulation

attributes. Device developers and manufacturers were also encountering situations where devices interacted with the body in unanticipated ways or deficiencies in the production process led to patient injuries and deaths. In the United States, this history was the driving force behind the 1976 Medical Device Amendments to the Food, Drug, and Cosmetic Act of 1938. By 1978, when the regulations required by this new law came into full effect, the production and clinical testing of medical devices were subject to the Food and Drug Administration (FDA) review. Many new devices entering the US market had to undergo the FDA review, either through the 510(k) premarket notification process or the premarket approval (PMA) process. The 1976 Amendments have been modified several times over the years and now also cover the device development process. This chapter provides an introduction to medical device classification, the preparation of premarket submissions, medical device clinical research, and manufacturing regulations.

The regulations developed as a result of the 1976 Medical Device Amendments sharing a common goal with the existing pharmaceutical regulations. They both strive to protect the public health; however, they approach this goal in different ways. The device regulations recognize differences between medical devices and pharmaceuticals and between the medical device industry and the pharmaceutical industry. In general, therapeutic medical devices exert their effects locally by cutting tissue, covering a wound, or propping open a clogged artery; therefore, both preclinical and clinical testing can be simplified as compared with the pharmaceutical approach. Many diagnostic devices do not even contact the patient; so, in these cases, pharmaceutical safety testing is entirely inappropriate. Differences in the structure of the medical device industry as compared with the pharmaceutical industry do not have a direct effect on regulation, but they do affect the pace of innovation. There are a relatively small number of very large pharmaceutical companies with large experienced regulatory staffs. There are a large number of very small medical device companies with few or no dedicated regulatory staff. In addition, the product life cycle time for a medical device might be as short as two or three years, or approximately one-tenth the time for a pharmaceutical product. In general, many medical device marketing applications such as 510(k)s and PMA supplements are submitted for incremental changes in medical devices. All of these factors make it essential that medical device professionals have an adequate understanding of both the technology underlying their company's products and of the applicable regulations. Development timelines in this industry are very short, and inappropriate strategic decisions can generate substantial delays or even preclude the introduction of a potentially lifesaving technology.

The objective of this chapter is to provide the reader with a step-by-step introduction to the regulatory issues associated with the medical device development process in the United States. This information will enable the reader to identify the major steps in that process. References are provided, throughout the text, for more detailed information.

IS IT A DEVICE?

PRODUCT JURISDICTION

When preparing the regulatory strategy for a product or technology, it is important to first determine which regulations apply. Is the product a device? A drug? A biologic? Two factors must be considered to make this determination. First, the indication for use of the product must be determined by management and clearly stated. Then, the primary mode of action for the product should be identified. And only then can the developer determine if that action is achieved through chemical action and metabolism (a drug) or by a physical action (device). If an alginate wound dressing contains an antibacterial agent, it is regulated as a device, so long as its primary intended purpose is to act as a (physical) barrier between the wound and the environment and the antibacterial agent only functions to enhance that device function. On the other hand, if the indication for use is to deliver the antibacterial agent (chemical) to the wound to treat an existing infection, then the alginate dressing might be considered as an inactive component of a drug product. In order to make this determination, one must carefully review the definition of a medical device contained in the 1976 Medical Device Amendments of the Food, Drug, and Cosmetic Act.

> An instrument, apparatus, implement, machine, contrivance, implant, in vitro reagent, or other similar or related article, including any component, part or accessory, which is
>
> 1. Recognized in the official National Formulary, or the USP, or any supplement to them,
> 2. Intended for use in the diagnosis of disease or other conditions, or in the cure, mitigation, treatment, or prevention of disease, in man or other animals, or
> 3. Intended to affect the structure or any function of the body of man or other animals, and which does not achieve its primary intended purposes through chemical action within or on the body of man or other animals and which is not dependent upon being metabolized for the achievement of any of its principal intended purposes.

In addition to this definition, there are also intercenter agreements[1] between the Center for Devices and Radiological Health (CDRH), Center for Drug Evaluation and Research (CDER), and Center for Biologics Evaluation and Research (CBER) that discuss jurisdictional issues. The Medical Device User Fee Act of 2002 (MDUFA02) established the Office of Combination Products. This office is an excellent source of information on these issues. See http://www.fda.gov/CombinationProducts/default.htm for more information on the Office of Combination Products and its functions. The FDA Office of Combination Products

TABLE 5.1
Medical Device Types

Function	Form
Therapeutic	Durable
Monitoring	Implantable
Diagnostic	Disposable

defined the primary mode of action in an August 25, 2005, federal register notice (available at http://www.fda.gov/OHRMS/DOCKETS/98fr/05-16527.pdf). If a sponsor requires an official product jurisdiction determination, he/she can file a request for designation with the Office of Combination Products. Assuming that the product is regulated as a medical device, we can consider the type of device and level of regulation.

TYPES OF MEDICAL DEVICES

There are a wide variety of medical devices in use today. They range from room-sized imaging systems that weigh several tons to ophthalmic implants that are less than 2-mm long and weigh only a few grams. Most *in vitro* diagnostic products (blood and urine tests) are also regulated as medical devices. Table 5.1 describes most devices using two of their characteristics.

Using this table, one can easily characterize most medical devices by determining the function of the device from the left column, then its form from the right column. For instance, a lithotripter that uses sound waves to break up stones in the kidney would be considered a durable therapeutic device, a pacemaker would be considered an implantable therapeutic device, and so on. Issues such as reuse, shelf life, and device tracking impact different types of devices in different ways.

MEDICAL DEVICE CLASSIFICATION

Once a determination has been made that a product is a medical device, the next issue that must be addressed is medical device classification. In simpler terms, "What kind of submission do I need to commercialize this device? Is it exempt from 510(k) notification requirements, subject to those requirements, or must we file a PMA application?" In order to answer this question, we need to know the class of the device.

There are three classes of medical devices. class I devices are the simplest devices, posing the fewest risks and subject to general controls. Most of them are exempt from premarket notification requirements [510(k)], and some are also exempt from compliance with the Quality System Regulation (QSR). As seen in Figure 5.1, the majority of types of medical devices are class I devices. Examples of class I devices include toothbrushes, oxygen masks, and irrigating syringes.

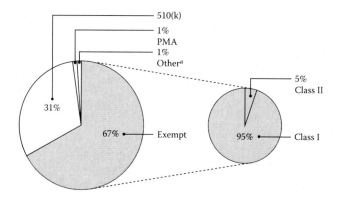

FIGURE 5.1 Device classification database search. (From GAO analysis of FDA data, http://www.gao.gov/assets/290/284882.pdf)
Notes: Data are for the 50,189 devices listed with the FDA by device manufacturers during the period October 1, 2002, through September 30, 2007. Even if their devices are exempt from premarket notification requirements, manufacturers must still comply with other FDA requirements, such as good manufacturing practice requirements specified in the FDA's quality system regulation. See 21 C.F.R. pt. 820 (2008).

Class II devices include many moderate risk devices. In order to market a class II device in the United States, the manufacturer must obtain clearance of a 510(k) premarket notification prior to commercialization. The purpose of this notification is to demonstrate that the new device is *substantially equivalent* to another device that has already gone through the 510(k) process or to a device that was on the market before the Medical Device Amendments were signed on May 28, 1976. Class II devices are subject to special controls, that is, Office of Device Evaluation (ODE) guidance documents, FDA-accepted international standards, and the QSR. Ultrasound imaging systems, Holter cardiac monitors, pregnancy test kits, and central line catheters are all class II devices. Nearly 4000 class II devices are cleared on to the US market each year.[2]

Most class III devices require PMA approval prior to marketing in the United States. These are devices that are not substantially equivalent to any class II device. They are usually technologically innovative devices. There are a small number of class III preamendments 510(k) devices; however, the FDA has been working diligently to either downclassify them to class II, or if their risk profile does not justify down classification, call for PMAs. Thirty-seven PMAs were approved in CY20011.[3]

Determining Device Classification

If the product in development is similar to other medical devices already on the US market with respect to its indication for use and its technological characteristics, then our classification determination becomes a search of the regulations. Title 21 of the Code of Federal Regulations (CFR), Parts 862–892 contains descriptions of a wide variety of medical devices arranged by medical practice area. The classifications and exemptions from 510(k) or the QSR, if any, are listed in this section of the regulations. The classification database in the CDRH Website can also be a useful tool for determining device classification (Table 5.2; Figure 5.2).

If a description in the CFR is consistent with the characteristics of the new device, then the device classification listed in that section of the CFR should apply. Precedents can be identified in another manner as well. If one is aware of other competing devices that are already on the market, one can search the 510(k)[2] or PMA[4] databases, within the CDRH Website, for those products and determine how they were classified. Figure 5.3 illustrates the process one can follow to identify possible predicate devices when only the name of a competitor is known.

When there is no obvious precedent to follow, it can be difficult to determine the appropriate device classification. The question can be explored informally via telephone calls with the appropriate branch chief within the ODE, but no binding decision will result from such discussions. Device developers can obtain a formal classification decision using the 513(g) request for classification process. The sponsor submits a brief document to the ODE describing the device, how it works, materials used, and similar devices, if any. The indication for use and draft labeling is also included along with a suggested classification and supporting rationale.

TABLE 5.2
Medical Device Classification

Device Classification Panel or Specialty Group	21 CFR Part
Anesthesiology	868
Cardiovascular	870
Clinical chemistry and clinical toxicology	862
Dental	872
Ear, nose, and throat	874
Hematology and pathology	864
Immunology and microbiology	866
Gastroenterology and urology	876
General and plastic surgery	878
General hospital and personal use	880
Neurology	882
Obstetrical and gynecological	884
Ophthalmic	886
Orthopedic	888
Physical medicine	890
Radiology	892

CFR, Code of Federal Regulations.

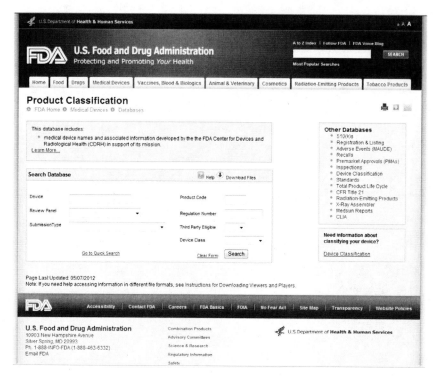

FIGURE 5.2 Selected pathways for marketing medical devices in the United States. (From http://www.accessdata.fda.gov/scripts/cdrh/cfdocs/cfPCD/classification.cfm)

There is a user fee of US$2971 for a 513(g) (FY 2012). In 60 days, the sponsor should receive a letter from the ODE, either confirming the sponsor's classification rationale or stating an alternate classification.

Reclassification

Once the FDA determines that a device is a class III PMA device, that type of device will always be a class III device, no matter how many other competitors follow with similar products. All the competitors that develop similar products will have to follow the PMA process to market their devices in the United States. The only way that situation can change is if the FDA approves a reclassification petition and downclassifies the device to class II or I. This type of reexamination can be initiated by either the FDA or the industry. In recent years, the FDA has examined many device types, their overall risk, and actual frequency of problems in the field and downclassified significant numbers of devices either from class III to II or from class II to I. These actions enable the FDA to focus more of its resources on the higher risk products. Industry groups have also submitted their own reclassification petitions and succeeded in downclassifying devices. There

FDA Medical Device Regulation

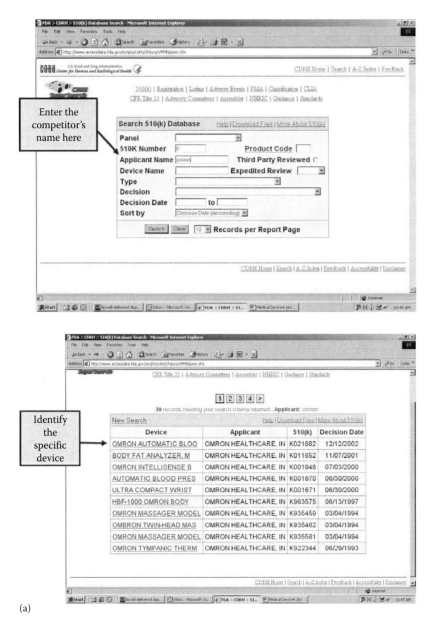

(a)

FIGURE 5.3 510(k) database search for a predicate device.

have also been recent efforts to increase regulatory oversight of some class II devices. For example, in 2011, surgical mesh used to treat pelvic organ prolapse was subject to a FDA public meeting and discussion of reclassifying it from class II to class III (see http://www.fda.gov/MedicalDevices/ProductsandMedicalProcedures/ImplantsandProsthetics/UroGynSurgicalMesh/default.htm).

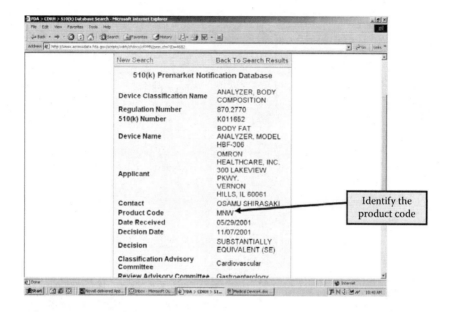

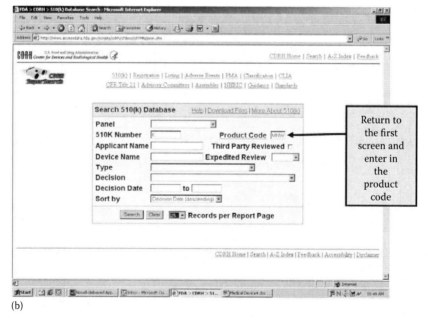

(b)

FIGURE 5.3 (Continued) 510(k) database search for a predicate device.

FDA Medical Device Regulation

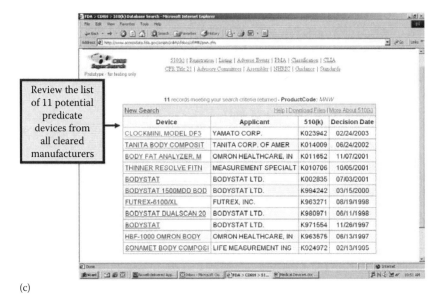

(c)

FIGURE 5.3 (Continued) 510(k) database search for a predicate device.

AN INTRODUCTION TO THE MEDICAL DEVICE APPROVAL PROCESS

STRATEGIC CHOICES

Now that the classification of the device is known, an appropriate regulatory pathway can be identified. Unlike the pharmaceutical regulatory process, a medical device developer is frequently presented with more than one regulatory path to the US market. A device such as software that analyzes magnetic resonance imaging (MRI) images is designated as a class II 510(k) product if it only measures the size or volume of anatomical structures. However, if the software detects abnormalities or provides diagnostic information, it would be considered a class III PMA device, so the indication for use is critical to determination of the regulatory path. A device developer may choose to "start small" and begin the FDA interactions with a 510(k) for simpler intended use and then, after gaining experience, move to the more challenging PMA once a revenue stream is established with the initial device. Generally, both the industry and the FDA would prefer to review medical devices as 510(k)s. This process provides the industry with timely reviews and conserves reviewing resources for the FDA. So, when speed to market is the prime consideration, one always attempts to follow the 510(k) path. Even within the 510(k) pathway, there are branches. If the FDA-recognized standards apply to the new device, the sponsor may choose to submit either an abbreviated 510(k) or a traditional 510(k). The review time is the same, but instead of containing the complete testing reports, an abbreviated 510(k) will contain a list of the recognized standards followed during device testing and a summary of the test results. This results in a smaller submission. Of course, a sponsor may choose an alternate test method, in which case the test protocol would need to be included in a traditional 510(k). In some cases, device developers may choose to propose a more complex PMA indication for use or, in a situation where the device classification is not clear, suggest the more complicated class III PMA designation. This can make sense when the developer may not have a strong intellectual property position but does have sufficient resources to conduct clinical trials. This strategy can result in the erection of a regulatory barrier of entry for other, less well-funded organizations. This strategy is often called creation of a "regulatory patent." Another consideration when deciding on a regulatory path is user fees. Since October 2002, the ODE has been authorized to charge fees for reviewing 510(k)s, PMAs, and PMA supplements (Table 5.3; Figure 5.2).

Submission Type	FY 2006	FY 2007	FY 2008	FY 2009	FY 2010	FY 2011	FY 2012
510(k)	$3833	$4158	$3404	$3693	$4007	$4348	$4049
PMA	$259,600	$281,600	$185,000	$200,725	$217,787	$236,298	$220,050
PMA 180-Day Supplement	$55,814	$60,544	$27,750	$30,109	$32,668	$35,445	$33,008

All the submission types mentioned in this section are discussed in more detail in later sections of this chapter.

MODIFICATION OF MARKETED DEVICES

Many changes can be made to 510(k) devices by following the design control provisions of the QSR, rather than submission of a new 510(k). Even when

TABLE 5.3
Selected the FDA Safety and Innovation Act—FDASIA Standard User Fees FY 2013

Application Type	Fee (US$)
PMA	248,000
Panel track PMA supplement	186,000
180-day PMA supplement	37,200
Real-time PMA supplement	17,360
30-day notice	3521
513(g)s	3348
510(k)—all types	4960
IDEs	No charge

IDE, investigational device exemption; PMA, premarket approval.

a new 510(k) is necessary, in many cases, a sponsor can choose to submit a special 510(k). If the sponsor is modifying its own device, the intended use is not changing and the fundamental scientific technology is not changing. The review period for a special 510(k) is 30 days. Changes to PMA products follow a more rigid process. Most changes require advance approval via the PMA supplement process. There are several types of PMA supplements with approval times ranging from 30 days to 180 days. The sponsor must also submit PMA annual reports that update ODE on all device changes and any new clinical data. Both the premarket and postmarket obligations must be considered when determining the preferred route to market. More information on postmarketing issues can be found in "Postmarketing Issues" section. The ease of modifying devices and other postmarket considerations also factor into the strategic regulatory planning process. It is far easier to update a 510(k) device than a PMA device.

DESIGN CONTROLS

Once the product definition and regulatory strategy have been prepared, class II and III device developers must work to comply with the design control provisions of the QSR (21 CFR 820) as the device development process moves forward. The QSR is the medical device equivalent of the pharmaceutical current good manufacturing practices (cGMPs). The QSR, unlike cGMPs, also regulates the device development process via its design control provisions (21 CFR 820.30). This section describes the device developer's obligations under the design control provisions of the QSR. Other sections of the QSR are discussed in "The Quality System Regulation" section.

THE DIFFERENCE BETWEEN RESEARCH AND DEVELOPMENT

The preamble to the QSR[5] states that research activities are not regulated by the QSR, but development activities are regulated. The regulation does not provide guidance for distinguishing between the two activities; however, the preamble does add, "The design control requirements are not intended to apply to the development of concepts and feasibility studies. However, once it decides that a design will be developed, a plan must be established...." Most device developers categorize investigations of a general technology as research and application of that technology to a particular product's development. For example, if a device developer creates a new laser technology, that effort would be considered research. Once the developer begins to apply that technology to a particular model of device with specific indications for use and user requirements, then they have begun the development phase and design controls must be applied. A device developer's design control standard operating procedures (SOPs) should clearly describe the point in the development process when design controls apply, and that definition should be consistently followed for all design projects.

DESIGN CONTROL COMPONENTS

There are eight components of design controls that stretch from planning for the development effort through design transfer (from development to manufacturing) and maintenance of existing designs. These controls apply to all class II and III medical devices and a small number of class I devices. The purpose of these controls is to ensure that devices are developed in a rational manner, in compliance with the firm's existing design control SOPs. Table 5.4 summarizes these components.

If a company is just starting to develop a medical device for the first time, the design control process must be fully described in SOPs and fully implemented before the development planning begins. Design controls are closely linked to many other QSR components and the entire system must work together to produce good product. Refer to "The Quality System Regulation" section for discussion of the other components of the QSR.

The design control regulation sets requirements for the development process. Firms must prepare and follow SOPs that comply with the regulations and that fully describe how the firm will meet all relevant regulatory requirements. All the relevant activities must be fully documented in the firm's design history file (DHF). For example, the regulation requires device developers to prepare a list of design inputs. Just like every other design control-related document, this list cannot be considered a static document. As the design process progresses, inputs are modified, added, or subtracted. The design input file must be maintained as a current document throughout the development process. Another important design control function is the design review. At least once during the design process, and more frequently for a complex design effort, the design must be reviewed to ensure that the design satisfies the design input requirements for the intended use of the device and the needs of the user. All other sources of design information, including design output reports, design verification documentation, and even actual prototypes should be part of this review. Most importantly, for regulatory compliance, a report must document all the design review activities and their results and list the individuals that participated in the review. The regulation requires that at least one member of the review team be an independent reviewer who has not been directly involved with the design effort.

Two other tasks must be considered throughout the device development process. Risk analysis should be conducted early in the design process and continually updated as the design evolves. The purpose of risk analysis is to identify potential risks associated with the device, and evaluate their effect. If a significant risk is identified, it must be mitigated (reduced), preferably by modifying the design. At the end of the design process, all substantial risks should be mitigated.

One method used to identify user-related risks is human factors evaluation. The FDA has been using human factors evaluations for an increasing number of devices. For example, in 2009 and 2010 the FDA began efforts to address the high number of infusion pump recalls and patient deaths (see http://www.fda.gov/MedicalDevices/ProductsandMedicalProcedures/General HospitalDevicesandSupplies/InfusionPumps/default.htm). The FDA currently

TABLE 5.4
Design Control Components

Design Activity	Personnel Involved	Examples of Issues
Design and development planning	Development, marketing, and management	• Determine timing for design reviews • Determine documentation requirements and departmental documentation responsibilities • Determine overall project timelines and budget
Design input	Development, management, sales and marketing, quality, and regulatory	• Identify users of the new device • Specify where the new device will be used • Describe the operating environment for the device • Document how long the new device will be used • Determine the user/patients requirements • Comply with regulations and standards • Develop specifications for the device • Develop, select, and evaluate components and suppliers • Develop and approve labels, user instructions, and training materials • Develop packaging • Document the processes and details of the device design • If applicable, develop a service program
Design output	Development	• Execute the design • Update risk analysis during the design process
Design review	Development, management, and others, as needed, including one person not directly involved in the design effort	• Determine if the design meets customer needs • Confirm that manufacturability and reliability issues are adequately addressed • Establish that human factors' issues are adequately addressed • Determine frequency of design reviews. At least one during the development program. Complex programs may have more, depending upon the developer's procedures

(Continued)

TABLE 5.4
(Continued) Design Control Components

Design Activity	Personnel Involved	Examples of Issues
Design verification	Development	• Confirm that the design outputs meet the design input requirements by reviewing data from tests, inspections, analysis, and formative human factors studies
Design validation	Development, management, quality, and clinical	• Perform under defined operating conditions on initial production units, or its equivalent • Include software validation and risk analysis, here as appropriate • Ensure that devices conform to defined user needs and intended uses • Include testing under actual or simulated use conditions • Validation plans, methods, reports, and review must be conducted according to approved SOPs • Include actual use clinical trials, simulated use testing, and other evaluations
Design transfer	Development, management, quality, and manufacturing	• Prepare a plan for the transfer of all the design components to manufacturing • Develop manufacturing facilities and utilities • Develop and validate manufacturing processes • Assure that all affected personnel are adequately trained • Assure that all manufacturing and quality systems function according to specifications • Transfer portions of the design in an incremental manner, rather than all at once
Design changes	Development and management	• Assure that design changes are tracked, verified, and validated • Assure that corrective actions are completed • Assure that the DHF is kept current and includes all design revisions

DHF, design history file; SOPs, standard operating procedures.

FDA Medical Device Regulation

expects other devices, especially those that are used by patients and lay caregivers to undergo rigorous human factors evaluations (see http://www.fda.gov/MedicalDevices/DeviceRegulationandGuidance/HumanFactors/default.htm). Formative human factor studies are frequently relatively small studies where potential users are observed as they use prototype devices. Lessons learned from these studies stimulate design changes as potential risks are identified. Once a final design has been established, a summative human factors study, or in some cases, a clinical trial is conducted to evaluate device performance in the hands of the expected user population. This last type of testing is considered design validation.

MEDICAL DEVICE CLINICAL RESEARCH

Once the regulatory pathway has been determined and development is underway, clinical data may be necessary. Keep in mind that the vast majority of 510(k) notifications do not contain clinical data. Figure 5.4 pictorially depicts the pathways for medical device clinical research. Unlike the pharmaceutical model, there are three levels of regulation of medical device clinical research. Some research is exempted from the investigational device exemption (IDE) regulation; some research is subject to just some sections of the IDE regulation; and other types of

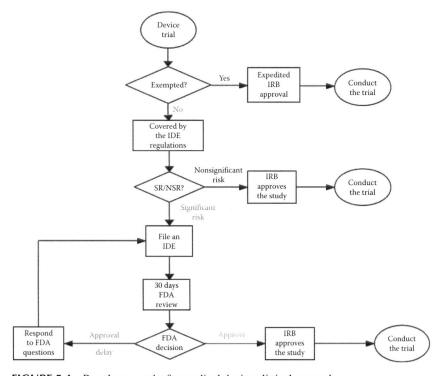

FIGURE 5.4 Regulatory paths for medical device clinical research.

research are subject to all sections of the IDE regulation. See http://www.fda.gov/downloads/RegulatoryInformation/Guidances/UCM126418.pdf for more information of risk determinations.

EXEMPTED STUDIES

Most exempted studies involve either previously cleared or approved devices or investigational *in vitro* diagnostic devices. If a sponsor wishes to conduct a study that, for example, compares the performance of their own previously cleared device with the performance of their competitor's previously cleared device, that study would be exempt from the IDE regulations, so long as both devices are used for their cleared indications. No prior FDA review or approval of the study is necessary.

Of course, due to privacy concerns and institutional regulations, any human clinical trial should use an informed consent form and be reviewed and approved by the appropriate institutional review board (IRB). Most *in vitro* diagnostic field trials are also exempt, so long as invasive means are not used to collect samples, performing the investigational assay does not consume sample material needed for medically necessary approved assays, and the data obtained from the investigational assay are not used to make treatment decisions. Also, in some cases when archived de-identified samples are used for *in vitro* diagnostic field trials, informed consent may not be necessary. See http://www.fda.gov/MedicalDevices/DeviceRegulationandGuidance/GuidanceDocuments/ucm078384.htm for more information. Animal studies and custom device studies are also exempt from the IDE regulation.

NONSIGNIFICANT RISK STUDIES

Many studies that do not involve highly invasive devices, risky procedures, and/or frail patients can be conducted under the nonsignificant risk (NSR) provisions of the IDE regulation (see http://www.fda.gov/downloads/RegulatoryInformation/Guidances/UCM126418.pdf). These provisions provide an intermediate level of control for the study without requiring the study sponsor to prepare and file an IDE. See Table 5.5 for a comparison of sponsor and investigator responsibilities. Areas where the requirements for NSR and significant risk studies differ are shaded. When a sponsor determines that a study is NSR, no FDA involvement is required, although many sponsors will consult with the FDA to confirm that the study is indeed NSR and that its design is consistent with the FDA expectations. Each IRB that reviews an NSR study must document three conclusions. First, that they concur with the sponsor's NSR determination; second, that the study protocol is approved; and third, that the consent form is approved. If just one IRB formally determines that a study is not NSR, then the sponsor must report this to the ODE. If all IRBs approve the study, it may proceed. In this case, the local IRBs monitor the progress of the study according to their own SOPs, and the FDA is not involved in the process (Table 5.5).

TABLE 5.5
NSR/SR Comparison Chart

Task	NSR Sponsor	NSR PI	SR Sponsor	SR PI
Submit an IDE to the FDA	−	−	+	−
Report ADEs to sponsor	−	+	−	+
Report ADEs to reviewing IRBs	+	+	+	+
Report ADEs to the FDA	+	−	+	−
Report withdrawal of IRB approval to sponsor	−	+	−	+
Submit progress reports to sponsor, monitor, and reviewing IRB	−	+	−	+
Report deviations from the investigational plan to sponsor and reviewing IRB	−	−[a]	−	+
Obtain and document informed consent from all study subjects prior to use of the investigational device	−	I	−	I
Maintain informed consent records	−	+	−	+
Report any use of the device without prior informed consent to sponsor and reviewing IRB				
Compile records of all anticipated and unanticipated ADEs and complaints	+	−	+	−
Maintain correspondence with PIs, IRBs, monitors, and the FDA	−[a]	−[a]	+	+
Maintain shipment, use, and disposal records for the investigational device	−[a]	−[a]	+	+
Document date and time of day for each use of the IDE device	−	−	−	+
Maintain signed investigator agreements for each PI	−[a]	−	+	+
Provide a current investigator list to the FDA every six months	−	−	+	−
Submit progress reports to the IRB, at least yearly	+	−	+	−
Submit a progress report to the FDA, at least yearly	−	−	+	−
Submit final study report to the FDA	−	−	+	−
Submit final study report to all reviewing IRBs	+	−	+	−
Monitor the study and secure compliance with the protocol	+	−	+	−
Notify the FDA and all reviewing IRBs if an investigational device has been recalled	+	−	+	−
Comply with IDE advertising, promotion, and sale regulations	+	+	+	+
Comply with IDE-labeling regulations	+	+	+	+

[a] Compliance with IDE regulations is recommended.

ADEs, adverse drug events; FDA, Food and Drug Administration; IDE, investigational device exemption; IRB, Institutional Review Board; NSR, nonsignificant risk; PI, principal investigator; SR, significant risk.

SIGNIFICANT RISK STUDIES

Significant risk studies require an approved IDE to treat patients in the United States. Typical significant risk studies involve implantable devices or devices that introduce significant quantities of energy into the body. Studies with devices that sustain or support life are nearly always considered significant risk. If a study sponsor is unsure of the risk status of a study, consultation with the appropriate branch within the ODE should be considered.

THE INVESTIGATIONAL DEVICE EXEMPTION

The IDE serves the same function for a significant risk medical device clinical trial as the investigational new drug (IND), described in Chapter 2, does for pharmaceutical clinical trials. The submission contains data that are similar in many respects to data contained in an IND. There are, however, some significant differences between the two submission types due to the differences in regulatory requirements between devices and drugs. First, although preclinical testing data are included in both the submissions, the data in an IDE conforms to the ISO 10993 biocompatability testing standard as modified by the FDA (see http://www.fda.gov/MedicalDevices/DeviceRegulationandGuidance/GuidanceDocuments/ucm080735.htm), rather than the "International Conference on Harmonization (ICH)" guidance. Relevant FDA guidance documents (special controls) may also list additional data expectations. The IDE regulation requires an investigational plan, but does not specify an investigator brochure. The international ISO 14155 medical device clinical research standard does require an investigator brochure. The IDE regulation also requires that the sponsor include a clinical monitoring SOP in the submission. Under the cost recovery provision of the IDE regulation, the sponsor may charge for the investigational device, so long as only research and development and manufacturing costs are recovered. An investigator agreement serves the function of the Form FDA 1572, used for pharmaceutical studies. More detailed information regarding IDEs can be found at http://www.fda.gov/MedicalDevices/DeviceRegulationandGuidance/HowtoMarketYourDevice/InvestigationalDeviceExemptionIDE/default.htm

UNIQUE ASPECTS OF MEDICAL DEVICE STUDIES

The informed consent, financial disclosure, and IRB regulations described in Chapter 9 apply equally for medical device studies. Provisions of the IND regulation and ICH guidelines do not apply to medical device studies. This section describes some of the unique features of medical device studies. Before we consider the regulatory differences between pharmaceutical and device studies, we need to review the procedural differences. Test article administration is frequently a prime concern in trials of therapeutic devices. In most drug trials, IV, IM, or PO administration of the test article is a trivial concern that is hardly discussed. The manner in which a surgical device is used or the technique by which an implantable device is placed in the body can mean the difference between success and failure in the trial. Because of this, investigator training is a critical aspect of many device trials. Protocol compliance while using the device and while recording data is also a critical issue. In addition, the clinical research associate (CRA) is called upon to transmit technical data between the technical development staff and investigators. Another global issue involves overall study design. Unlike most pharmaceutical studies that are both masked and randomized, the vast majority of device studies are not masked. Most of the time, it is not possible or ethical to mask the device, especially if the device is an implant or a surgical device. Often it is possible to mask a patient assessor to reduce bias. There are also several key regulatory differences between pharmaceutical clinical trials and medical device clinical trials. First, the ICH guidelines only apply to pharmaceutical studies, not to medical device studies. The greatest effect is seen on adverse device effects analysis and reporting (Figure 5.5). The IDE regulations permit an investigator to analyze a potential adverse device effect for 10 days before reporting it to the local IRB and the sponsor (most sponsors impose a 24-hour reporting period). The sponsor then has another 10 days to evaluate the event to determine if it should be reported to the ODE, all reviewing IRBs, and all participating investigators. The IDE regulations do require the sponsor to directly communicate this information to the IRBs. This responsibility cannot be delegated to the investigators. While ICH guidances do not apply, some, such as those that describe format and contents of clinical study reports, may offer device companies good suggestions for organizing their study reports. The IDE regulation also does not require the preparation of an investigator brochure. In some cases, especially for multinational studies, a sponsor may choose to prepare such a document, even though it is not required. The Form FDA 1572 is another inapplicable document. It requires the investigator to comply with key provisions of the IND regulation, so it is not relevant to device studies. In its place, we have the investigator agreement. It serves roughly the same purpose as the Form FDA 1572. Its contents are specified in 21 CFR 812.43(c). Although not required by the regulation, many sponsors ask that the principal investigator list the subinvestigators in the agreement, as this list will simplify the gathering of financial disclosure information. There is usually a second investigator agreement, not subject to the FDA review, that covers financial compensation, publishing priorities, and other unregulated activities. Lastly, the cost recovery provision of the IDE regulation [21 CFR 812.20(b)(8)] permits the sponsor to charge for the device. The sponsor can charge enough to recover research and development costs. This provision cannot be used to commercialize an investigational device.

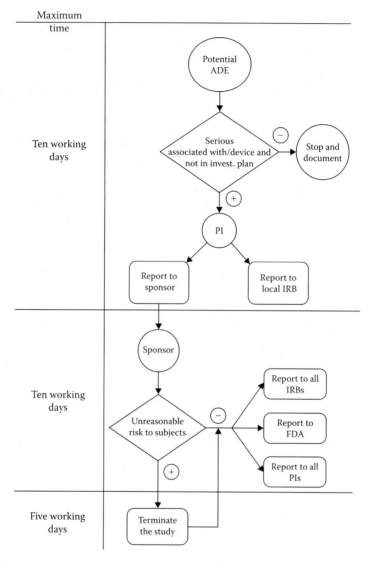

FIGURE 5.5 Adverse device effect reporting.

THE 510(k) PREMARKET NOTIFICATION

More than 3000 medical devices are cleared on to the US market every year through the 510(k) premarket notification process. This represents approximately half the new devices that appear in the US market in a given year. The 510(k) process is relatively rapid, flexible, and adaptable to many different device types and risk levels.

FDA Medical Device Regulation

The goal of the 510(k) process is the following:

> Demonstration of substantial equivalence to a device that was on the US market prior to May 28, 1976 or to a device that has *already gone through the 510(k) clearance process.*

Devices that have successfully gone through the 510(k) process are described as "510(k) cleared." A distinction is made between those devices that have been reviewed according to the substantial equivalence standard from those that have been reviewed according to the PMA application, safety, and effectiveness standard. PMA devices are "approved."

The previously cleared device included for comparison purposes in a 510(k) is called the predicate device. A 510(k) may contain multiple predicate devices that address various features of the device. The device designers should be able to provide regulatory personnel with assistance, identifying key technological characteristics that demonstrate substantial equivalence. These data should already be part of the design inputs required as part of design controls. Generally, little, if any, manufacturing data are included in a 510(k). Sterile devices will include information on the sterilization process, including sterilization process validation activities and the sterilization assurance level. *In vitro* diagnostic products will frequently include data on the production of key reagents such as antibodies or nucleic acid probes. The other part of substantial equivalence relates to the indication for use. Frequently, one medical device can be used for many indications in a variety of medical specialties. When new indications are added, those indications must be cleared in a traditional or abbreviated 510(k). The 510(k) must cite a predicate device with the same indication for use.

When searching for potential predicate devices, several information sources are useful. Two FDA databases, the 510(k) database[2] and the classification database,[6] can be very helpful. The 510(k) database is especially useful when one knows either the name of potential predicate devices or the manufacturer of the device. The classification database can be used to identify a particular device type and its corresponding product code. One can then transfer the product code to the 510(k) database and generate a listing of all similar devices. Sales and marketing staffs and competitor Websites are also excellent sources of predicate device information (Figure 5.3).

SUBSTANTIAL EQUIVALENCE

The two pillars of substantial equivalence are "intended use" and "technological characteristics." The sponsor must demonstrate that the new device has an intended use that is substantially equivalent to a predicate device and that the technological characteristics of the new device are substantially equivalent to a predicate device. The predicate device must be a device that has already been cleared through the 510(k) process or a device that was in commercial distribution prior to May 28, 1976, when the FDA was first able to regulate medical devices. A PMA device cannot serve as a 510(k) predicate device. There is some flexibility in ODE's approach to the 510(k) process, especially with respect to technology. The devices do not have to be identical. An acceptable predicate device can have different technological characteristics from the new device, so long as they do not raise new questions of safety and effectiveness and the sponsor demonstrates that the device is as safe and as effective as the legally marketed device. Different technological characteristics might include changes in materials, control mechanisms, overall design, energy sources, and principles of operation. Safety and effectiveness can be demonstrated through engineering analysis, bench or animal testing, or human clinical testing. Recent 510(k) reform efforts have criticized the practice of "split predicates" where the sponsor cites one predicate device with respect to intended use and another predicate device for technology. While this practice is not specifically prohibited in the regulations, it is increasingly more difficult to obtain clearance when utilizing this approach. If it is not possible to identify a suitable predicate device, or devices, the sponsor may have to consider filing a PMA or a *de novo* 510(k), if appropriate.

TYPES OF 510(k)s

There are four types of 510(k) premarket notifications. They are briefly described below. Figure 5.6 describes the decision process used to determine which type of 510(k) should be submitted. Each type of 510(k) is briefly described in the following sections. See http://www.fda.gov/MedicalDevices/DeviceRegulationandGuidance/HowtoMarketYourDevice/PremarketSubmissions/PremarketNotification510k/default.htm for more information on these types of 510(k)s.

Traditional 510(k)

The traditional 510(k) is filed when the sponsor has developed a device that they believe is substantially equivalent to one or more devices that have already been cleared through the 510(k) process, or were already on the market before the 1976 Medical Device Amendments were signed on May 26, 1976. In addition, the subject device is not a modification of one of the manufacturer's cleared devices nor does the application contain any declarations of conformance with the FDA-recognized standards.[7] Once this 510(k) is submitted, the targeted review period is 90 days.

Abbreviated 510(k)

This 510(k) is similar to the traditional 510(k) in function. A sponsor can choose to comply with the FDA-accepted standards during the testing process. A declaration of conformance is included in the 510(k), stating that the device meets the specifications in the referenced standards. Unlike a traditional 510(k), entire test reports do not need to be included. This simplifies both the 510(k) preparation and review processes. Once this 510(k) is submitted, ODE has 90 days

FDA Medical Device Regulation

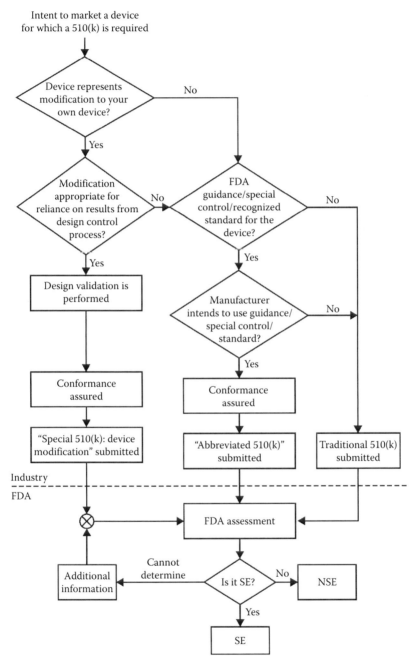

FIGURE 5.6 The new 510(k) paradigm. (From A New 510(k) paradigm—Alternate approaches to demonstrating substantial equivalence in premarket notifications, US FDA.)

to review the document. More information on the format for both traditional and abbreviated 510(k)s can be found at "Guidance for Industry and FDA Staff: Format for Traditional and Abbreviated 510(k)s," http://www.fda.gov/downloads/MedicalDevices/DeviceRegulationandGuidance/GuidanceDocuments/ucm084396.pdf

Special 510(k)

A special 510(k) is submitted when a sponsor has modified his/her own device, has not added a new indication for use, and has not altered the fundamental scientific technology of the device. Design controls, including a risk analysis, must be conducted. Reviews for special 510(k)s are generally processed within 30 days.

De Novo 510(k)

The *de novo* 510(k) is a 510(k) without a predicate device. It is not a commonly used path (13 clearances in 2012); but in some circumstances, it is appropriate: where the sponsor can demonstrate that the product has few risks and the extensive PMA safety and effectiveness review is not warranted. The device should be discussed with ODE in advance, before embarking on this path. The *de novo* process has been under review as part of 510(k) reform efforts and may be enhanced in the future. A draft guidance titled "Draft Guidance for Industry and Food and Drug Administration Staff—De Novo Classification Process (Evaluation of Automatic Class III Designation)" found at http://www.fda.gov/downloads/MedicalDevices/DeviceRegulationandGuidance/GuidanceDocuments/UCM273903.pdf describes potential changes to the *de novo* process.

510(k) COMPONENTS

The most common sections of a traditional 510(k) are described later. Many of these sections are also present in the other types of 510(k)s.

The Cover Sheet (Form FDA 3514)

This five-page form provides the ODE with general information related to the submission in a standardized format. Completion of this document is not mandatory. Only relevant data fields should be completed. The applicant signature is not required. Indications should be taken word for word from the body of the 510(k). A sample cover sheet can be found at http://www.fda.gov/downloads/AboutFDA/ReportsManualsForms/Forms/UCM080872.pdf

The Cover Letter

This letter should be no more than one or two pages long and identify the device, very briefly summarize the contents of the application, and provide the name, address, telephone, e-mail address, and fax numbers of the contact person. The type of 510(k) should also be specified.

The Table of Contents

This section helps to create a "reviewer-friendly" document by making it easy for the reviewer to locate each key section. Although it is not specifically required in the regulations, it is an expected component of any 510(k). Key sections of the 510(k) should be listed in the order they appear in the 510(k) along with the page number of the section. Index tabs, used selectively, can also aid the reviewer during the review process. All pages of the 510(k) should be numbered consecutively. This numbering facilitates communication between the reviewer and the sponsor during the review process.

User Fee Information

A copy of the completed medical device user fee cover sheet (available at http://www.fda.gov/oc/mdufma/coversheet.html) must be included in this section. The unique payment identification number present in this form enables the ODE to confirm that the user fee payment has been received. The actual user fee payment is not included in the 510(k). The information at the preceding URL describes the user fee payment process in detail. The FY 2013 user fee for all types of 510(k)s is US$4960.

Form FDA 3654

This form is used to describe the level of compliance with the FDA-recognized standards. It is expected in traditional and abbreviated 510(k)s.

Statement of Substantial Equivalence

This optional section can "sell" the 510(k) to the ODE by providing a well-reasoned rationale for a substantial equivalence determination. This section may not be necessary when there is a very simple comparison between a single predicate device and the new device. When a traditional or abbreviated 510(k) involves multiple predicate devices and complex technological comparisons, this type of statement can help communicate the sponsor's rationale. It contains a brief summary of device background information, along with a list of the predicate device(s) and, most importantly, a narrative description of the sponsor's substantial equivalence claim. If appropriate, cross-references to other sections of the 510(k) may be included.

Labeling

This section must provide the ODE with all printed material associated with the device, including printing fixed to the outside of the device, its packaging, operator's manual or in the case of software-controlled devices, text programmed into the electronics for display. Frequently, information displayed on video display screens is also reproduced in the operator's manual; so it does not have to be included twice. Patient information brochures, if used, should also be included.

Advertising and Promotional Material

If provided, the ODE will review the documentation and inform the sponsor of areas of noncompliance. This is optional information. If included, material should be clearly copied. Copies of actual brochures, especially if they are not on standard-size paper or include foldouts, are difficult for ODE document control personnel to handle. Advertising copy must be consistent with the indications for use mentioned in the 510(k).

Comparative Information

This is the heart of the 510(k). This section must contain data that demonstrate that the 510(k) device is "substantially equivalent" to the predicate device(s). Careful selection of comparative parameters is essential. Comparison charts listing parameters and values for the predicate device and the 510(k) device are common. Bench and clinical testing data may also be included. Advertising for the predicate device may also be included to support statements describing the predicate device. This section must clearly demonstrate that the new device is substantially equivalent to one or more predicate devices with respect to indication for use and technological features such as materials used and operating principle.

Biocompatability Assessment (If Necessary)

Medical devices contain a wide variety of materials, from stainless steel and titanium in orthopedic implants to plastics in catheters or even living cells in wound care products.[8] The data in this section must demonstrate that the device materials do not cause toxicity. Toxicity can occur through direct contact between the device and the body, such as a wound care product or an implantable device. Toxicity can also occur if materials such as plasticizers or mold-release agents leach from polymers, such as the tubing and components of a heart–lung bypass circuit, which carry blood out of and back into the body. Adverse effects are often localized, but can be systemic, or even carcinogenic effects can occur; so the standard requires more extensive testing when the device is implanted, rather than contacting intact skin and for permanent implants, as opposed to devices that contact the body for less than 24 hours.

An FDA-modified version of ISO 10993 is used to determine testing appropriate for a specific device. See http://www.fda.gov/MedicalDevices/DeviceRegulationandGuidance/GuidanceDocuments/ucm080735.htm for more information on the use of ISO 10993. The FDA document includes a testing matrix that uses the length of exposure and type of exposure to determine which tests are appropriate. Before conducting recommended testing, it is advisable to confirm the testing plan with the ODE, as requirements may vary for some devices.

Full reports of each required test are included in a traditional 510(k), especially if the test protocols have been modified from those specified in the ISO 10993. A summary table of all biocompatibility testing and summary of results are often useful. If the medical device does not contact the patient, biocompatibility data are generally not necessary.

Truthful and Accurate Statement

This statement identifies a person who takes legal responsibility for the accuracy of the 510(k). It follows the requirements of 21 CFR 807.87(j).

> I certify that, in my capacity as (*the position held in company*) of (*company name*), I believe to the best of my knowledge, that all data and information submitted in the premarket notification are truthful and accurate and that no material fact has been omitted.

The statement must be signed and dated by a responsible person employed at the submitting company. A consultant cannot sign it.

Clinical Data

ODE may request clinical data to demonstrate substantial equivalence to a predicate device. It may also be necessary when, as described in "Substantial Equivalence" subsection, the sponsor must demonstrate that the new device does not raise new questions of safety and effectiveness. At some point, ODE reviewers will become more familiar with the device technology and indication for use and will only require engineering data. This can occur once the first three or four 510(k)s for that generic type of device have successfully gone through the review process. Clinical data requirements for other 510(k) devices are specified in guidance documents and do not change over time. Generally, 510(k) clinical trials are smaller and simpler than most PMA clinical trials. Depending on the risk level of the trial, an approved IDE may be necessary to conduct the trial.

Shelf Life (If Necessary)

Stability of device components and packaging integrity (for sterile devices) must be demonstrated. The "Shelf Life of Medical Devices" guidance document (http://www.fda.gov/downloads/medicaldevices/deviceregulationandguidance/guidancedocuments/ucm081366.pdf) offers general advice. The useful life of *in vitro* diagnostic products must be determined. A full report of real-time or, where appropriate, accelerated aging studies must be included.

Indication for Use Form

This form clarifies, for any interested party, the device's cleared indication(s) for use. The sponsor lists the indications for use on an ODE form. If the sponsor wishes to promote the device for a new indication, another traditional or abbreviated 510(k) must be cleared. Once a 510(k) is cleared, this form, the clearance letter, and the 510(k) summary are available from the FDA via its Website.

510(k) Summary

Summaries are released to the public via the FDA's Website. They provide interested parties with a brief description of the device and some of the data included in the 510(k).

The content of the summary is described in 21 CFR 807.92. All relevant items must be present, or the ODE will request clarification, potentially delaying 510(k) clearance.

Practical Aspects for 510(k)s

It is important to conduct research early in the 510(k) process and become aware of the cleared indications and technologies for competitive products. It is possible to request a competitor's 510(k) under the Freedom of Information Act, although processing times can often exceed 12 months, so this is not usually a practical option. For older 510(k)s, commercial information brokers may offer considerably faster response times.

Several recent guidance documents can be useful when preparing a 510(k). They include the following: "Draft Guidance for Industry and Food and Drug Administration Staff—Refuse to Accept Policy for 510(k)s" (http://www.fda.gov/downloads/MedicalDevices/DeviceRegulationandGuidance/GuidanceDocuments/UCM315014.pdf), which contains useful checklists and "Guidance for Industry, FDA Staff—Factors to Consider When Making Benefit-Risk Determinations in Medical Device Premarket Approval" (http://www.fda.gov/downloads/MedicalDevices/DeviceRegulationandGuidance/GuidanceDocuments/UCM296379.pdf) and "De Novo Classifications and Draft Guidance for Industry and Food and Drug Administration Staff—The 510(k) Program: Evaluating Substantial Equivalence in Premarket Notifications [510(k)]" (http://www.fda.gov/downloads/MedicalDevices/DeviceRegulationandGuidance/GuidanceDocuments/UCM284443.pdf).

ODE encourages sponsors to submit an electronic version of the 510(k) instead of one of the paper copies. The electronic version must be identical to the paper copy. Generally, it consists of a pdf file and may contain some internal navigation aids. Sponsors wishing to utilize more sophisticated technology should consult, in advance, with ODE.

Once a 510(k) is filed, ODE will e-mail the sponsor a letter acknowledging receipt of the submission and including the "K" number used for internal tracking. It is important to keep a copy of every document sent to or received from the FDA. Sponsors should also designate one company FDA contact person. That individual should document all phone conversations with reviewers. The FDA contact people should keep in mind that when ODE reviewers call with questions, they should listen carefully, but not leap to unsupported conclusions. If an ODE reviewer asks for specific data, confirm the data with experts in your company if you have any doubts. In most circumstances, a delay by a day or two will not be significant compared with the risk of misstatement. Increasingly, communications with reviewers occur via e-mail. Additional data can be officially submitted via fax or e-mail if the reviewer concurs. Once the reviewer's questions have all been answered, the reviewer's conclusions are reviewed prior to generating a clearance letter. A copy of the clearance letter is usually faxed to the sponsor shortly after it is signed. Commercial distribution can then begin. The official copy of the letter is mailed to the sponsor.

POSTSUBMISSION CONSIDERATIONS FOR 510(k)s

Manufacturers must comply with the QSR with respect to device modifications, production, and quality operations. Injuries or deaths (to patients or medical personnel) must also be reported to the FDA in accordance with the medical device reporting (MDR) regulation (21 CFR 803). Manufacturers are subject to inspection by the FDA investigators who review the QSR and the MDR compliance. Manufacturers must also register and list with the FDA. Refer to "Postmarketing Issues" section for a more detailed discussion of postmarketing responsibilities.

THE PMA APPLICATION

INTRODUCTION TO THE PMA

PMAs are necessary when the device developer wishes to market an innovative device in the United States that is not substantially equivalent to any other device that has been cleared through the 510(k) process. The PMA must demonstrate that the device is safe and effective. The PMA process is considerably more complex than the 510(k) process. Typical review times are approximately one year. Unlike most 510(k)s, a detailed manufacturing section describing the methods for building and testing the device must be included. Prior to final approval of the PMA, the CDRH office of compliance must review and approve the results of a preapproval inspection of the device manufacturing and development facilities. The sponsor of the clinical trial and two or three of the clinical investigation sites are also often subject to CDRH bioresearch monitoring (BIMO) inspections to confirm compliance with relevant sections of 21 CFR 812. Lastly, the postmarket requirements of a PMA are considerably more complex than those related to a 510(k). Specifically, a PMA annual report must be filed with the ODE each year and changes in labeling, materials, manufacturing, and quality methods, and specifications as well as changes in manufacturing location must all be reported to, and approved by, the ODE, in advance. This is done through the PMA supplement process.

THE PMA PROCESS

PMAs are large and complex documents, often considerably greater than 2000 pages. It can frequently take several years to obtain all the preclinical, clinical, and manufacturing data necessary for the PMA. It is essential that the PMA preparation effort be well planned, with good coordination between all functional areas involved in the development process. Advance research before a regulatory strategy is prepared should include a wide variety of sources. Shortly after a PMA device is approved, the approval letter, summary of safety and effectiveness, and official labeling are placed on the CDRH Website. These documents provide greater technical and regulatory detail than a 510(k) summary. The PMA submission itself is not available via the Freedom of Information process.

Once the indication for use and the device description have been established, it is important to confirm the key elements of the development plan with the appropriate reviewing branch within the ODE. The device developer may choose to obtain this information via an informal telephone call, an informal pre-IDE meeting, a formal designation meeting, or a formal agreement meeting. See "Draft Guidance for Industry and FDA Staff Medical Devices: The Pre-Submission Program and Meetings with FDA Staff," available at http://www.fda.gov/downloads/MedicalDevices/DeviceRegulationandGuidance/GuidanceDocuments/UCM311176.pdf for a more detailed description of these meetings. Generally, the more formal the meeting, the less interactive the discussion. Less formal meetings, while not generating binding agreements, can encourage very productive technical exchanges. The choice of meeting type involves balancing business, regulatory, and clinical needs.

Once a PMA development plan has been established and reviewed by the ODE, it is time to execute it. Generally, multiple activities such as manufacturing development and validation, preclinical functional and biocompatibility testing, and clinical testing proceed along parallel, often simultaneous tracks. In some cases, it may be clear during the planning phase that some data, such as manufacturing process information or preclinical testing data, may be available long before the clinical trial has ended. In these cases, it may be advantageous to submit the pieces of the PMA to the ODE as they are completed, rather than to send all the data at the very end. This process is called a modular PMA. If a PMA sponsor chooses to submit a modular PMA, a PMA shell or outline of the PMA must be prepared and approved by the ODE. The shell describes the contents of each module. As the modules are submitted, the ODE reviews them independently. Once review of a module has been successfully completed, the ODE sends the sponsor a letter stating that the module is "locked" and will not be reopened unless some portion of data already submitted changes in later stages of the development process. When the last module is submitted, the ODE considers the PMA complete.

User Fee

The user fee for the review of a PMA is payable either when the entire traditional PMA is submitted or when the first module of a modular PMA is submitted. The FY 2013 user fee for PMA review is US$248,000.00.

Advisory Panels

When a PMA device raises questions that the ODE reviewers have not previously addressed, they may choose to refer those questions to one of the advisory panels maintained for this purpose. Advisory panels are made up of experts in the field, who are not the FDA employees or from the industry. Many panel members are in academic medicine. The panel has one nonvoting industry representative and one nonvoting consumer representative. An executive secretary, usually a senior ODE reviewer, coordinates administrative details. The conclusions of the advisory panel are not binding on the FDA although they are almost always followed. Transcripts of advisory panel meetings are available via the CDRH Website. Videotapes of these meetings are also available from private sources. If competitive products have gone through the panel process, these meeting minutes can provide a great deal of valuable information on the types of data and analysis expected. If such a panel meeting occurs during your development process, it is very helpful if regulatory, medical, and technical development personnel attend in person. This can make preparation for your own panel meeting easier. More information on these panels can be found at http://www.fda.gov/AdvisoryCommittees/CommitteesMeetingMaterials/MedicalDevices/MedicalDevicesAdvisoryCommittee/default.htm

Clinical Data

According to Section 515 of the Food, Drug, and Cosmetic Act, a PMA must provide valid scientific evidence that there is a "reasonable assurance" that a device is both safe and effective. CFR 21, 860.7(c)(2) states that this evidence can come from

- Well-controlled investigations.
- Partially-controlled investigations.
- Objective trials without matched controls.
- Well-documented case histories conducted by qualified experts, and reports of significant human experience with a marketed device from which it can fairly and responsibly be concluded by qualified experts that there is reasonable assurance of safety and effectiveness of a device under its conditions of use.

In practice, the vast majority of PMA studies are designed as well-controlled studies where patients are randomized to either a treatment or a control group. Less frequently, studies can be designed to compare the investigational device to a historical control group, provided that the historical control group accurately reflects current US medical practice and the demographics of the US population. Data from other types of studies must always be reported to the ODE; however, they generally cannot stand as the sole source of performance data.

Use of International Data

Due to the international nature of the medical device industry, human clinical data may be available from ex-US studies before US development efforts have begun. How should these data be treated? Can they be used to support the PMA? Does the FDA require US clinical data?

There are no FDA requirements that a PMA must contain US clinical data. Good, credible, and ethical data will be accepted from any location. The ODE suggests that sponsors planning to submit international data in a PMA discuss their plans with them early in the development process. As with any clinical study, it is critical to assure that the study meets the ODE's expectations regarding medical and scientific issues such as the endpoints selected and the comparators used. If all of these parameters are consistent with ODE expectations, then there is one last set of tests before the data can be accepted. According to 21 CFR 814.15, the study must

- Be conducted in accordance with the Declaration of Helsinki or local ethical procedures, whichever is stricter.
- Use a patient population similar to the US patient population.
- Use a standard of care and medical practice similar to that in the United States.
- Be performed by competent investigators.
- Generate data, including source documentation, that are available for audit by the FDA.

Sponsors must be especially careful that study patients are not treated with drugs or procedures that are not available in the United States.

Components of the PMA

The PMA regulation (21 CFR 814) contains a description of the components of a PMA. ODE has produced numerous guidance documents that describe various PMA sections. Many of these guidance documents are product specific. Two of the more generic guidance documents can be found at http://www.fda.gov/downloads/MedicalDevices/DeviceRegulationandGuidance/GuidanceDocuments/UCM313368.pdf and http://www.fda.gov/MedicalDevices/DeviceRegulationandGuidance/GuidanceDocuments/ucm089764.htm

The items listed below include the major sections of a PMA. The length and complexity of each section will vary according to the technical details and regulatory issues associated with the product:

- Cover page
- Table of contents
- Summary of safety and effectiveness
- Device description and manufacturing data
- Performance standards referenced
- Technical data (nonclinical and clinical)
- Justification for a single investigator
- Bibliography
- Device sample (if requested)
- Labeling
- Environmental assessment (if necessary)

THE QUALITY SYSTEM REGULATION

The QSR regulates both the device development and the manufacturing process for all class II and III devices from the beginning of the development phase until the device is no longer supported by the manufacturer. It also covers the manufacturing process for many class I devices. It does not cover the research process for any medical devices. The goal of the QSR is to create a self-correcting system that reliably produces robust device designs and production methods, ensuring that devices perform in a manner consistent with their intended use. In many ways, the QSR has evolved into the glue that holds the medical device regulatory process together from development through end of use. Much of the information that is included in a 510(k) or PMA is taken from the DHF, prepared as a result of the Design Control requirements of the QSR. As discussed earlier, the existence of the QSR makes the special 510(k) possible. Once a device is marketed, the corrective and preventive action (CAPA) provisions of the QSR are closely related to compliance with the MDR regulation. An additional advantage of the QSR is that it follows the philosophy of the international medical device standard, ISO 13485, which helps to enable device companies that sell their product internationally to maintain common systems for most design- and production-related activities. In most cases, the QSR requires more extensive documentation than ISO 13485.[9] The system works by requiring specific activities and documentation beginning during the development process. The manufacturing and quality processes also require specific evaluations and procedures, all of which must be documented. Frequently, the FDA field investigators will follow the quality system inspection techniques (QSITs) approach[10] when inspecting a device facility. This process breaks the QSR compliance into four main modules and four satellite modules, some of which may not be applicable to all device firms. The FDA investigator will choose a subset of those modules and determine the firm's compliance with the QSR. This means that not every system is reviewed during a QSIT inspection; however, this process does yield a general assessment of the QSR compliance. Many firms consider the QSR requirements to be only a beginning and build on them, adding various customer-oriented feedback loops and financial accountability to the process. These integrated business systems can generate significant returns on the investment by reducing time to market, reducing the number of field corrections and recalls, and increasing customer satisfaction and device safety. The remaining portions of this section describe some of the provisions of the QSR. Design controls were already discussed in "An Introduction to the Medical Device Approval Process" section. Although these sections of the QSR are discussed separately, Figure 5.7 pictorially demonstrates how these functions are connected to each other. Readers should refer to the regulation[11] for complete information.

DESIGN CONTROLS

Design Controls (21 CFR 820.30) regulate the device development process. The research process is not regulated by the FDA. Additional information is included in Table 5.4.

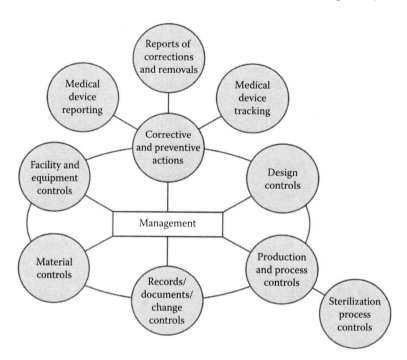

FIGURE 5.7 The seven primary QSR subsystems. (From FDA guide to inspections of quality systems, Washington, DC, August 1999.)

MANAGEMENT CONTROLS

Device firms need to demonstrate that they have management systems in place that can adequately control all the processes that take place in the life cycle of their products from the development phase onward. As Figure 5.8 illustrates, management is at the center of the quality system. The QSR holds "management with executive responsibility" ultimately responsible for the tasks specified in the regulation. Clearly, a device manufacturer with six employees will have less complex systems than a manufacturer with 600 employees. One SOP or a single organizational structure would not be appropriate for all device manufacturers.

One key provision of the QSR involves the controls that management places on the regulated system. First, there must be a quality policy in place, implemented and understood by all levels of employees. A quality plan and quality system procedures must also be in place. Next, management has the responsibility to assure that there are adequate resources and organizational structures to carry out all the activities specified in the regulation. A management representative must be formally appointed, must be actively involved in maintaining the quality system, and must regularly report those efforts to management with executive responsibility. Part of maintaining the quality system involves testing the system with prescheduled audits conducted by company staff that is not directly involved in the function audited. These audits must be conducted according to an SOP and recorded in an audit log

FDA Medical Device Regulation

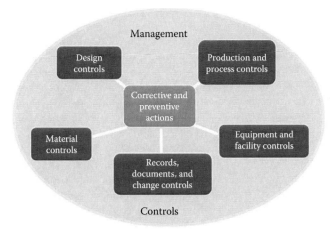

FIGURE 5.8 CAPA diagram. (From the FDA QSIT workshop presentation. The Quality System Inspection Technique: "QSIT"; http://temenet.silesia.pl/wp-content/uploads/2013/07/FDA-Quality-System-Inspection-Technique-QSIT-presentation.pps; http://www.fda.gov/ohrms/dockets/98fr/091099e.txt.)

and audit results documented. (FDA investigators do not generally have the authority to request copies of these audit reports.) The function of these audits is for the company itself to identify and then correct any quality system problems detected in the audits. Management reviews of a wide variety of quality data must be conducted at regular intervals and documented. These data include, but are not limited to, audit reports. Other sources of quality data include rework records from the manufacturing floor, incoming quality control (QC) testing summaries, service records, customer complaints and inquiries, and final inspection records. All of these data sources combine to paint a picture of the status of the company's products. It is critically important that the firm can demonstrate that action is taken as a result of these data. Identification of quality issues is important, but correction of problems and confirmation of the effectiveness of such corrections must also be documented.

CORRECTIVE AND PREVENTIVE ACTION

The CAPA portion of the regulation makes the firm's quality system self-correcting and self-improving. The five functional areas depicted in the boxes in Figure 5.9 feed information in the CAPA system. Under the supervision of the management, these data are processed and initiatives developed and executed that are intended to identify the causes of the problems and correct them. Data sources for the CAPA system include internal audits; incoming, in-process, and final QC testing results; service and repair records; and customer feedback. A variety of statistical tools may be used to better evaluate these data. Failure investigations should be conducted, according to a predetermined SOP, to determine the root cause of device failures. Once this has been done, a corrective action plan must be prepared and the corrective actions verified, or in appropriate instances, validated.

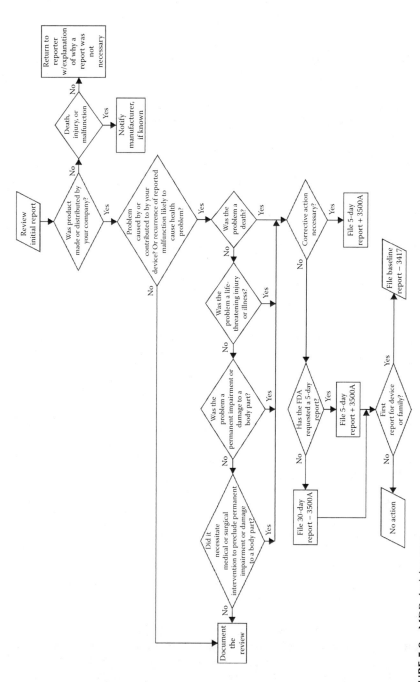

FIGURE 5.9 MDR decision tree.

PRODUCTION AND PROCESS CONTROLS

Production and process controls are the systems at the heart of the manufacturing process. Documentation is a major part of the control process. The device master record (DMR), a compilation of records containing the procedures and specifications for a finished device, is a key document for this functional area. Rather than existing as a discrete document, it is frequently an index that directs the reader to other documents where the necessary information is located. The device history record (DHR) is a compilation of records containing the production history of a finished device or a production run of devices. It usually contains manufacturing documentation, testing results, labeling documentation, and release/approval documentation. A single DHR may be generated for a large, expensive durable medical device, while another DHR may describe a production run of 10,000 disposable devices. Validation documentation, when necessary, is also a key part of production and process controls. Any production process whose output cannot be 100% checked once it is completed must be validated to establish, by objective evidence, that a process consistently produces a result or product meeting its predetermined specifications. Typically, processes such as sterilization or molding of plastic parts are validated. Other activities such as calibration, servicing and maintenance of production and testing equipment, and cleaning and maintenance of buildings must also be documented.

The goal of the QSR is to weave a web of systems that closely monitor development efforts to assure that a high-quality design is created, that the production of that device occurs in a controlled and predictable manner, and that various streams of quality data are appropriately analyzed and used to effect CAPA, when necessary.

POSTMARKETING ISSUES

REGISTRATION AND LISTING

Within 30 days of placing a medical device into interstate commerce, the manufacturer must register and list with the FDA. The device registration fee in FY 2013 is US$2575 per year. All device manufacturers, US and international, must register. The FDA Safety and Innovation Act (FDASIA) has imposed additional requirements on some facilities, especially contract manufacturers and sterilizers. Refer to http://www.fda.gov/MedicalDevices/ResourcesforYou/Industry/ucm314844.htm?source=govdelivery for more information. The purpose of registration and listing is to inform the FDA of the existence of the manufacturer. At some point after registration, the FDA may choose to inspect the device development and manufacturing facilities to ensure compliance with the QSR.

MEDICAL DEVICE MODIFICATIONS

Medical device technology evolves at a very rapid rate. Often the version of the device that receives initial PMA is a version or two older than the one sold outside the United States or sold by their competitors at the time of the PMA. 510(k) devices also change quickly. In both cases, sponsors need to understand how the FDA process will affect their product upgrade timelines and budgets. Modifications for all class II and III devices must be developed in accordance with the design control provisions of the QSR. Design controls have added enough extra confidence to the system so that, since 1998, the FDA has created new processes such as special 510(k)s and 30-day notices for PMAs that permit sponsors to rapidly implement some device modifications, as long as they comply with the design control provisions of the QSR.

Modifications to 510(k) Devices

There are three main classifications of 510(k) device modifications. They include those that require a documented review and a determination by the company that a new 510(k) is not needed, those that require a special 510(k), and those that require a new traditional or abbreviated 510(k). A cleared 510(k) cannot be supplemented.

A useful source of more detailed information on changes to 510(k) devices can be found at http://www.fda.gov/downloads/MedicalDevices/DeviceRegulation andGuidance/GuidanceDocuments/ucm080243.pdf. There are no annual reports required for 510(k) products (Table 5.6).

Modifications to PMA Devices

Modifications to PMA devices are more closely controlled than modifications to 510(k) devices. Table 5.7 briefly summarizes the various types of PMA supplements (see http://www.fda.gov/medicaldevices/deviceregulationandguidance/guidancedocuments/ucm089274.htm). Fees for PMA supplements, in FY 2013, range from US$3,968.00 to US$186,000.00, depending on the supplement type.

TABLE 5.6
Modifications to 510(k) Devices

Regulatory Action	Examples of Modifications
Review document in a memo to the file	Redesign the external case of a durable medical device so that it consists of few pieces to reduce production costs
File a special 510(k)	Add a feature that has already been incorporated in another device of the same type[a]
File a traditional or abbreviated 510(k)	Add a new indication, significant change in technology

[a] Modification to firm's own device and no change in intended use or fundamental scientific technology.

TABLE 5.7
PMA Supplement Types

PMA Supplement Type	Examples of Modifications
180-days supplement	Introduce a major change in the design of the device or in manufacturing or QC methods
180-days panel track supplement	Add a new indication for use where clinical data are required to support the application
Special PMA supplement changes being effected[a]	Introduce a change, which enhances the safety of a device, such as labeling changes that add or strengthen a contraindication, warning, precaution, or information about an adverse reaction.
30-days' notice[a]	Change the type of process used (e.g., machining a part to injection molding the part)
Real-time supplement[b]	Introduce minor design modifications that would otherwise require a 180-day supplement
Annual report	Update the microprocessor for the device when equivalence test has previously been approved by the ODE.

[a] The sponsor may choose either submission type.
[b] With the prior approval of the responsible ODE branch chief.
ODE, Office of Device Evaluation; PMA, premarket approval; QC, quality control.

PMA sponsors must also submit a PMA annual report to the ODE and pay a US$6680.00 filing fee (FY 2013) every year. This report contains updates on ongoing clinical trials, device modifications, adverse device effects, and MDR reports. More information on PMA annual reports can be found at http://www.fda.gov/MedicalDevices/DeviceRegulationandGuidance/HowtoMarketYourDevice/PremarketSubmissions/PremarketApprovalPMA/ucm050422.htm#annual

Medical Device Reporting

Significant problems with marketed medical devices must be reported to the FDA, using the Form FDA 3500A (MedWatch). While this same form is used to report pharmaceutical adverse events, Section D, suspect medical device; Section F, for use by user facility/distributor devices only; and section H, device manufacturers only are specific to devices. The process for evaluating and reporting device incidents is described in 21 CFR 803 and is not related to the ICH procedures employed for pharmaceuticals. The MDR regulation was originally implemented in 1984, and the final regulation was published in December 1995 and effective from July 31, 1996. An MDR SOP must be in place for every device manufacturer, regardless of device class. This applies even if the firm has never made an MDR report. MDR reports are available on the CDRH Website: http://www.accessdata.fda.gov/scripts/cdrh/cfdocs/cfMAUDE/search.CFM. A flowchart that summarizes the MDR process is included (Figure 5.9).

MDR reporting time frames. The manufacturer must report incidents to the FDA five working days after becoming aware of events requiring remedial action to prevent an unreasonable risk of substantial harm or events for which the FDA has required five days of reporting. This type of notification commonly occurs when a recall or field correction is necessary. The manufacturer must report incidents to the FDA 30 working days after becoming aware of the information that reasonably suggests that a device may have caused or contributed to a death or serious injury or if the device malfunctions in a manner likely to cause or contribute to death or serious injury. It is important to note that the regulation does not differentiate between injuries to patients, medical professionals, or family members. An injury to anyone that is caused by the device can be reportable.

Key MDR definitions. Serious injury: life-threatening, permanent impairment, or damage, or medical/surgical intervention necessary to preclude such damage. Cosmetic or trivial irreversible damage is not serious.

> Malfunction: the failure of the device to meet its performance specifications or otherwise perform as intended.
> "Becomes aware": when any employee, at any level of the company becomes aware of a reportable event.
> "Reasonably suggests": a professional medical opinion relating to the causal relationship between the adverse event and the medical device. If a physician, working for the manufacturer, concludes that an event is not related to the device, no report is necessary. This decision must be documented.
> "Caused or contributed to": causation can be attributed to device failures or malfunctions due to improper design, manufacturing methods, labeling, or operator error. Remedial action: any action that is not routine maintenance, routine servicing, and is intended to prevent the recurrence of the event.

Other MDR requirements. Manufacturers must retain all MDR records for two years or for the expected life of the device, whichever is longer. The types of records that must be retained include all MDR-related forms submitted to the FDA, explanations why reports were not submitted for specific events that were not reported, and documentation relating to all events investigated. Written procedures must be present for identification and evaluation of events, a standardized review process to determine reportability, and for procedures to assure that adequate reports are submitted to the FDA in a timely manner. Additional information on the MDR regulation can be found at http://www.fda.gov/medicaldevices/safety/reportaproblem/default.htm

UNIQUE DEVICE IDENTIFICATION

Congress authorized the FDA to require unique identifiers for medical devices in 2007. A proposed rule (see http://www.regulations.gov/#!documentDetail;D=FDA-2011-N-0090-0001) was published in the July 10, 2012, *Federal Register.* The goals of the Unique Device Identifier (UDI) effort include enhancement postmarket surveillance of medical devices, better management of device recalls, and potentially, reduction of medical errors. At present, there is no universal system to identify medical devices. Once removed from primary packaging, it is often impossible for health care providers to identify the model or manufacturer of implantable or single use devices. Consequently, many device failures are likely not reported to manufacturers or the FDA. The UDI system, combined with electronic medical records, is expected to improve reporting and vastly increase the postmarket device performance data available to the FDA and industry. Implementation of this proposed regulation will affect device labeling, complaint handling, MDR reporting, and other device safety systems.

Advertising and Promotion

Unlike pharmaceuticals, no preclearance of advertising copy is required for medical devices, even PMA devices. Promotional material must conform with cleared or approved indications for use. If a device is cleared for a general indication, more specific indications cannot be promoted, unless they are specifically cleared.

REFERENCES

1. Food and Drug Administration, Intercenter Agreements, October 1991. Available at http://www.fda.gov/oc/combination/intercenter.html
2. Food and Drug Administration, 510(k) database. Available at http://www.accessdata.fda.gov/scripts/cdrh/cfdocs/cfPMN/pmn.cfm
3. Food and Drug Administration, PMA database. Available at http://www.accessdata.fda.gov/scripts/cdrh/cfdocs/cfPMA/pma.cfm
4. Food and Drug Administration. Available at http://www.accessdata.fda.gov/scripts/cdrh/cfdocs/cfPMA/pma.cfm
5. Food and Drug Administration. Final Rule. Medical Devices; Current Good Manufacturing Practice (cGMP) Final Rule; Quality System Regulation. *Federal Register*, 1996; 61:195, 52602–52662.
6. Food and Drug Administration. Available at http://www.accessdata.fda.gov/scripts/cdrh/cfdocs/cfPCD/classification.cfm
7. Food and Drug Administration. Available at http://www.accessdata.fda.gov/scripts/cdrh/cfdocs/cfStandards/search.cfm
8. Helmus, M.N. (ed.). (2003). *Biomaterials in the Design and Reliability of Medical Devices*. New York: Kluwer Academic/Plenum Publishers.
9. Trautman, K. (1997). *The FDA and Worldwide Quality System Requirements Guidebook for Medical Devices*. Milwaukee, WI: ASQC Quality Press.
10. Food and Drug Administration. (1999). *Guide to Inspections of Quality Systems*. Washington, DC: Food and Drug Administration, August.
11. Code of Federal Regulations, Title 21, Section 820.
12. US Government Accountability Office. Report to Congressional Addressees, January 2009. Available at http://www.gao.gov/assets/290/284882.pdf
13. US Food and Drug Administration. Product Classification. Available at http://www.accessdata.fda.gov/scripts/cdrh/cfdocs/cfPCD/classification.cfm

14. US Food and Drug Administration. A New 510(k) paradigm—Alternate approaches to demonstrating substantial equivalence in premarket notifications. Available at http://www.fda.gov/MedicalDevices/DeviceRegulationandGuidance/GuidanceDocuments/ucm080187.htm
15. The Quality System Inspection Technique: "QSIT", http://temenet.silesia.pl/wp-content/uploads/2013/07/FDA-Quality-System-Inspection-Technique-QSIT-presentation.pps; http://www.fda.gov/ohrms/dockets/98fr/091099e.txt

6 A Primer of Drug/Device Law: What Is the Law and How Do I Find It?

Josephine C. Babiarz

CONTENTS

Introduction	170
What Is a "Law"?	170
Who Makes Laws?	171
Who Interprets Laws?	171
Who Enforces Laws?	171
What Is the Difference between a Federal Law and a State Law?	172
Which Laws Are More Important—The State or the Federal? What Is "Preemption"?	172
Are There Any Times When State Laws Control Medical Products?	175
Where Do I Find Laws?	176
How Do I Find Current Laws?	178
What Do All These Numbers Mean? What Is a Citation?	179
What Is the Difference between the USC and the Public Laws? How Are Laws Published?	181
What Is the Difference between the FDCA and the USC?	181
What Is an "Amendment"?	182
Why Can't I Find Section 510(k) in the USC?	183
Where Is the Biologics/Biosimilars Law? I Can't Find It in 21 USC Anywhere	183
What Is a "Regulation"?	184
What Is the Difference between a Law and a Regulation?	184
Which Is More Important—A Law or a Regulation?	185
What Is the Difference between the USC and the CFR?	185
How Do I Find a Current Regulation?	185
How Do I Find Older Regulations?	186
What Is "Guidance"?	186
Conclusion	187

INTRODUCTION

Working in a "regulated" environment has many connotations, but to those of us in medical products, the "regulators" always include the Food Drug and Cosmetic Act (FDCA), with its tangle of amendments and the Food and Drug Administration (FDA) itself, which issues innumerable regulations and guidances. You cannot be "in compliance" with regulations you have never read or laws you cannot find. Hence, this chapter.

In regulatory affairs, success often lies in the details, and the details are found in the regulations. As an example—the Human Protection Regulation at 21 Code of Federal Regulations (CFR) Part 50 requires that a participant sign and date in the informed consent form. The participant is to fill in the date in the form—not the principal investigator or the Clinical Research Associate (CRA). Unless you have read the regulation, you may think, like a lot of nonregulatory people do, that having the participant sign the form was all you needed—that making the participant fill in the date in the form was irrelevant or clearly less important. After all, the form is documenting that the participant is in the trial at a certain date. So, some folks—woe to them—might use a date stamp to memorialize when a participant signs or have the CRA date it. These folks are surprised when an FDA inspector writes up a site report, leading to a 483. After all, did not the participant sign the form? Yes, but the regulation requires that the participant *date* the form. This requirement is very clear *if you read the regulation*.

This requirement provides insight into the FDA's approach to documenting evidence—when the participant dates the form, the participant verifies that he/she was there on that date. That act of dating is evidence that the participant agreed to the informed consent prior to the commencement of the trial. However, if you are looking for informed consent in 21 United States Code (USC), or cannot distinguish between the different versions of informed consent in the CFR, you do not have a chance of finding the applicable regulation, much less reading it.

This chapter will help you know the difference between a law and a regulation and how to find the current ones. Wherever possible, you will be given Internet addresses. While the appearance of the Web pages themselves may change, the fundamentals of what you are searching for and the rules and techniques you use to search are the same. Once you understand the basics, you will be able to use this chapter to locate the specific information you need to succeed.

This chapter is organized as a list of frequently asked questions, to help you find the information efficiently. Intrepid regulator—forge on! You can find it!

WHAT IS A "LAW"?

A law is a rule you have to follow. The laws can also be called "statutes," "public laws," "acts," or "codes." New laws are "enacted" (meaning they are suddenly there and you have to follow them). Old laws are "repealed" (meaning that they really go away and you do not have to worry about them anymore, except maybe to prove that you complied when they were in effect) or "amended" (meaning that they say something a little different now, and you probably need to know what changed and comply with that change).

Most of the laws that concern us will be amended from time to time; we will have examples of amendments and see how they work in "What Is an 'Amendment'?" section.

WHO MAKES LAWS?

In the United States, laws are passed in two steps: first, the law is written and voted by an elected group of people (the Congress, if it is a federal law, or a legislature, if it is a state law); second, it is signed by the president, on the federal level, or a governor, on a state level. If the president vetoes a law, that is, refuses to sign it, then the Congress can override the veto by a vote of at least two-thirds of its members. Governors' vetoes can also be overridden.

WHO INTERPRETS LAWS?

There are two branches of government that interpret laws—the executive and the judicial. Because these branches have different jobs, their interpretations are also different.

The executive branch (meaning the president and all agencies controlled by the president), which includes the Department of Health and Human Services (HHS) and inside that department, the FDA, interprets the laws by issuing regulations, which tell people what to do in order to comply with the laws. The executive branch is also charged with enforcing the laws, and is in charge of the prosecutors and police.

The judicial branch (meaning the courts, including the Supreme Court of the United States) also interprets laws. The primary role of the judiciary is to decide whether a certain law is constitutional (usually whether Congress had the power to enact it, as with the court case on health care reform), to resolve conflicts between two or more federal laws or between a federal law and state law, and also to determine whether the regulations issued by an agency are consistent with the laws and the Constitution. A court can void a regulation if it decides that the regulation is too vague to be followed, contradicts the law, or gives the executive more power than Congress wrote in the law.

WHO ENFORCES LAWS?

At times, it seems like everybody does. The fact is that the executive branch of government is charged with enforcing laws. This list includes the Department of Justice, the Drug Enforcement Administration (DEA), the Federal Bureau of Investigation (FBI), and the state police for criminal matters. An FDA inspector may uncover something troublesome and refer a matter out for criminal investigation, so treat all folks carefully.

WHAT IS THE DIFFERENCE BETWEEN A FEDERAL LAW AND A STATE LAW?

There are two levels of government in the United States—federal and state. On the federal level, the US Constitution establishes the three branches of the government: the Congress, which passes the laws; the executive, which enforces the laws; and the judiciary, which interprets the laws and decides on conflicts between the branches.

The Constitution recognizes that the states each have their own, independent government. The states also have three branches of government: the legislative, the executive, and the judiciary, which function in the same way as their federal counterparts. Under the states, there is a third level of government, the local (city, town, or county).

You must comply with all applicable laws on each level. The Constitution and the courts prevent conflicts between the two levels of government, and each level has its own "turf" so to speak.

WHICH LAWS ARE MORE IMPORTANT—THE STATE OR THE FEDERAL? WHAT IS "PREEMPTION"?

"Preemption" means that the federal laws are superior to, have control over, trump all others, and simply just must be obeyed. Any state laws that contradict or conflict with the federal laws are invalid. This *does not* mean that you do not have to worry about any state laws; in fact, you must obey all the state laws that do not contradict or conflict with federal laws. Here are some general preemption concepts and rules.

A perfect example of preemption is the new drug application (NDA). In order to market a prescription drug in the United States, you need an approved NDA issued by the FDA. You go through the approval system one time only, at the federal level, and when you get the NDA approved, you can market the product in all 50 states. The good news is that you need not get separate approvals from each state, that is, California, Idaho, Texas, and Mississippi, to market the product there. The other news is that you cannot get approval in just one state—say Nevada—and have that approval give you permission to market in all other 49 states. So, from a regulatory and economic perspective, the federal approval evens the playing field and eliminates the various different types of approval testing 50 states could require. If you did try to market a prescription drug on the basis of state-level approval, you would be in violation of the FDCA and be marketing a product illegally, and claims of "off-label," "misbranded" would apply.

A Primer of Drug/Device Law 173

Preemption does not mean you can ignore the states, and there are recent decisions from the US Supreme Court which show the limitations of the preemption doctrine. The question the Court decided in several recent cases is whether the FDA approval of a product meant that the manufacturer was free from any product liability claims on a state level. (A product liability claim is one that argues the product sold was defective, because of a defect in its design or manufacture, or that its warnings were inadequate, and because of that defect, a person was injured. These claims are very expensive for drug, device and biologics manufacturers alike.) The preemption rule is different for brand name drugs and generic drugs, and for PMA-approved devices and 510(k) devices. (The reader should refer to earlier chapters in this book for discussions of these terms.) To make sense out of these different conclusions, we look to the policy and the actual language of the laws.

The Congressional policy for and scope of authority of the FDA makes it a different type of federal agency. You can get a refund from the Internal Revenue Service (IRS) if you have overpaid a tax; an individual consumer never gets a "refund" or payment from the FDA if a drug did not work for the indicated condition. Congress determined it was not economically possible for patients to get refunds; and, consequently, the FDCA does not allow people to bring personal injury actions. The enforcement of the FDCA is limited to the FDA's authority over product manufacturers. The FDCA does not control physicians or other prescribers. The policy question Congress had to address is how do people injured by defective drugs and devices get relief? Some are injured because of the product and require additional medical care. The additional medical care can be covered by private insurance, funded by premiums paid by fellow employees in most cases, and some of these injuries are covered by public health insurance, such as Medicare and Medicaid. Congress had to consider the costs of medical product injuries due to defects on many sides. Congress decided that the FDA would not be financially responsible to individuals for personal injury. This policy decision squares with the system of review established in the FDCA—manufacturers design and conduct clinical trials, and the FDA rules on whether the product is marketable based on the evidence submitted. The system has become very intricate over the decades, and consequently, the rules for compensating a patient for personal injury caused by a defective medical product—a case usually brought in a state court—are complex. The Supreme Court decisions seem contradictory if read superficially; you have to think deeply about the system to harmonize these conclusions.

The leading case on preemption of brand name or innovator drug claims decided by the US Supreme Court is *Wyeth v. Levine* [129 S. Ct. 1187 (2009); 555 US 555]. Wyeth sold the drug Phenergan under an approved NDA; the Phenergan label warned about significant risks of administering Phenergan by the IV push method. A patient received the drug via IV push, and lost her arm to gangrene attributed to the IV push method used to administer the drug. The precise language that warned against the IV push was approved by the

FDA as part of the product label. The Supreme Court decided that Congress did not write express language in the statute, which said that the FDA labeling controlled state cases, and therefore, the FDCA did *not* preempt state law failure to warn cases. Accordingly, Wyeth is responsible to the patient and must pay damages for failing to adequately warn against the adverse effects, again despite the fact that the warning label was approved by the FDA. The Wyeth case is controversial, and it is easy to predict that many will lobby Congress to change the law.

The Wyeth case governs state preemption of brand name or innovator drug claims; the rule for generic drugs is different. The rule for generic drugs is that FDA approval of an Abbreviated NDA (ANDA), generic drug submission, preempts state cases based on (1) product defect; and (2) design defect (including the claim that the product is unreasonably dangerous). The product defect case, *Pliva v. Mensing* [131 S.Ct. 2567; 180 L. Ed. 2nd 580 (2011)] was based on a failure-to-warn claim, just as Wyeth was. Two patients took metoclopramide, a generic of Reglan (the innovator or brand name drug) which was approved for digestive tract disorders. After taking metoclopramide over an extended period, the patients developed a severe neurological disorder, tardive dyskinesia. The Supreme Court looked at the ANDA approval process, and noted the differences between the NDA and ANDA processes. In the NDA approval process, the Court emphasized that a manufacturer had to prove the drug is safe and effective and that the proposed label is accurate and effective. However, the generic approval process enacted by the Hatch–Waxman Amendments allows manufacturers to develop generic drugs inexpensively, without duplicating the clinical trials needed for the innovator drug (slip opinion, p. 5–6; additional citations omitted). Further, under the FDA regulations, the emphasis on generic approval is the demonstration that the active ingredient in the generic drug is identical to that in the innovator drug. The Court concluded that the brand/innovator manufacturers and generic manufacturers have different responsibilities under the FDCA, and that the generic manufacturer is responsible for ensuring that the label on the generic product is the same as the brand label. The Court then concluded that the generic manufacturer had no authority to change the label for the generic drug, because that change would result in different labels for the brand/innovator drug and the generic drug. The Court accepted the FDA's policy that the label for the two drugs must be identical and the state product liability claim was preempted. In 2013, the Supreme Court decided that the FDCA preempted a decision by a state court that a drug was unreasonably dangerous by design. In the case, *Mutual Pharmaceutical Co. Inc. v. Bartlett* [133 S. Ct. 2466 (2013)], the FDA-approved labeling did not warn about toxic epidermal necrolysis at the time the prescription was written. In addition, the physician prescribed the brand name drug but the pharmacist dispensed the generic form. The Court held that the claim was preempted, despite the fact that FDA required an explicit warning about toxic epidermal necrolysis after the product was prescribed and dispensed.

Case law concerning devices also has two seemingly contradictory conclusions, and the decision for devices rests on what type of device is the subject of the court case. This difference is due to specific language in the FDCA. In the Supreme Court case, *Reigel v. Medtronic* 552 US 312 (2008), the Court had to decide whether a balloon catheter had adequate warnings against misuse. The facts were that the balloon catheter was contraindicated for the patient Reigel; the case was based on whether the warning, as stated in the FDA-approved labeling, was in fact adequate and preempted state law claims. The Reigel decision is the opposite of the Wyeth decision because of the language in the FDCA. The FDCA allows the FDA to preempt state requirements for a class III, PMA-approved device. The balloon catheter is a class III device, and went through a premarket approval (PMA) application. Based on this language, the Supreme Court decided that once class III devices were approved under a PMA, the FDA's approval preempted any state personal injury claims. The preemption protection only applies to approved class III devices. In an earlier case, involving a class II device, cleared for market under a 510(k), the Supreme Court decided there is no preemption of state personal injury claims, and the cleared, class II device is still subject to state personal injury claims [*Medtronic v. Lohr* 518 US 470 (1996)].

Federal preemption does not do away with all state law requirements of a company that makes devices. Federal law does *not* preempt the ability of states to require any company or person doing business in the state to register with the state, pay a corporate excise tax to the state, a property tax to the locality where the business operates, or from collecting sales tax on the sales of any devices in that state.

ARE THERE ANY TIMES WHEN STATE LAWS CONTROL MEDICAL PRODUCTS?

Yes, there are areas where the states exert control over medical products and their development, manufacture, clinical investigations, and other activities, which impact your ability to manufacture and distribute medical products.

Federal law does not address certain really important things like how old you need to be before you can sign a contract (or give informed consent) or what medical data privacy rights you have apart from the Health Insurance Portability and Accountability Act (HIPAA). Legal requirements for disclosures in informed consent are also found on a state level, and the FDA's regulations clearly indicate that federal and state laws *both* govern the document and the process.

In these cases, the federal government adopts the state's definition in enforcing the federal regulation. So, let us say your state says a person can legally sign a contract at 18 years of age. Another state says you have to be 21 years old. Even if your company is located in the state with the 18 years age requirement, and you submit your FDA application from your state, and the FDA has reviewed and approved your informed consent form (contract), when you use subjects in a state that says you have to be 21 years to sign a contract, you need to have a parent or guardian sign for the 18-year-olds. The FDA does not preempt the local requirement of 21 years. You can have 18-year-olds sign only in a state that allows 18-year-olds to be bound to contracts. You cannot argue that the federal government preempted the age of consent, where a state says you have to be 21 years old to be bound to a contract.

Pharmacy laws are one of the biggest examples of how each state can and does exert control over medical products. The FDCA provides that certain medical products are available only by prescription from a physician. The FDCA goes on to say that it does not regulate physicians. The states regulate physicians and pharmacists, as well as how certain products are stored, dispensed, and used. As discussed earlier, there are many instances where the federal government is pleased to let the states "work out the details," so to speak.

You should also be on the lookout for certain state laws governing biologics, which can sometimes be cloaked as privacy statutes. These have significant impact on genetic testing, data collection, and product development, which relies on such data.

In conclusion, there is interplay between federal and state laws. The correct answer to the question is that you must comply with all laws that apply to your product. Just complying on a federal level, or on a state level, is not enough.

WHERE DO I FIND LAWS?

You can find them in a lot of places; one of the key responsibilities of the US government is to publish laws, so that the public can read and comply with them. Laws are available in printed books and online, 24/7; there is no excuse for not finding them.

First, if you are going to make a career in regulating medical products, it is worth your while to buy a copy of the relevant FDA-enforced laws. You can buy just the FDA volume, *Title 21*, from a number of publishers, including the Government Printing Office. One is available from the USC Service, Lawyers' Edition, Lexus Law Publishing; *be sure to get the most current version, including updates*. The Lawyers' editions or Annotated Editions contain not only the laws but also key court cases and the amendment history of the sections, which can save you considerable time. You should buy updates pocket parts each year.

Students balk at the cost of the book when the information is available online, but most quickly decide the book is worth the cost in convenience and organization. The Internet gives you access to information in little pieces; it can be very frustrating to use. When you have a text, you can read the entire law that you are researching; you do not have to scroll through sections or have pages reload. You can find the US laws on the Internet, using the Library of Congress Website,

A Primer of Drug/Device Law 177

at http://www.loc.gov/law/help/guide/federal/uscode.php (accessed on August 19, 2012), which looks like this:

```
Research & Reports | Guide to Law Online | How Do I Find...? | Current Legal Topics | Guides to Our Collections
International | Nations | U.S. Federal | U.S. States & Territories | About the Guide | Guide Index

    ⊚ U.S. Code                                    ⊚ Public Laws
    ⊚ Statutes at Large

U.S. Code
  ›  United States Code (U.S. House of Representatives, Office of the Law Revision Counsel) offers introductory material to the
     Code and access to a search engine. Titles 1 through 51, the Table of Popular Names, and Tables I to VI are based on
     Supplement IV of the 2006 edition (January 7, 2011) of the Code. The Organic Laws are based on the 2006 edition (January 3,
     2007) of the Code.
       ›  Download the United States Code
       ›  Search the United States Code
       ›  Search Prior Versions of the U.S. Code - beginning with 1988 ed., Suppl. II (January 2, 1991)
       ›  United States Code Classification Tables - sorted by Congress, then by either Public Law or U.S. Code section; includes
          prior tables extending to 1995
       ›  USCprelim - provides an advance posting of the next online U.S. Code
       ›  U.S. Code (Beta) - for testing and feedback
  ›  United States Code (U.S. Government Printing Office, FDsys) provides a search engine for searching the 2006 edition,
     Supplement IV of the Code (as of January 7, 2011)
  ›  United States Code (Cornell Legal Information Institute) provides an updated interface to the U.S. Code, with links to notes
     and legislative activity through THOMAS, various search options, and a popular name table

                                                                                              ⊚ Back to Top

Statutes at Large
  ›  United States Statutes at Large (Law Library / National Digital Library Program)
     v. 1-18 (1789-1875)
  ›  United States Statutes at Large (Digitized) (U.S. Government Printing Office / FDsys) v. 65-116 (1951-2002) authenticated,
     available in large PDF files
  ›  United States Statutes at Large (U.S. Government Printing Office, FDsys) v. 117-122 (2003-2008) authenticated

                                                                                              ⊚ Back to Top
```

This site links to Thomas, the US Congressional source for information. The link to "download the code" does not take the place of the code book in my opinion; it organizes information by date and Congressional action.

A number of law schools have Websites with useful information, and you can access most of them without paying tuition. One that I like is the Legal Information Institute of Cornell Law School; the Website is www.law.cornell.edu. This site frequently connects sections with hypertext links, which is convenient. There are also Websites such as Findlaw.com® that can help you locate the specific law you want, but since Findlaw links to the federal government and Cornell Law School sites, you might as well go directly to them. For current information on Congressional actions, status of bills, and a history of Congressional intent, the Library of Congress Website is a treasure of useful information. The site is named after Thomas Jefferson, and the Web address is http://thomas.loc.gov/. Note that the Thomas URL does not start with "www."

There is an important trick to finding the precise section of the law you are seeking—and that is to understand the two systems under which the laws governing the Food and Drug Administration are organized. The answer to this question can be found in the "What Is the Difference between the USC and the Public Laws? How Are Laws Published?" section.

HOW DO I FIND CURRENT LAWS?

The key to finding current laws is to know the code numbers, the Congressional Session, the popular name of the law, or its formal title. Any one of these can lead you to the correct place. As an example, let us outline how you would really find "Obamacare," if you wanted to read the law. "Obamacare" is the pundit name for the law; the law as enacted is not called "Obamacare," so doing a text search in one of the legal Websites is not going to work. When you only know a provision or a name, try a general search engine. I used "obama care law text" as a search in Google® (accessed on August 19, 2012); it gave me several good options. The first three hits were the following:

1. Read the Law/Health Care.gov
   www.healthcare.gov/law/full/
   Mar 23, 2010—The Affordable *Care Act* puts in place comprehensive health reforms that enhance the...Read the *law* at HealthCare.gov to learn more....The first link listed above contains the full *text* of the Affordable *Care Act* and the Health...
2. Text of the Obamacare ruling and bill
   brainshavings.com/obamacare/
   You want the full *text* of the *bill*? Just the facts? Here you go. The *bill* everyone knows as *Obamacare* is officially called the Patient Protection and Affordable...
3. Patient Protection and Affordable Care Act—Wikipedia, the free...
   en.wikipedia.org/wiki/Patient_Protection_and_Affordable_Care_Act
   4.3.1 Public opinion; 4.3.2 Term *"Obamacare"*...The *text* of the *law* expands Medicaid eligibility to include all individuals and families with incomes up to 133%...

The first hit, a ".gov" address, is to the federal government, and this actually gives me the correct name of the law: "The Patient Protection and Affordable Care Act." The third hit is to Wikipedia, an ".org" site; ".org" denotes a not-for-profit, but not necessarily unbiased Website. Since I am only looking for the official name, to find the actual law myself, this works as well. When I have the true name of the law, I can use any of the previously mentioned methods to read the law.

Remember to check the publication date and understand the source's policy for obtaining current laws. Basically, you want to be sure that any source you check

A Primer of Drug/Device Law 179

is updated, so you are not reading something that is old and may not have been updated or is missing the most recent enactments.

It sounds obvious to check the date, but this is a detail that many overlook. With a book, check the title page, with the date of the copyright. If it is this year's date, it is almost current, except for any laws that have passed since the book was published. For an earlier copyright date, check if there is a back "pocket part." If the book is more than a year old, there should be an update, known as the pocket part inserted in the back flap or "pocket" of the book. Because there is an additional cost for the pocket part, and there are recent library budget cuts, some libraries have cancelled their pocket part subscriptions. But it is always worth a check.

For Internet resources, you still have to check the publication date. This is not the date of your search, at the bottom of the page you print, nor is it the date the page was updated; it is the date that this compilation of laws was last enacted and updated. The compilation date tells you that the compilation does *not* include any laws passed after that date.

Some of the libraries require you to insert a date before you can access the material. One of these is the version of the USC available on the FDsys page of the US Government Printing Office's Website (http://www.gpo.gov/fdsys/browse/collectionUScode.action?collectionCode=USCODE&searchPath=Title+42&oldPath=&isCollapsed=true&selectedYearFrom=2011&ycord=1596). The pull-down menu that asks you to choose the year before it displays the USC indicates the most current version available there is dated 2011. Consequently, the latest amendments to the FDCA, known as "PDUFA V," Prescription Drug User Fee Act, to some, and more formally as the "FDA Safety and Innovation Act (FDASIA)," signed into law on July 9, 2012, is not included in the version on the Website. You can access the language of the act via the FDA Website, which links you back to the GPO to a specific link, http://www.gpo.gov/fdsys/pkg/BILLS-112s3187enr/pdf/BILLS-112s3187enr.pdf, or you go to Thomas, type in "safety and innovation" in the text search, and select the correct law from the listed search results.

When you retrieve the updated versions of the code, you are in fact accessing and using the Public Laws as well. The section "What Is the Difference between the USC and the Public Laws? How Are Laws Published?" explains the similarities and differences between the public laws and the USC.

WHAT DO ALL THESE NUMBERS MEAN? WHAT IS A CITATION?

Most laws are organized in outline form; the numbers identify sections and paragraphs of the law. The "citation" is the standard reference tool that identifies where specific text can be found. For the USC, there is a title number, which comes first, and then the sections and paragraphs. Happily, Google and many other search engines think of numbers as text, so if you type in the citation, within quotation marks, you can often go right to the actual law. This search technique also works for regulations.

The US Office of Patents and Trademarks has published a wonderful graphic and a definition set, accessed August 17, 2012, at http://www.uspto.gov/main/glossary/lawsandrules.htm

The GPO says it best, and that page is reprinted here for you:

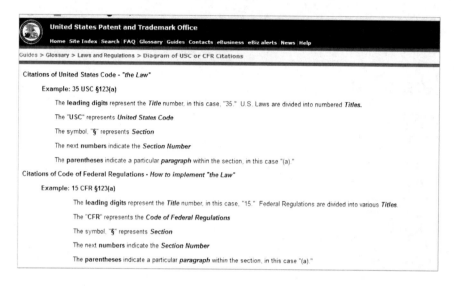

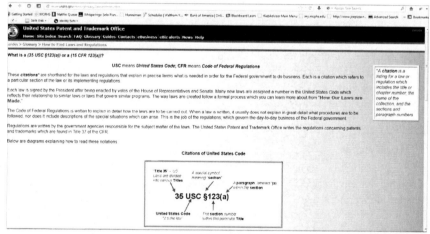

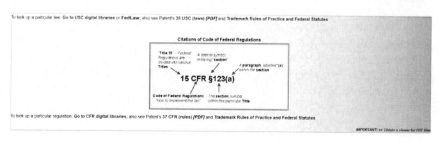

WHAT IS THE DIFFERENCE BETWEEN THE USC AND THE PUBLIC LAWS? HOW ARE LAWS PUBLISHED?

Really and truly, the USC and the public laws are *both* laws passed by Congress and they both say the same things. They are just organized in a different way. The permanent public laws are those laws that Congress works on and passes on a daily basis. The public laws are referenced by the identity number of the Congress working on them. The Congress in session on January 1, 2012, is the 112th Congress, so all of its laws begin with the number 112. Remember that the House of Representatives is elected for a term of two years, so the number of each Congress covers two years. The FDASIA was passed by the 112th Congress, and it became Public Law 112-144, which means it was the 144th law passed by the 112th Congress.

As Afshin Shamooni, one of my students, phrased it, the public laws are like a diary, where each law is recorded on the day it is passed. Congress can pass different laws on the same subject in the same year and in different years. The Public Laws reflect Congressional thinking at a particular time. When Congress decides to enact or change a law, that enactment or change is described in a new Public Law. The Public Law contains both the content of the new law Congress wants to pass, and the instructions on the way to change the existing laws or insert the new laws. The only way to find out all the laws passed on a subject using the public laws is to read and search all of them. This is not particularly efficient, which is why the Congress ordered a consolidated format, which we call the USC.

The USC puts all of the public laws passed on any one subject into one big chapter, or title. The USC changes the specific language Congress wanted to change in the earlier versions, and then reprints the entire document with all the changes and additions in one place. It is current up to the date of its publication with all the revisions that Congress has enacted, all in one place. Because it is one document, the numbering system of the USC is different from the numbers used in the Public Laws, which are simply shorter. However, the substance, the content of the actual law, is the same. An example of the similarities and differences between the Code and Public Law, with a sample and instruction, is found in the "What Is an 'Amendment'?" section.

WHAT IS THE DIFFERENCE BETWEEN THE FDCA AND THE USC?

The main difference is the section numbering; the actual substance, the language, is the same. The FDCA is the name of a public law, originally passed decades ago and updated regularly. The USC is the name of the compiled law, and Title 21 contains the FDCA, with all its amendments that regulators normally use.

The section "What Is an 'Amendment'?" explains what an amendment is and how public laws are integrated into the code. The section "Why Can't I Find Section 510(k) in the USC?" has the Websites that correlate the section numbers of the FDCA with the USC.

WHAT IS AN "AMENDMENT"?

An "amendment" is a change to a law or regulation. An amendment is formally approved—that is, passed just as a law, or in the case of a regulation, issued. As discussed earlier, the Public Law version contains the new language of the law, and the instructions on where this new language goes—its position in the actual wording.

Many of the Public Law provisions are simply instructions to the editors of the USC, telling the editors what words and punctuation marks to change. One can only understand the intent and operation of the new law by making these changes and reading the now-edited text. Notice that each public law has its own table of contents and section numbers, and that these section numbers are not the same as those in the code. Each public law follows its own outline numbering system, and the code, because it incorporates all changes from all public laws, has a much larger outline and many more numbers.

The words that make up the laws in both the FDCA and the USC are the same; it is the classification and numbering system that changes with the source you are using. Only the public laws contain the instructions for actually editing the main body of law, in our case, the FDCA, and by making those changes, impacting the USC itself.

This is easier to understand using an actual example. We will use the effective date of the new PDUFA fees for this example; these changes are included in the FDASIA, which was cited above. The pre-FDASIA 2011 version of Title 21, Chapter 9, Section 379h, gives the FDA the authority to assess certain types of drug fees "beginning in fiscal year 2008." This date reflects PDUFA IV, which started in that fiscal year. The actual language is shown below:

§ 379h. Authority to assess and use drug fees
(a) Types of fees
Beginning in fiscal year 2008, the Secretary shall assess and collect fees in accordance with this section as follows:
   (1) Human drug application

The new law, FDASIA enacted in 2012, increased the fees, and the new amounts are due in fiscal year 2013, which starts October 1, 2012. The law which empowers the FDA to charge the increased fees, as well as the instruction on where to put the new language, is in Section 103 of FDASIA, and is shown below:

SEC. 103. AUTHORITY TO ASSESS AND USE DRUG FEES.
   Section 736 (21 USC 379h) is amended—
      (1) in subsection (a)—
         (A) in the matter preceding paragraph (1), by striking
"fiscal year 2008" and inserting "fiscal year 2013"

In this way, Congress has clearly instructed the Law Revision Counsel to change the date from 2008 to 2013. The Public Law follows the outline form and at Section 103 of Public Law 112-144, Congress has authorized a change in date, in a particular section of the law. The USC section amended is 379h; the FDCA section amended is Section 736.

WHY CAN'T I FIND SECTION 510(K) IN THE USC?

Do not panic. "They" have not eliminated that wonderful loophole, known as the "same-as" or "me-too" exemption for devices. The section number by which the provision is known refers to the FDCA. In the FDCA, the numbering system places that section at 510(k). When the amendments were incorporated into the USC, the appropriate numbering system for premarket notification was section 360(k). You can find the link from the FDA Website, http://www.fda.gov/RegulatoryInformation/Legislation/FederalFoodDrugandCosmeticActFDCAct/FDCActChapterVDrugsandDevices/default.htm, accessed August 20, 2012. This page does include a cross-reference of FDCA and USC sections, but the chart uses the FDCA sections as a reference, and you have to hunt for the appropriate USC numbers. The FDA has removed a page which assisted with these cross-references; when you knew one or the other of the citations, you could find the cross-reference. The page is no longer updated, but still of use, and can be accessed at http://crawls.archive.org/katrina/20050908025833/http://www.fda.gov/opacom/laws/crossref.html

For the disbelievers, here is the infamous section 510(k); the number in brackets indicates the USC section, 21 USC 360. The actual provision reads as follows:

510(k) REGISTRATION OF PRODUCERS OF DRUGS AND DEVICES

SECTION 510 [21 USC 360]

(a) As used in this section....
(k) Each person who is required to register under this section and who proposes to begin the introduction or delivery for introduction into interstate commerce for commercial distribution of a device intended for human use shall, at least ninety days before making such introduction or delivery, report to the Secretary (in such form and manner as the Secretary shall by regulation prescribe)—
(1) the class in which the device is classified under section 513 (360c) or if such person determines that the device is not classified under such section, a statement of that determination and the basis for such person's determination that the device is or is not so classified, and
(2) action taken by such person to comply with requirements under section 514 (360d) or 515 (360e) which are applicable to the device.

WHERE IS THE BIOLOGICS/BIOSIMILARS LAW? I CAN'T FIND IT IN 21 USC ANYWHERE

The correct name of the "biosimilars" law is the Biologics Price Competition and Innovation Act (BPCIA) of 2009. This is a section of the Patient Protection and Affordable Care Act (PPAC Act, discussed earlier). The BPCI did not amend the FDCA and is not under Title 21, which is where the FDCA is located. The BPCI amends the Public Health Service Act (PHSA), which controls biologics and is under Title 42 of the USC. The FDA uses the same definition for biologics as in the PHSA, and the FDA has authority and jurisdiction over biologics that are marketed as drugs. To locate this law, use the reference to the PPAC itself, starting at Title VII, Section 7001 (see http://www.gpo.gov/fdsys/pkg/PLAW-111publ148/pdf/PLAW-111publ148.pdf). As of the date of publication of this chapter, the FDA Website link to the PHSA was not updated with this BPCI text. (Author's search was to select the link to "Other Laws Affecting FDA" from the Regulatory Information Page, http://www.fda.gov/RegulatoryInformation/Legislation/default.htm, accessed on August 20, 2012.)

WHAT IS A "REGULATION"?

A regulation is a binding instruction issued by an agency (in our case, the Secretary of HHS for the FDA) that tells you how to interpret and comply with a law. You must follow a regulation—that is, if you fail to follow a regulation, and you have an inspection, the FDA inspector must write up your failure on a 483; failures to follow regulations usually end up in the "issued warning letter" section of the FDA Website, not a good place to be.

WHAT IS THE DIFFERENCE BETWEEN A LAW AND A REGULATION?

Laws and Regulations are issued by different branches of government and have different functions. However, each must be obeyed.

Laws come from legislative bodies, like Congress, and set policy in broad terms. Regulations come from the executive branch, and provide details on how laws are to be implemented, or obeyed. The FDA is part of the executive branch of government and is under HHS. HHS is a cabinet agency, whose secretary reports to the president.

Congress sometimes directs the executive branch to issue regulations. That was the case with FDAMA, where Congress decreed that regulations concerning dissemination of information on unapproved products be issued. The FDA did promulgate an initial set of regulations, which restricted the amounts and types of information manufacturers could publish concerning unapproved products. Litigation ensued over the breadth of the regulation, and the courts ultimately decided that the regulation was overly broad, in that it infringed constitutional rights of commercial free speech, and so struck down the existing regulation.

To update the status of that regulation, the law which directed the FDA to issue regulations on the dissemination of information on off-label products expired. In place of that regulation, the FDA has issued guidance, which can be found at http://www.fda.gov/RegulatoryInformation/Guidances/ucm125126.htm

While the courts have the power to nullify regulations that are not consistent with the statutes, have been improperly issued (usually meaning that there have been inadequate public hearings), or exceed the Agency's authority, these cases are really few and far between. Most of the time, the FDA's regulations are given great deference by a court, and are upheld.

WHICH IS MORE IMPORTANT—A LAW OR A REGULATION?

The problem for regulators is that both are equally important. Violation of laws can result in criminal penalties, but hopefully no one is reading this chapter with an eye to "cutting it close on the out-of-jail" end of things. Violation of regulations results in warning letters, which is why a lot of "old-timers" in the industry insist that "a regulation is a law you follow."

WHAT IS THE DIFFERENCE BETWEEN THE USC AND THE CFR?

The USC stands for the United States Code and contains *laws*. The CFR stands for the Code of Federal Regulations and contains *regulations*. The CFR does *not* have laws, and the USC does *not* have regulations. The USC is enacted by Congress and the CFR is the domain of the executive branch, in our case, HHS.

A CRF has nothing to do with either one of them. CRF stands for "Case Report Form."

HOW DO I FIND A CURRENT REGULATION?

You can find current regulations by checking the GPO Access electronic CFR page, found at http://ecfr.gpoaccess.gov/cgi/t/text/text-idx?c=ecfr&tpl=%2Findex.tpl. This site is updated within 48 hours of the issuance of a change; this is actually the easiest way to find a current regulation.

If you use the GPO Access main page, for the CFR, note that these are updated only once a year. The GPO updates the FDA regulations once a year, April 1. This means that a change in the regulations which was effective before April 1 will be reflected in the current CFR, and any that happen later, will not be reflected in that version; changes made after April 1 will not be included until the next year. To update this version, use the *Federal Register* link on the GPO Access main page. All regulations must be published in the *Federal Register* in order to be effective.

The FDA Website itself may not always have the most current version of a regulation, and it usually lists the revision date of the regulation you have accessed. As a note of caution, *always* check the revision date of the Web page you are reading. On the PDF version of the regulation, the date is shown at the top of the right-hand page as 21 CFR Ch. I (4–1–12 Edition). In other text versions, the date is shown as follows:

    Volume: 1
    Date: 2012-04-01
    Original Date: 2012-04-01
    Title: PART 50—PROTECTION OF HUMAN SUBJECTS

Remember that for the FDA regulations, the 2012 date means that the version you are looking at contains regulations published and in effect up to April 1; there are still eight months left in the year, and any changes made after April 1 will not be included in this version until April 2013. To determine whether or not there are any changes to a particular regulation, search the *Federal Register* for any published changes to the regulation, from April 1 through the date that is important to you. This *Federal Register* page is one of my favorite haunts. I use it to see if there are any regulation changes in the offing, find older versions of regulations (you will need these to establish what the rules were when you started a trial, process, etc.), and also keep abreast of what guidances the FDA has contemplated or issued that may be relevant. This is one of the best government sites around. Take some time to learn the advanced search features; you will not be sorry.

HOW DO I FIND OLDER REGULATIONS?

If you need to reference older regulations, you can use the main CFR page from GPO Access; the Annual Edition allows you to select the year you search. For example, say you want to know what regulation was in effect on June 30, 2003. The current edition of 21 CFR goes up to April 1, 2003. You can go to the 2004 version of the CFR and do a comparison, line by line to find any changes, and then do more research to determine when that change, if any, took place. Or, you can simply use the *Federal Register*, searching for that regulation as a text, for example, "21 CFR Part 50." You can insert the specific days for your search in the advanced search feature. Generally, I add one or two days to each end of my search, just to remove all doubt.

You cannot use the e-CFR site described above to research older versions of a regulation; it accesses only the current version.

WHAT IS "GUIDANCE"?

The FDA issues guidance on a number of subjects. As the lead paragraph says, the guidance represents the Agency's thinking, but is not binding. That means you should read it to determine what the Agency's view on a subject is or was at a particular time. The disclaimer also means that following a guidance does not guarantee that your application will be filed. Some guidances are, by the Agency's own admission, hopelessly out of date, but it just has not gotten around to revising them yet.

You should discuss what guidances to follow at the preliminary meetings you hold with your FDA reviewers. This removes doubt about what you are expected to do, and hopefully makes the job easier. You should always read and understand a relevant guidance before your preliminary meeting, so you can ask intelligent questions about how the guidance impacts your application. "Official" guidances are published in the *Federal Register* and on the FDA Website. "Draft" guidances are only available on the FDA Website; draft guidances are still valuable, giving insight into the FDA thinking on a particular subject, even though they are not formally issued.

There is a central listing of guidances available from the FDA's Website at http://www.fda.gov/Drugs/GuidanceComplianceRegulatoryInformation/Guidances/ucm121568.htm, accessed August 18, 2012.

CONCLUSION

As a regulator, you are responsible for knowing and abiding by the applicable regulations; in order to do this, you must be able to read and understand both the laws and regulations; it goes with the territory. With a few guidelines and experience, you can certainly get the information you need to be compliant.

The sheer appearance of dense text—the quantity of words, the jumble of section numbers and parentheses can be daunting. There are suggestions to simplify this activity. First, remember that laws are written in outline format. The section numbers, letters, and parentheses are there to help you organize the law. Using the outline format helps you keep the subject, the main point of the section you are reading, in mind. Second, the laws are frequently cross-referenced, meaning that one law contains the section numbers of another law. This means that the two sections are related; and in certain law libraries, the sections are hyperlinked, so you can refer to each of them easily. Finally, be on the lookout for ordinary words—"and," "or"—and look carefully at what these conjunctions are joining. Finding an "or" offers you choices between conditions, opportunities, while the "and" requires compliance with all conditions.

This chapter is intended to help you find the information you will need to succeed as a regulator. Bear in mind that the FDA is at heart, a scientific agency, not a law school, and you have, in the other chapters of this book, resources that will guide you on the policy and process of regulatory affairs.

In conclusion, time spent learning these terms and citations should give you a great start in getting and keeping your products in regulatory compliance!

7 The Development of Orphan Drugs

Scott N. Freeman

CONTENTS

Introduction 189
The Orphan Drug Act of 1983 and Its Amendments 190
Incentives for Orphan Drug Development 191
    Orphan Drug Exclusive Approval 191
    Tax Credit for Clinical Development 191
    Orphan Products Grants 192
    Exemption from Marketing Application Fee 192
    Written Recommendations for Orphan Drug Development 193
Orphan Drug Designation 193
    Patient Population Determination 193
    Orphan Subsets of Nonrare Diseases or Conditions 194
    Medically Plausible Basis of Effectiveness 194
    Determination of Orphan Drug Sameness 194
    Clinically Superior Orphan Drugs 196
    Orphan Drug Experience in the United States 196
Conclusion 197
Acknowledgments 197
References 197

INTRODUCTION

Millions of people in the United States and many more throughout the world are afflicted by one of the estimated 5000 to 8000 rare diseases or conditions that have been identified to date.[1] While some diseases such as cystic fibrosis or hemophilia are well characterized, little is known about the vast majority of rare diseases. Despite the great need for novel safe and effective treatments for these diseases or conditions, the small patient populations may not be perceived by sponsors as representing viable markets to make a profit or even recoup the costs of research and development. Sponsors may also be wary in instances where the drug may not be patentable. Thus, even with the existence of drugs with potential uses for rare diseases or conditions, there may be reluctance to "adopt" them for commercial pursuit. As such, these drugs have come to be known as "orphan" drugs.[2] As described in this chapter, legislation has addressed many of the aforementioned disincentives—making the prospect of developing drugs for rare diseases or conditions more commercially appealing.

THE ORPHAN DRUG ACT OF 1983 AND ITS AMENDMENTS

Prompted by the urging of a small grassroots coalition of patient advocacy groups largely spearheaded by Abbey Meyers, as well as public sentiment brought about by actor Jack Klugman of the television series *Quincy, M.E.*, the Orphan Drug Act was legislated to address barriers impeding drug development for rare diseases or conditions. Both the Senate and the House passed the Act in December 1982, and it was signed into law on January 4, 1983, by then President Ronald Reagan.[3]

The Act, as amended, established financial and regulatory incentives to promote potentially promising drugs for the treatment, prevention, or diagnosis of rare diseases or conditions.[4] These incentives include (1) the potential for seven years of market exclusivity, (2) a tax credit covering 50% of qualified clinical testing expenses, (3) the possibility of grants for clinical investigation to help defray expenses, (4) an exemption from marketing application fees, and (5) written recommendations for nonclinical and clinical investigations for marketing approval purposes.

To be eligible for these benefits, a sponsor must request and obtain an orphan drug designation from the Food and Drug Administration (FDA). Prior to amendment of the Act in 1984, the FDA would provide orphan drug designation when the disease or condition subject to the request occurred so infrequently in the United States that there was no reasonable expectation that the costs of development and production would be recovered by sales in the United States.[3] This placed a burden on sponsors to provide detailed financial information to prove an anticipated lack of profitability regardless of how small the patient population would be.

The initial response to the Act was less than enthusiastic given the reluctance of sponsors to disclose financial information to the FDA as well as the difficulties associated with estimating development costs and anticipated sales. To rectify this, the Act was amended in 1984 to alternatively allow a drug to receive orphan drug designation if the proposed disease or condition affected less than 200,000 persons in the United States.[5] A drug for a disease or condition exceeding this patient population limitation may still be eligible for orphan drug designation if the sponsor meets the original financial-based criterion.

Initially, the Act only stipulated seven years of market exclusivity for orphan-designated drugs that could not be patented. However in 1985, the Act was amended to generally allow orphan-designated drugs to be eligible for the market exclusivity incentive regardless of their patentability.[6] The Act was again amended in 1988 to include the requirement that an orphan drug designation request be filed prior to the submission of a marketing application for the drug for the rare disease or condition.[7] Before this amendment, an orphan drug

The Development of Orphan Drugs

designation request could be filed at any point prior to the FDA's marketing approval.

To implement the Act, the FDA was tasked with promulgating standards and procedures for determining eligibility criteria for orphan drug incentives.[8] In 1991, the FDA issued a notice of proposed rule making titled "Orphan Drug Regulations."[9] After reviewing public comments, the regulations were finalized in 1992.[10] The FDA issued proposed revisions to the regulations in 2011 to clarify certain language and propose areas of minor improvement.[11] In June 2013, the FDA issued a final rule which amended the original regulations.[12]

INCENTIVES FOR ORPHAN DRUG DEVELOPMENT

Orphan Drug Exclusive Approval

Perhaps the most significant incentive provided by the Act is market exclusivity.[13] This allows the first sponsor to obtain marketing approval to market their orphan-designated drug for the approved use for a period of seven years without competition from other sponsors with the same drug for the same use. Once the FDA approves a marketing application for an orphan drug from the first sponsor for a rare disease or condition, and that drug has orphan designation, the FDA may not approve a marketing application for the same drug for the same use for seven years unless the sponsor holding exclusivity provides consent or cannot assure the availability of sufficient quantities of the approved drug.[14] This exclusivity ensures predictable and often significant revenue from sales due to the lack of competition from other sponsors.

Suspension or withdrawal of exclusivity occurs if the sponsor's orphan drug designation is revoked.[15] An orphan drug designation may be revoked if the application contained an untrue statement of material fact, omitted required material information, or if the drug was not eligible for orphan designation at the time of the request.[16] The withdrawal or suspension of exclusivity does not affect the marketing approval status of the drug. The sponsor must also notify the FDA at least one year in advance of any discontinuation of drug production after the drug is approved for marketing.[17] This requirement is necessary for the FDA to attempt to find another sponsor prior to discontinuation to ensure that the drug remains on the market for patients who are in need of it.

Orphan drug marketing exclusivity is limited to the approved indication for use of the drug that is covered by the designation. During the period of exclusivity, another sponsor may receive market exclusivity for the same drug for a different use.

Tax Credit for Clinical Development

The Act allows sponsors of a drug with orphan designation to claim a tax credit equal to 50% of the expenses incurred by clinical testing for the designated rare disease or condition against Federal taxes owed. This tax credit only applies to clinical studies conducted between the date of orphan drug designation and the date of marketing approval. These clinical studies must be conducted under an active investigational new drug (IND) application. The sponsor cannot claim the tax credit for any expenditure funded by a grant, contract, another person, or government entity.[18] Currently, the tax credit is a component of the general business credit that can be carried back for one year and forward for 20 years.

Orphan Products Grants

The goal of the Orphan Products Grants Program is to support the clinical development of products (drugs, biologics, medical devices, and medical foods) for use in rare diseases or conditions where no current therapy exists or where the proposed product will be superior to the existing therapy. The FDA provides grants for clinical studies on safety and/or effectiveness that will either result in or substantially contribute to market approval of these products. The disease or condition being studied must be considered rare as defined by the Act (i.e., those that affect less than 200,000 persons in the United States).

Grants are available for foreign or domestic, public or private, and for-profit or nonprofit entities, as well as state and local governments and Federal agencies that are not part of the Department of Health and Human Services (HHS). Organizations that engage in lobbying activities are not eligible to receive grant awards.

Grants are awarded on a competitive basis contingent on the ranking of priority review scores. Currently, grants of up to US$200,000 in total (direct and indirect) costs may be awarded for phase 1 clinical studies for up to three years. Phase 2 and 3 studies may qualify for grants of up to US$400,000 per year for up to four years. On average, about 100 new grant applications are received per year of which 10–15 are funded. Over 500 grants have been funded through the Orphan Products Grants Program that has contributed to the marketing approval of over 40 products.

Clinical investigations that are supported by the Orphan Products Grants Program must be conducted under an active IND (drugs) or an active investigational device exemption (medical devices). They must also comply with applicable human research subject protection regulations and good clinical practice (GCP) guidelines, as well as have Institutional Review Board (IRB) approval. Additionally, there must be evidence that sufficient product is available for the study. Grant availability is published as a request for applications (RFAs) in the *Federal Register*.

Exemption from Marketing Application Fee

The Prescription Drug User Fee Act (PDUFA) was promulgated in 1992 and authorized the FDA to collect fees (application, establishment, and product fees) from sponsors submitting marketing applications for new drugs and certain biological products to help expedite the review process.[19] The FDA Modernization Act (FDAMA) of 1997 provided legislation that exempted sponsors of orphan-designated drugs from the application fee for the designated rare disease or condition.[20] It also stipulated that establishment and product fees remain in place but may be waived on a case-by-case basis. These fee exemptions represent a significant cost-saving benefit. For fiscal year 2013, the application fee was set at US$1,958,800 for a standard marketing application requiring full review of clinical data and US$979,400 for applications not requiring full review of clinical data or supplements requiring the review of clinical data. Additionally, the product fee was set at US$98,380 and the establishment fee at US$526,500.[21]

WRITTEN RECOMMENDATIONS FOR ORPHAN DRUG DEVELOPMENT

The Act also includes a regulatory incentive in the form of written recommendations from the FDA for preclinical and clinical investigations to support marketing approval of an orphan drug.[22] This provision is intended to foster early communications between sponsors and review divisions in order to provide guidance on the data necessary for an IND submission. After obtaining an IND, there are multiple points in the clinical development process where meetings with the FDA are warranted to ensure that studies are being conducted in line with regulatory requirements.

ORPHAN DRUG DESIGNATION

To be eligible for the incentives set forth in the Act, a sponsor must obtain an orphan drug designation. Orphan designation is for drugs intended to treat, prevent, or diagnose a rare disease or condition. What follows are the primary criteria for whether a drug qualifies for orphan designation: (1) the drug must be for a disease or condition that affects less than 200,000 persons in the United States (i.e., for a rare disease or condition or for an orphan subset of a nonrare disease or condition as described as follows), (2) there must be a medically plausible basis to expect the drug to be effective for the given rare disease or condition, and (3) the drug may not be the same drug for the same use as one that already has marketing approval unless a plausible hypothesis of clinical superiority is provided.

A sponsor may obtain an orphan drug designation for the same drug for multiple rare diseases or conditions; conversely, numerous sponsors may acquire an orphan designation for the same drug for the same rare disease or condition. An orphan drug designation request may be filed at any point during development but must be submitted to the Office of Orphan Products Development (OOPD) in the FDA prior to the submission of a marketing application for the designated rare disease or condition.

PATIENT POPULATION DETERMINATION

As per the Act and implementing regulations, the patient population is defined as the number of people in the United States affected by the disease or condition for which the drug is indicated—which must be fewer than 200,000 persons in the

United States. When the drug is a therapeutic, the patient population is typically the prevalence of the disease or condition. If the disease or condition has an average duration of less than one year, the annual incidence is generally considered to be an acceptable estimate. In instances where the drug is a diagnostic, preventative, or vaccine, the patient population consists of the number of people who would receive the drug annually.[23]

ORPHAN SUBSETS OF NONRARE DISEASES OR CONDITIONS

In general, a drug receives orphan designation for the diagnosis, treatment, or prevention of a rare disease or condition (i.e., a disease or condition that affects less than 200,000 persons in the United States). However, a drug may also be designated for use in what is referred to as an "orphan subset" of a nonrare disease or condition (i.e., a disease or condition that affects 200,000 or more persons in the United States). An "orphan subset" is a regulatory term specific to the Orphan Drug Regulations. To establish an orphan subset, sponsors must show that the remaining persons with the disease or condition would not be appropriate candidates for the drug owing to some property(ies) of the drug (e.g., toxicity profile, mechanism of action, previous clinical experience with the drug).[24] For example, a drug may be too toxic for use in treating a disease or condition except in patients refractory to or intolerant of other less toxic treatments; the refractory and intolerant patients may be an appropriate orphan subset for the purpose of designating this drug.

MEDICALLY PLAUSIBLE BASIS OF EFFECTIVENESS

In line with the intent of the Act, orphan drug designation may be granted for drugs that show promise in the diagnosis, treatment, or prevention of rare diseases or condition. Whether a drug is "promising" depends on whether there is a medically plausible basis for expecting the drug to be effective for the treatment, prevention, or diagnosis of a given rare disease or condition. Demonstration of this medical plausibility is contingent on the scientific rationale provided within the orphan drug designation request.

DETERMINATION OF ORPHAN DRUG SAMENESS

The Act and implementing provisions stipulate that the first sponsor to obtain marketing approval for an orphan drug for a rare disease or condition will receive seven years of market exclusivity for the approved use if the sponsor has obtained orphan drug designation for the relevant disease or condition. During this seven-year period, the same drug for the same orphan use may not be granted marketing approval without demonstrating clinical superiority (except in limited circumstances).[25] It is therefore important to delineate how the FDA determines whether two drugs would be considered the same.

For drugs considered to be small molecules, a second drug intended for the same use as the first drug would be considered the same if the active moiety is the same as that of the first drug. The active moiety is defined as the part of the drug, other than components that make it a salt, ester, or other noncovalent derivative (such as a complex, chelate, or clathrate), that is responsible for the physiological or pharmacological action of the drug.[26] Thus, changes to the chemical structure of a drug's active moiety could render that drug to be considered a different drug for purposes of orphan designation and exclusivity.

For large molecule (macromolecule) drugs such as proteins, polysaccharides, and polynucleotides, there often exists a certain degree of heterogeneity, and it is possible to make minor structural modifications without significantly altering the activity of the drug. Were the FDA to allow such insignificant structural differences or changes to render a second drug different from the first drug for purposes of determining orphan drug designation and exclusivity, it would undermine the exclusivity incentive.[27] The Orphan Drug Regulations therefore define the sameness of two large molecules intended for the same use on the basis of the principal molecular structural features; not all such features.[28] Examples include the following:

- Two protein drugs would be considered the same if their structural differences were due to post-translational events, infidelity in transcription or translation, or contained minor differences in amino acid sequence. Additionally, two protein drugs would be considered the same if the only differences were glycosylation patterns or tertiary structures.
- Two polysaccharide drugs would be considered the same in instances where their saccharide repeating units were identical, even if the number of units were to vary or if structural differences were due to postpolymerization modifications.
- Two polynucleotide drugs that consist of two or more distinct nucleotides would be considered the same if they had an identical sequence of purine and pyrimidine bases bound to an identical sugar backbone.
- Two closely related, complex, partially definable drugs for the same therapeutic intent—such as two live viral vaccines for the same indication—would be considered the same.

The regulatory definitions for the sameness of two orphan drugs protects the sponsor of an orphan-designated drug with exclusive marketing approval from attempts at overcoming exclusivity by making only minor modifications to a second drug. However, if a second drug that is otherwise considered to be the same drug for the same use demonstrates clinical superiority (discussed later), it would then be considered a different drug and not blocked from marketing by the first sponsor's exclusivity. It would also earn its own period of exclusivity upon approval.

CLINICALLY SUPERIOR ORPHAN DRUGS

Orphan drug exclusive approval ensures that sponsors will not be subject to competition from other sponsors for that same drug for that same approved use for a period of seven years. However, to ensure that this exclusivity does not interfere with the prompt availability of superior follow-on drugs, a second drug with the same active moiety or principal molecular structural features as that of a previously approved orphan drug for the same use would be considered a different drug if it is shown to be clinically superior.[9] To incentivize the development of potentially superior drugs that would otherwise be considered the same drug as one with approval, only a plausible hypothesis of clinical superiority is required for orphan drug designation. However, to obtain orphan drug exclusivity, the sponsor must demonstrate that its drug is in fact clinically superior to the previously approved "same drug" at the time of marketing approval. If such clinical superiority is shown and marketing approval is obtained, the follow-on drug will have orphan drug exclusivity independent of that of the first drug.

Clinically superior means that the follow-on drug has a significant therapeutic advantage over that of the approved orphan drug that is otherwise the same drug. This may be demonstrated in one or more of the following ways:

- The drug must provide greater effectiveness as assessed by the drug's effect on a clinically meaningful endpoint from adequate and well-controlled clinical studies. In general, this necessitates a similar level of evidence that would be required to support a comparative effectiveness claim for two different drugs; in most cases, direct comparative clinical trials would be necessary.
- The drug must provide greater safety, for example, by eliminating an ingredient or contaminant that is associated with relatively frequent adverse events, in a substantial portion of the target population. In some cases, comparative clinical trials will be necessary.
- In rare instances, where neither greater effectiveness nor greater safety has been shown, a second drug may be considered clinically superior if it is determined to otherwise provide a major contribution to patient care. This basis for clinical superiority is quite narrow and is not intended to allow for the approval of any drug that may confer a minor convenience over the previously approved drug.[29]

ORPHAN DRUG EXPERIENCE IN THE UNITED STATES

From the promulgation of the Act in 1983 through 2011, the OOPD received 3659 orphan drug designation requests. Through the same time period, 2539 orphan drug designations were granted resulting in an average of approximately 69% of all requests receiving designation. Prior to the passage of the Act, only a handful of drugs had been approved for rare diseases or conditions in the United States. Since the passage of the Act through 2011, there have been 397 marketing approvals of orphan-designated drugs.

CONCLUSION

The Act provides financial and regulatory incentives to promote the development of drugs for rare diseases and conditions, and it has been generally heralded as a success. These incentives have led to the marketing approval of drugs for rare diseases or condition that may not have otherwise come to fruition. Many of these approved products are for rare diseases or conditions for which there were no effective treatments or the treatments available were of very limited effectiveness. This has impacted the lives of countless people in the United States. While much has been accomplished, a great need still remains. The Act remains an important incentive to help bring to market novel treatment options for those still in need.

ACKNOWLEDGMENTS

The author thanks Gayatri R. Rao, MD, JD; Debra Y. Lewis, OD, MBA; Henry H. Startzman III, MD; and Margaret W. Renner, JD, for their expert assistance.

REFERENCES

1. IOM (Institute of Medicine). 2010. *Rare Diseases and Orphan Products: Accelerating Research and Development*. Washington, DC: The National Academies Press.
2. Congress used the term "orphan drug" in enacting the Orphan Drug Act to refer to drugs intended for use by such a small population that they have little to no commercial value, and hence generally lack sponsors (i.e., are "orphans"). H.R. Rep. 97-840, Pt. 1, at 6 (1982).
3. See Public Law No. 97-414 (1983).
4. The term "drug" in this chapter refers to both small molecules and large molecules (i.e., chemical and biological drugs).
5. See Public Law No. 98-551 (1984).
6. See Public Law No. 99-91 (1985).
7. See Public Law No. 100-290 (1988); a marketing application refers to a new drug application (NDA) or biologics license application (BLA).
8. See Section 526(d) of the Federal Food, Drug, and Cosmetic Act (FDCA) [21 USC 360bb(d)].
9. See 56 Federal Register 3338 (1991).
10. See Code of Federal Regulations, Title 21, Part 316 (1992).
11. See 76 Federal Register 64868 (2011).
12. See 78 Federal Register 35117 (2013).
13. See Section 527 of the FDCA (21 USC Section 360cc); 56 Federal Register 3341 (1991).
14. See Section 527(b) of the FDCA [21 USC Section 360cc(b)].
15. See Code of Federal Regulations, Title 21, Section 316.29(b).
16. See Code of Federal Regulations, Title 21, Section 316.29.
17. See Section 526(b) of the FDCA [21 USC Section 360bb(b)].

18. See Code of Federal Regulations, Title 21, Section 1.28-1(b)(3)(i).
19. See Public Law No. 102-571 (1992).
20. See Section A; Public Law No. 105-115 (1997).
21. See 77 Federal Register 45639 (2012).
22. See Code of Federal Regulations, Title 21, Subpart B.
23. Code of Federal Regulations, Title 21, Sections 316.20 and 316.21.
24. See Code of Federal Regulations, Title 21, Section 316.3(b)(13).
25. As noted, these exceptions involve the sponsor's inability to assure availability of sufficient quantities of the approved drug and the sponsor's written consent for other applications to be approved during the exclusivity period. Section 527(b) of the FDCA [21 USC 360cc(b)].
26. See Code of Federal Regulations, Title 21, Section 316.3(b)(2).
27. See 57 Federal Register 62077 (1992).
28. See Code of Federal Regulations, Title 21, Section 316.3(b)(13)(ii).
29. Code of Federal Regulations, Title 21, Section 316.3(b)(3); 56 Federal Register 3343 (1991).

8 CMC Sections of Regulatory Filings and CMC Regulatory Compliance during Investigational and Postapproval Stages

Prabu Nambiar, Steven R. Koepke, and Kevin Swiss

CONTENTS

Introduction 200
Pharmaceutical Quality 200
    Drug Substance 201
    Manufacture 202
    Characterization (Structure Elucidation).. 204
    Control of Drug Substance (Specifications)... 205
    Reference Standards or Materials 206
    Container Closure System 206
    Stability 207
    Drug Product 207
        Description and Composition of the Drug Product 207
        Pharmaceutical Development 207
        Components of the Drug Product 208
        Drug Product Development 208
    Manufacture (Manufacturer(s)/Method of Manufacture) 210
    Control of Excipients (Specifications) 211
    Control of Drug Product (Specifications) 211
    Justification of Specification(s) 214
    Reference Standards or Materials 214
    Container Closure System 214
    Stability 215

    Additional Information for Biotechnology Products 215
    Additional Information for Combination Products 217
CMC Regulatory Compliance 218
    Managing CMC Changes and Maintaining Regulatory Compliance 218
    Managing Changes during IND Stages 219
    Managing Changes during Postapproval Stages 223
CMC-Specific Meetings with the FDA 226
    Pre-IND Meeting 226
    EOP2 Meeting 227
    Pre-NDA Meeting 227
    Other Meetings (Type C) 228
FDA Initiative on "Pharmaceutical cGMP for the 21st Century—A Risk-Based Approach" and Its Impact on CMC Drug Development and Change Management 228
Conclusion 230
References 231

INTRODUCTION

The chemistry, manufacturing, and controls (CMC) section of a regulatory filing [investigational new drug (IND), IND amendments, IND annual reports, new drug application (NDA) or biologics license application (BLA), postapproval CMC supplements, and NDA annual reports] contains detailed information pertaining to the characteristics, manufacturing, and quality aspects of the drug substance and drug product. Under the International Conference on Harmonization (ICH) Common Technical Document (CTD) format,[1] the CMC section is referred to as the quality section and the structure is outlined in the ICH CTD guidance.[2] This chapter first discusses the details of the quality section of a CTD, followed by how CMC changes are managed during the IND development phases and postapproval stages. As this book addresses the Food and Drug Administration (FDA) regulatory affairs, the focus of the discussions is primarily based on the US FDA expectations and requirements. Other health authority requirements such as those found in Europe, Canada, Japan, or China could vary significantly from this chapter with respect to CMC, and the reader is encouraged to seek other reference materials as appropriate.

PHARMACEUTICAL QUALITY

The quality section, which is the module 3 in a CTD (ICH M4Q), describes how the drug substance and the drug product are manufactured and how the consistency of their quality will be assured from batch to batch. The contents of these quality sections will evolve with time and experience in both quantity and detail of information. The quality section of a marketing application (CTD/NDA) describes the CMC processes for commercial product, and therefore these sections of a marketing application are very detailed. However, for an IND application, while the same basic information is required, it may be supplied in much less detail because of the preliminary stage of development. The CMC information filed in an IND or NDA/CTD is reviewed by the Agency to ensure that the drug substance and the drug product meet the "quality standards" and do not pose any significant safety risk or compromise efficacy during the intended use in the targeted patient population.

Although in each phase of the clinical investigation sufficient information should be submitted to assure the proper identification, quality, purity, and strength of the investigational drug, the amount of information needed to make that assurance will vary with the phase of the investigation, the proposed duration of the investigation, the dosage form, and the amount of information otherwise available. For example, although stability data are required in all phases of the IND to demonstrate that the new drug substance and drug product are within acceptable chemical and physical limits for the planned duration of the proposed clinical investigations, if very short-term tests are proposed, the supporting stability data can be correspondingly very limited. It is expected that with the progression of a product through the phases of the IND, additional information will be provided. The process of updating/ amending the CMC information for an IND through the development phases is outlined in later sections of this chapter. The final application should contain the information necessary to ensure the identity, strength, quality, and purity of the product. The information to be provided in the quality module (module 3) of a marketing application should include the following information about the drug substance and drug product.

Drug Substance

General Information (Nomenclature/Structure/Physicochemical Properties)

Nomenclature. All appropriate names or designations for the drug substance should be provided along with any codes, abbreviations, or nicknames used in the application to identify the drug substance. Any "official" US names (USAN, CAS, etc.) or foreign names (BAN, JAN, INN, etc.) that have not yet been finalized should be identified as proposed.

Structure. This section contains only summary information relating to structure and other characteristics. More detailed information concerning proof of structure is to be provided in the characterization section. Information that is expected to be provided here includes

1. One or more drawings to show the overall chemical structure of the drug substance, including regiochemistry and where appropriate, absolute stereochemistry.
2. Molecular formula, including any hydrates/solvates/salts, etc.
3. Molecular weight of total structure (formula weight).

For a naturally derived protein drug substances, the information should include

1. The number of amino acid residues.
2. The amino acid sequence indicating glycosylation sites or any other posttranslational modifications.
3. A general description of the molecule (e.g., shape, disulfide bonds, and subunit composition).

General properties. A list of the general physicochemical properties of the drug substance should be provided. Relevant properties are those physical, chemical, biological, and microbiological attributes relating to the identity, strength, quality, purity, and/or potency of the drug substance, and, as appropriate, drug product. The information should include, as appropriate, the following:

1. A general description of the drug substance (e.g., appearance, color, and physical state)
2. Melting or boiling points
3. Optical rotation
4. Solubility profile (aqueous and nonaqueous, as applicable follow United States Pharmacopeia or USP general notice on solubility definition)
5. Solution pH
6. Partition coefficients
7. Dissociation constants (pK_a or pK_b)
8. Identification of the physical form
9. Biological activities

For a naturally derived protein drug substances, additional information should be included, such as the following:

1. Isoelectric point
2. Extinction coefficient
3. Any unique spectral characteristics

Manufacture

This section is divided into several subsections that as a whole describe the manufacturing process and its controls on both process and materials.

Manufacturer(s). The name, address, and manufacturing responsibility should be provided for each firm (including contract manufacturers and testing laboratories) and each site (i.e., facility) that will be involved in the manufacturing or testing of the drug substance. If the current good manufacturing practice (cGMP) activities leading to the production of the drug substance (manufacturing, release testing, stability testing, etc.) are outsourced, this section should provide the required details for each of the outsourced contract provider. If the site is outside of the United States, then a US agent with a US address must be used which could be a subsidiary of the company or even the IND or NDA applicant.

Description of the manufacturing process. This section should include a schematic flow diagram that gives the steps of the manufacturing process and shows where each material enters the process. The entire manufacturing process from the starting materials to the final drug substance released for testing should be depicted. This schematic flow diagram should be accompanied by a narrative description of the manufacturing process that represents the sequence of manufacturing steps undertaken and should include the scale of production. The narrative provided should include more details than that provided in the flow diagram. The description should identify all process controls along with any associated numeric ranges, limits, or acceptance criteria. Any process controls that are considered critical process controls should be highlighted. All critical operating parameters, environmental controls, process tests and all tests performed on intermediates, postsynthesis materials, and unfinished drug substance should be listed along with their associated numeric ranges, limits, or acceptance criteria. The noncritical controls should be listed separately from the critical tests to distinguish them from the critical tests that constitute the specification for the intermediate, postsynthesis material, or unfinished drug substance. For justifications of the criticality of process controls, a cross-reference should be provided to the relevant drug substance process development sections.

Materials from biological origin should include additional detailed information on isolation procedures, preparation procedures, and procedures to maintain traceability of biological materials. A discussion regarding the risk of adventitious agents and a statement that any bovine-derived materials originate in bovine spongiform encephalopathy (BSE)-free countries should also be provided. For materials that may have biological origin, specific tests and acceptance criteria to control microbial contamination should also be included in the specification. A discussion assessing the risk with respect to potential contamination with adventitious agents should be provided when appropriate.

Control of materials. Materials used in the manufacture of the drug substance (e.g., starting materials, solvents, reagents, and catalysts) should be listed along with quality and control (specifications) for these materials. If advanced intermediates are used as starting materials, a detailed description should be provided for each starting material with appropriate justifications. Generally, definition of such advanced intermediates as starting materials are discussed and agreed upon with the Agency during phase-2 or end-of-phase 2 (EOP2) stages of drug development. Particular attention should be paid to potential impurities that could originate from the starting material and the possibility of them resulting in the drug substance. Starting material specifications with appropriate justifications and also the sponsor's change-control strategy to manage future CMC changes to starting materials should be included. Additional details can be found in the draft guidance ICH Q11: Development and Manufacture of Drug Substances, available at http://www.fda.gov/downloads/Drugs/GuidanceComplianceRegulatoryInformation/Guidances/UCM261078.pdf

If the drug substance is to be sterile, validation information relating to any sterilization process (e.g., drug substance and packaging components) should be submitted.

The application should also contain a process development section in the drug product section in which a description and history of the manufacturing process for the drug substance throughout the various development phases should be provided, and this will be described later. For development activities that include quality-by-design (QbD) approaches (discussed in the later part of this chapter), description of those experiments and summary of results should be provided here. Any unique QbD definition(s) designated by the sponsor and used in defining the product and process should be clearly explained. Typical QbD-related information provided in this section includes design of experiments (DoEs) and other studies carried out to understand the criticality of input or intermediate material attributes and process parameters. Particular attention should be paid on impurities (in starting materials and/or intermediates) and demonstration of purging capabilities (e.g., spiking studies), where appropriate. Data-driven justification should be provided where the QbD activities resulted in defining the criticality of input material attributes (e.g., impurity specification) and/or process parameters and defining the design space (multidimensional combination and interaction of input variables and process parameters that have been demonstrated to provide quality assurance) and control strategy. See ICH Q8 and Q11 (draft) guidance documents for additional details.

CHARACTERIZATION (STRUCTURE ELUCIDATION)

Data and analysis to support the determination of the structure of the drug substance should be provided. The chemical structure of the drug substance should be confirmed using physical and chemical techniques such as elemental analysis, mass spectrometry (MS), proton and carbon nuclear magnetic resonance (NMR) spectroscopy, ultraviolet (UV) spectroscopy, infrared (IR) spectroscopy, single-crystal X-ray crystallography, X-ray powder diffraction spectroscopy, differential scanning calorimetry, thermogravimetric analysis, and any other relevant tests (e.g., functional group analysis, derivatization, and complex formation). For naturally derived proteins, the primary, secondary, tertiary, and, if applicable, quaternary structures should be confirmed using appropriate techniques such as amino acid compositional analysis, full amino acid sequencing, peptide mapping, and MS. Additional tests (e.g., isoform analysis and carbohydrate composition or sequence) may be warranted for glycoproteins.

Information on drug substance impurities should also be provided as part of characterization. A discussion of the actual and potential impurities most likely to arise during manufacture, purification, and storage of the drug substance should be provided. Impurities of all kinds (e.g., organic, inorganic, and residual solvents) should be included in the discussion. Additional information such as structural alerts of the impurities that could potentially cause unusual toxicity (e.g., genotoxicity) should be discussed. Summary of *in silico* studies evaluating the structural alerts and any following confirmatory studies (e.g., Ames test) should be provided and cross-referenced to appropriate control strategy section (e.g., process control, control of materials, or control of drug substance) of module 3 and safety section (module 4) of the e-CTD. For drug substances of biological origin and semisynthetic drug substances, the description of impurities should include, if appropriate, those related to the natural origin of the material [e.g., pesticide residues, heavy metals due to the concentration of metals by certain plant species, and related substances whose concentrations vary with changes in harvesting conditions (species, location, season, and organ)].

CONTROL OF DRUG SUBSTANCE (SPECIFICATIONS)

The proposed specifications for the drug substance used to produce the drug product should be provided. A specification is defined as a list of tests, references to analytical procedures, and appropriate acceptance criteria, which are numerical limits, ranges, or other criteria for the tests described. Guidance on setting specifications is outlined in guidance ICH Q6A; for biotechnology products, the guidance is outlined in ICH Q6B.[3] The specification establishes criteria to which each batch of drug substance should conform to be considered acceptable for its intended use. Conformance to specification means that the drug substance, when tested according to the listed analytical procedures, will meet the listed acceptance criteria. The specification sheet should list all tests to which each batch of a drug substance will conform and the associated acceptance criteria, and it should also include a reference to the analytical procedures that will be used to perform each test. The acceptance criteria are the associated numerical limits, ranges, or other criteria for the tests described (for further guidance, see ICH Q6A). For drug substances developed under QbD paradigm, end-product testing could be one element of overall control strategy (ICH Q8). In such cases, the overall control strategy should be summarized in this section.

To support the proposed specifications, a description of relevant batches manufactured and the results of these batch analyses should be provided. Batch analysis, which is a collation of analytical data for all the tests included in the specifications, should be provided for all drug substance batches used for (1) nonclinical studies; (2) drug product clinical efficacy and safety, bioavailability, and bioequivalence; and (3) primary stability studies. Batch analysis data should also be provided for any other batches that are being used to establish or justify specifications and/or evaluate consistency in manufacturing. It is recognized that analytical methods may change during the course of development, and the batch analysis reports should include information concerning the analytical method utilized. The information submitted on each of the batches should include a description of the batch. The description should include the following:

1. Batch identity (i.e., batch number) and size
2. Date of manufacture
3. Manufacturer and site of manufacture

4. Manufacturing process, where applicable
5. Use of batch (e.g., nonclinical, clinical, and stability)

The Agency will utilize all available information to evaluate the submitted application. A written justification for the proposed drug substance specifications based on the relevant development data, information on impurities, standards in an official compendium, batch analyses data, stability studies, toxicology data, and any other relevant data should be submitted along with the proposed specifications. Specification for impurities should include organic and inorganic impurities and residual solvents. Guidance on impurities in new drug substance and residual solvents are outlined in ICH Q3A and Q3C, respectively.[4]

Following ICH Q3A, some impurities will require structural identification. In addition, in the chemical synthesis of the drug substance, some intermediates, reagents, or solvents are reasonably assumed to be genotoxic. (Note: The assessment of individual impurity genotoxicities are beyond this chapter and the reader is encouraged to see "Genotoxic and Carcinogenic Impurities in Drug Substances and Products: Recommended Approaches.") Then these chemicals should be assessed as to which are expected to be unusually toxic, genotoxic, carcinogenic, or may have a functional group which has been linked to genotoxicity (e.g., epoxide, nitrosamine, and nitrophenyl). Should a genotoxic compound be present in the active pharmaceutical ingredient (API), it is recommended to quantitate the level to below 1.5 µg otherwise additional protracted animal testing is generally required.

REFERENCE STANDARDS OR MATERIALS

Information on any reference standards or reference materials used for testing the drug substance (API) should be provided. These should include any postulated or actual impurity or related substance reference standards. If the reference standard is obtained from an official source, this should be stated. When the reference standard is not from an official source, it should be fully characterized by the applicant. Additionally, information as to how another reference standard is qualified (e.g., purchase reference, API batch is further purified and tested, or a pure API batch is used) should be provided in this section.

CONTAINER CLOSURE SYSTEM

A description of the container closure system for the drug substance should be provided, including the identity of materials of construction of each primary packaging component (which has drug contact), its specifications, and reference to the applicable 21 Code of Federal Regulations (CFR) to which the component is based upon. The same type of information should be provided for any functional secondary packaging components. Only a brief description should be provided for any secondary packaging components that do not provide additional protection. The suitability of the container closure system should be discussed with respect to protection from moisture and light; compatibility of the materials of construction with the drug substance, including the potential for sorption to container; and leaching. (In addition, see FDA guidance on Container Closure Systems for Packaging Human Drugs and Biologics: http://www.fda.gov/downloads/Drugs/.../Guidances/ucm070551.pdf)

STABILITY

A summary of all relevant stability studies conducted, protocols used, and the results of the studies should be provided. ICH guidance on stability studies is outlined in ICH Q1A, Q1B, Q1D, and Q1E; guidance for biotechnology products is outlined in ICH Q5C.[5] (Note: The ICH Q1F guidance discussing stability studies for hot or humid zones has been withdrawn.) The discussion should include, for example, (1) a summary of stability batches tested, storage conditions used, attributes tested, shelf life acceptance criteria, test schedule, amount of data available, and analysis of data (including a summary of any statistical analyses performed) and (2) conclusions regarding the label storage conditions and retest or expiration dating period. A postapproval stability protocol and stability commitment should be provided for monitoring the drug substance over the course of the application's lifetime.

DRUG PRODUCT

Description and Composition of the Drug Product

This section should include a brief description of the dosage form, the container closure system, and a statement of composition of the drug product. The composition statement describes the qualitative and quantitative formulation of the drug product as intended for use. The composition statement must contain a list of all components used in the manufacture of the drug product regardless of whether or not they appear in the finished drug product. The composition statement should include (1) quality of the material used [i.e., USP, National Formulary (NF), American Chemical Society (ACS) reagent, technical grade, or drug master file (DMF), etc.]; (2) the function of the component; (3) the amount of the component on a per-unit basis; (4) the total weight, volume, or other appropriate measure of the unit on a lot basis; and (5) any additional explanatory information.

Pharmaceutical Development

The pharmaceutical development section should contain information on the development studies conducted to establish that the dosage form, formulation, manufacturing process, container closure system, microbiological attributes, and usage instructions are appropriate for the purpose specified in the application. The studies included in this section are in addition to those routine control tests conducted on a lot-by-lot basis according to specifications (e.g., release testing and stability testing). A brief description of each of the components of this section follows (see also the ICH Q8 guidance).[6]

Components of the Drug Product

Drug substance. Any key physicochemical characteristics [e.g., water content, solubility, particle-size distribution, polymorphic form, solvation or hydration state, pH, and dissociation constant (pK_a)] of the drug substance that can influence the performance or manufacturability of the drug product should be discussed. This includes the compatibility of the drug substance with the excipients used in the drug product. For combination drug products, the compatibility of the two (or more) drug substances with each other should also be discussed.

Excipients. The choice of excipients, their concentration, and the characteristics that may influence the drug product performance or manufacturability should be discussed in context to the respective role of each excipient. Any excipient ranges present in the batch formula should be justified with data in this section. The use of any functional excipients (e.g., antioxidants and penetration enhancers) to perform throughout the intended shelf life of a drug product should also be demonstrated. The use of any novel excipients (those that are used in the United States for the first time in a human drug product or by a new route of administration) should be discussed and justified. It should be noted that the manufacturing, chemistry, and controls information for any novel excipient should be provided in the same level of detail as that provided for a new drug substance. This information would be expected to be included in an appendix.

Drug Product Development

Formulation development. A brief summary describing the development of the drug product taking into consideration the proposed route of administration and usage should be provided. For modified release drug products, a detailed description of the release mechanism (e.g., erodible matrix system, barrier erosion, and diffusion) should be included. Any parameters relevant to the performance or manufacturability (e.g., powder flow characteristics) of the drug product should be addressed. Physicochemical and biological properties such as pH, osmolarity, dissolution, redispersion, reconstitution, particle-size distribution, aggregation, polymorphism, rheological properties, biological activity or potency, and/or immunological activity can be relevant.

Overages. An overage is a fixed amount of the drug substance in the dosage form that is added in excess of the label claim. Any overages included in the formulation should be justified. It should be noted that normally overages in drug products can only be justified by manufacturing process losses or the inability to dispense the total amount of product. With a single grandfathered exception, drug substance overages in the drug product to increase stability expiry periods are not acceptable.

Manufacturing process development. The selection and optimization of the manufacturing process, in particular critical aspects of the process, should be explained. It is important that in this section the differences between the manufacturing processes used to produce lots for the clinical safety and efficacy, bioavailability, bioequivalence, or primary stability batches and the process be identified. Information on the manufacturing process differences between the production of the clinical batches that support efficacy or bioequivalence and primary stability batches and the procedures and equipment proposed for production batches should be provided. The information should be presented in a way that facilitates comparison of the processes and the corresponding batch analyses information (e.g., tables). Differences in equipment (e.g., different design, operating principle, and size), manufacturing site, and batch size should be delineated for each submitted batch.

For development activities that include QbD approaches (discussed in the later part of this chapter), description of those experiments and summary of results should be provided here. Any unique QbD definition(s) designated by the sponsor and used in defining the product and process should be clearly explained. Typical QbD-related information provided in this section includes DoEs and other studies carried out to understand the criticality of input or intermediate material attributes and process parameters. Data-driven justification should be provided where the QbD activities resulted in defining the criticality of input or intermediate material attributes or process parameters and defining the design space (multidimensional combination and interaction of input variables and process parameters that have been demonstrated to provide assurance of quality) and control strategy. If there are mathematical models derived from the DOE activities, the specific nature of the application of those models should be explained. If the models are used as part of the overall control strategy (which have higher significance), the Agency expects the sponsor to describe how the model will be refined and maintained as the knowledge base continues to increase through the commercial production. See ICH Q8 for additional details.

Container closure system. Container closure system refers to the sum of packaging components that together contain and protect the dosage form. A brief description of the container closure systems and the container closure system used for storage and transportation of drug products should be provided. The suitability of the container closure systems should be discussed and taken into consideration, the choice of materials, protection from moisture and light, compatibility of the materials of construction with the dosage form (including sorption to container and leaching), safety of materials of construction, and performance (such as reproducibility of the dose delivery from the device when presented as part of the drug product). Any additional relevant information to support the appropriateness of the container closure system or its use should be provided as warranted.

Microbiological attributes. For sterile products in particular, the microbiological attributes of the drug product, drug substance, and excipients should be discussed. Should the drug substance, excipients, or both promote or prevent bacterial or fungal growth, information should also be provided here.

Compatibility. The compatibility of the drug product with any diluents (e.g., constitution, dilution of concentrates, and admixing) or dosage devices specified in the drug product labeling and the compatibility of the drug product with likely coadministered drug products should be addressed to provide appropriate and supportive information for the labeling. The information should be used to identify in the labeling of diluents and other drug products that are compatible with the drug product as well as those that are found to be incompatible. Compatibility studies should assess, for example, precipitation, sorption onto injection vessels or devices, leachables from containers and administration sets, and stability. The design and extent of the compatibility studies depend on the type of drug product and its anticipated usage.

Manufacture (Manufacturer(s)/Method of Manufacture)

Manufacturer. The name, address, and manufacturing responsibility should be provided for each firm (including contract manufacturers, packagers, and testing laboratories) and each site (i.e., facility) that will be involved in the manufacturing, packaging, or testing of the drug product. Each site should be identified by the street address, city, state, and the drug establishment registration number. If the site is outside of the United States, then a US agent with a US address must be used which could be a subsidiary of the company or even the IND or NDA applicant.

Method of manufacture. A batch formula should be provided that includes a list of all components used in the manufacturing process, their amounts on a per-batch basis, including overages, a reference to their quality standards, and any explanatory notes. Batch formulas should be provided for the intended validation batch sizes of each formulation.

A description of the manufacturing process and process controls should be provided, including a flow diagram of the manufacturing process. The submitted flow diagram should include the following:

- The entire manufacturing process giving the steps of the process and showing where materials enter the process. The diagram should identify each of the critical steps and any manufacturing step where, once the step is completed, the material might be held for a period of time before the next processing step is performed.
- The identity of the material being processed in each step.
- The identification of any critical process controls and the points at which they are conducted.
- The type of equipment used in each step of the process.

A manufacturing process description, including packaging stages, which represents the sequence of steps undertaken and the scale of production, should be provided. This description should provide more detail than that provided in the flow diagram. In lieu of the manufacturing process description, a master batch record containing all pertinent information may be provided. Steps in the manufacturing process should have the appropriate process controls identified and associated numeric values submitted.

All critical process controls and their associated numeric ranges, limits, or acceptance criteria should be identified and justified. Any research studies or information that supports the justification included in the "Pharmaceutical Development" section should be cross-referenced. This is important to achieve postapproval regulatory flexibility. For sterile products, validation information relating to the adequacy and efficacy of any sterilization process should be submitted. When applicable, validation information should be provided for processes used to control adventitious agents.

Control of Excipients (Specifications)

USP/NF compendial excipients. When a compendial excipient is tested according to the monograph standard, no additional testing needs to be submitted unless pertinent to the quality of the final product.

Non-USP/NF compendial—non-novel excipients. Information for each individual excipient should be submitted. Generally, additional CMC information for the excipient will be required and should be provided with reference to a DMF, if applicable.

Novel excipients. Full CMC information on novel excipients should be included in an appendix or referenced to a DMF, if applicable. This information include pharmacology–toxicology testing for safety.

The FDA inactive ingredients database should be consulted for any uncommon and all novel excipients. Justification for the use of these excipients may be found therein or in the USP Food Chemical Codex and should be reported in this section. If no justification can be found for the particular excipients and route of administration, then toxicological testing is generally required and agreement should be sought with the Agency.

Excipients of human or animal origin. Any excipient of human or animal origin should be identified and a specification submitted, regardless of whether or not the excipient appears in the finished drug product (e.g., processing agent). Should any material be imported into the United States, additional certification is generally required from the US Department of Agriculture prior to importation. These materials include gelatin, lactose, and nonvegetable-derived fatty acids including stearates.

Control of Drug Product (Specifications)

The proposed specifications for the drug product should be provided. The specifications establish criteria to which each batch of drug product should conform to be considered acceptable for its intended use. Conformance to specification means that the drug product, when tested according to the listed analytical procedures, will meet the listed acceptance criteria. A specification is one part of the strategy to control drug product quality. They are proposed and justified by the manufacturer and approved by the Agency. Specifications are established to confirm the quality of drug products rather than to establish full characterization and should focus on those characteristics found to be useful in ensuring product quality as it relates to safety and efficacy. ICH Q6A provides guidance for proposing acceptance criteria, which should be established for all new drug substances and new drug products, that is, universal acceptance criteria, and for those that are considered specific to individual drug substances and/or dosage forms.

- Description—a qualitative statement about the appearance of the dosage form and packaging including phase (e.g., solid, cream, and liquid), color, and packaging type.
- Identification—an identifying test should be specific for the new drug substance, for example, IR spectroscopy or two orthogonal chromatographic procedures (i.e., high performance liquid chromatography or HPLC and thin layer chromatography or TLC).
- Assay—a specific, stability-indicating procedure should be included to determine the content of the new drug substance.
- Impurities—organic and inorganic impurities and residual solvents are included in this category. For further information on impurities in new drug product, refer to ICH Q3B[7] (also refer to ICH Q3A and ICH Q3C).

In addition to the universal specifications for drug substance and drug product, it is expected that additional specifications specific to both the drug substance and drug product will be necessary to control the quality of the product. These additional specifications will be dependent on the properties of the drug substance, type of dosage form, the route of administration, and the patient population.

- Performance test—this is a measure of the rate of release of the drug substance from the drug product which could be a dissolution test for tablet or capsule or drug release test for a transdermal.
- Uniformity of dosage units—this term includes both the mass of the dosage form and the content of the active substance in the dosage form; a pharmacopeial procedure (e.g., USP <905>) should be used.
- Water content—a test for water should be included where appropriate.
- Sterility—all parenteral products should have a test procedure and acceptance criterion for evaluation of sterility.
- Extractables/leachables—generally, where development and stability data show evidence that extractables from the container or closure systems are consistently below levels that are demonstrated to be acceptable and safe, elimination of this test can normally be accepted. Extractables testing would also be expected from most liquids including parenterals, loaded syringes, transdermals, metered-dose inhalers (MDIs), and nasal sprays.
- Following ICH Q3B, some impurities, extractables, and leachables may require structural identification. (Note: The assessment of individual impurity genotoxicities are beyond this chapter, and the reader is encouraged to see "Genotoxic and Carcinogenic Impurities in Drug Substances and Products: Recommended Approaches.") Then these chemicals should be assessed as to which are expected to be unusually

toxic, genotoxic, carcinogenic, or may have a functional group which has been linked to genotoxicity (e.g., epoxide, nitrosamine, and nitrophenyl). Should a genotoxic compound be present in the API, it is recommended to quantitate the level to below 1.5 µg, otherwise additional protracted animal testing is generally required.

For complex products such as drug–device combinations (e.g., MDIs), additional specifications related to the performance of the device should be included. Information should be provided for all analytical procedures listed in the specification. Analytical validation information for all analytical procedures used in the specifications, including experimental data, should be provided. Validation of an analytical procedure is the process of demonstrating that analytical procedures are suitable for their intended use.

For drug products developed under QbD paradigm, end-product testing could be one element of the overall control strategy (ICH Q8). In such cases, the overall control strategy should be summarized in this section.

Batch analysis, which is a collation of analytical data for all tests included in the specifications, should be provided for all batches used for clinical efficacy and safety, bioavailability, bioequivalence, and primary stability studies. Batch analysis data should also be provided for any other batches that are being used to establish or justify specifications and/or evaluate consistency in manufacturing. It is recognized that analytical methods may change during the course of development, and the batch analysis reports should include information concerning the analytical method utilized. The information submitted on each of the batches should include a description of the batch. The description should include the following:

- Batch identity (i.e., batch number), strength, and size
- Date of manufacture
- Site of manufacture
- Manufacturing process, where applicable
- Container closure system
- Use of batch (e.g., bioavailability and stability)
- Batch number of the drug substance used in the drug product
- Batch number of any novel excipients that are not compendial

All expected drug product impurities should be listed in this section of the application, whether or not the impurities are included in the drug product specification. Drug substance process impurities that could carry over to the drug product should be listed here even if they are not degradants and are normally controlled during drug substance testing. It is expected that a cross-reference will be provided for the qualified level of an impurity.

Degradation products. Degradation products of the active ingredient can arise during drug product manufacture or as reaction products of the active ingredient with an excipient and/or immediate container closure system. Attempts should be made to identify all degradation products found at significant levels in the drug product (ICH Q3B).

Residual solvents. The level of residual solvents in a drug product should be controlled in the specifications. Because these are known compounds, the identity and presence of residual solvents in the finished drug product can usually be confirmed by using routine analytical techniques.

Miscellaneous drug product impurities. Any miscellaneous drug product impurity is an impurity other than (1) a degradation product, (2) a residual solvent, or (3) an extraneous contaminant that is more appropriately addressed as a GMP issue (e.g., metal shavings). Miscellaneous drug product impurities include, for example, container closure system leachables, excipient degradants, organic volatile impurities, aluminum, and ethylene oxide residuals.

Justification of Specification(s)

Justification for the proposed drug product specifications should be provided. The justification should be based on relevant development data, batch analyses, characterization and qualification of impurities, stability studies, and any other relevant data. Data from the clinical efficacy and safety, bioavailability, bioequivalence, and primary stability batches and, when available and relevant, development and process validation batches should be considered in justifying the specification. If multiple manufacturing sites are planned, data from these sites should be provided to help establish the relevant tests and acceptance criteria. This is particularly true when there is limited initial experience with the manufacture of the drug product at any particular site. Proposed acceptance criteria can include a reasonable allowance for analytical and manufacturing variability.

Reference Standards or Materials

Information on the reference standard or reference materials used in testing the drug product should be provided. The information on the reference standards for drug substance and drug substance impurities will be provided in "Drug Substance" section and need not be replicated here. A list of available reference standards should be provided in this section for any impurities that are unique to the drug product. The reference standards could be for impurities from drug substance and excipient interactions, impurities formed during drug product manufacturing, or an excipient impurity or leachable from the container closure system that is included in the drug product specification.

Container Closure System

A description of the container closure system for the drug product should be provided, including the identity of materials of construction of each primary packaging component and its specification. This information should include the composition, regulation, specifications, and architectural drawings of all primary packaging materials. The regulation requirement that pertains to the product contact or patient contact materials conform to applicable 21 CFR indirect food additive regulations. The same type of information should be provided for functional secondary packaging components as is provided for primary packaging components. For nonfunctional secondary packaging components (e.g., those that neither provide additional protection nor serve to deliver the product), only a brief description should be provided. (Also see the FDA guidance on Container Closure Systems for Packaging Human Drugs and Biologics, available at http://www.fda.gov/downloads/Drugs/.../Guidances/ucm070551.pdf)

STABILITY

The types of studies conducted, the protocols used in these studies, and results of the studies should be summarized. ICH guidances on stability studies are outlined in Q5C for biotechnology products and in ICH Q1 for other products, primarily small molecules.[5] This summary should include (1) a summary of stability batches tested, storage conditions used, product attributes tested, shelf life acceptance criteria, test schedule, amount of data available, and analysis of data (including a summary of statistical analyses, if performed); (2) conclusions regarding the labeled storage conditions and the proposed shelf life; and (3) conclusions regarding in-use storage conditions and shelf life, if applicable. Detailed results from these stability studies undertaken on primary stability batches should be included.

It is important that the analytical procedures used to generate the data in each of the stability studies be identified. A summary of any changes in the analytical procedures should be provided if the analytical procedure was changed over the course of generating the stability data. The summary should identify when an analytical procedure changed, the differences between the analytical procedures, and the impact of the differences with respect to the data being reported. A postapproval stability protocol and stability commitment should be provided for monitoring the drug product over the course of the application lifetime.

Constitution or dilution studies performed as part of formal stability studies to confirm product quality through shelf life should also be reported in this section of the application. This is in addition to the data submitted in the compatibility section of the drug product. The design and any results from drug product stress testing and thermal cycling studies should be provided here. The information should be used, as appropriate, to support the validation of analytical procedures.

ADDITIONAL INFORMATION FOR BIOTECHNOLOGY PRODUCTS

Viral adventitious agents and transmissible spongiform encephalopathy or TSE agents. All developmental or approved products manufactured or processed in the same areas as the applicant's products should be identified when there is potential for cross-contamination with TSE agents. For nonoral, nontopical products, this information should also be provided when there is potential for cross-contamination with viral adventitious agents. Information should be included on the design features of the facility and procedures to prevent cross-contamination of areas and equipment.

For protein products. Protein products due to their potential for contamination require additional production information be provided in the applications. A diagram should be provided illustrating the manufacturing flow, including movement of raw materials, personnel, waste, and intermediates in and out of the manufacturing areas. Information should be presented with respect to adjacent areas or rooms that may be of concern for maintaining integrity of the product. Information on all development or approved products manufactured or manipulated in the same areas as the applicant's product should be included. A summary description of the product-contact equipment and its use (dedicated or multiuse) should be provided. Information on preparation, cleaning, sterilization, and storage of specified equipment and materials should be included, as appropriate. Information should be included on procedures (e.g., cleaning and production scheduling) and design features of the facility (e.g., classifications) to prevent contamination or cross-contamination of areas and equipment where operations for the preparation of cell banks and product manufacturing are performed.

Therapeutic proteins can be produced in varying manners. These can be from microbial cell lines of microbial, human or animal origin, or plant or animal tissues. These production systems must be characterized in order to ensure reproducibility of the identity, quality, and potency of the protein produced. The product must be ensured to be free of any viral contamination (ICH Q5AR1).

If the production system is a recombinant cell line, the expression construct must be characterized using nucleic acid techniques (ICH Q5A). The analysis of the expression construct at the nucleic acid level should be considered as part of the overall evaluation of quality. It should be noted that this testing only evaluates the coding sequence of a recombinant gene, and not the translational fidelity nor other characteristics of the recombinant protein, such as secondary structure, tertiary structure, and posttranslational modifications.

It is important that the therapeutic protein be as well characterized as possible to ensure the final quality of the product. The three-dimensional conformation of a protein is an important factor in its biological function. Proteins generally exhibit complex three-dimensional conformations (tertiary structure and, in some cases, quaternary structure) due to their large size and the rotational characteristics of protein alpha carbons. Current analytical technology is capable of evaluating the three-dimensional structure of many proteins. Methods such as X-ray crystallography and multidimensional NMR spectroscopy can help define tertiary protein structure and, to varying extents, quaternary structure. A protein's three-dimensional conformation can often be difficult to define precisely using current physicochemical analytical technology. In evaluating the integrity of the three-dimensional structure and quality of a protein, functional assays are also critical tools (ICH Q6B).

Specifications derived from the characterization studies will include an identity test that is based on the unique aspects of the molecular structure. Specifications

CMC Sections of Regulatory Filings

are linked to a manufacturing process, and this is especially true for biotechnology products. Unlike simple molecules, purity/impurities and potency are normally separate methods since potency can be a function of the three-dimensional structure and may not be capable of being measured by the same systems. In addition, the drug substance can include several molecular entities or variants. When these molecular entities are derived from anticipated posttranslational modification, they are part of the desired product. When variants of the desired product are formed during the manufacturing process and/or storage and have properties comparable to the desired product, they are considered product-related substances and not impurities and be included in potency.

ADDITIONAL INFORMATION FOR COMBINATION PRODUCTS

Combination products are therapeutic and diagnostic products that combine drugs, devices, and/or biological products. Because combination products involve components that would normally be regulated under different types of regulatory authorities, they raise challenging regulatory, policy, and review management challenges. Differences in regulatory pathways for each component can impact the regulatory processes for all aspects of product development and management, including preclinical testing, clinical investigation, marketing applications, manufacturing and quality control, adverse event reporting, promotion and advertising, and postapproval modifications.

For combination products involving either drugs or biotechnology products as active ingredients, regardless of the jurisdiction designation, the active ingredients are reviewed by the regulatory body responsible for drugs or biotech. While there are still some legacy products and applications, it has generally been recognized that regulation of drugs and biotech requires specialized knowledge.

Combination products are subject to the standard ICH Quality Guidances. While there is some flexibility in the application of the guidances due to the nature of the products (low drug content, need for sterilization, difficulty with analytic levels especially impurities/degradants, requirement for potentially complex elution/dissolution test, etc.), generally combination products are expected to be of at least the quality of US generic drugs (assay of label claim 90%–110% or better, impurity/degradant levels of <5%). It should also be noted that despite their cost per individual unit and short development times for device products, combination products are expected to have the appropriate ICH-type stability data both real time and accelerated at the time of submission to regulatory authorities. There is liberal application of matrixing when there are multiple sizes or lengths involved, but it is expected that each basic design be represented in the stability matrix.

Combination products involving drugs or biotechnology are generally treated in a manner similar to a complex drug delivery device such as MDIs or dry powder inhalers (DPIs). It is necessary to characterize both amount of dose as well as the nature of the dose delivered. This can lead to the need for complex elution techniques or in the case of MDIs and DPIs, particle-size distributions of the dose.

CMC REGULATORY COMPLIANCE

After review and approval/acceptance of the CMC information by the Agency, the CMC processes and procedures described in the IND/CTD/NDA become a binding commitment. Thus, all future batches of that particular drug substance and drug product are expected to be manufactured by the processes and procedures described in the regulatory filing(s) so that they meet the quality criteria described in the application. Continuing to maintain this commitment is referred to as CMC Regulatory Compliance and it is a statutory requirement (see 21 CFR 211.100). The details of how the CMC procedures are followed by a firm in a compliant manner are governed by the firm's operating procedures defined under GMP. The QC/QA departments perform the compliance verification by QC release testing, batch record review, and product release. The Agency may also verify the CMC regulatory compliance during the GMP inspections. If, for any reason, the sponsor has to deviate from a filed or approved process or procedure, the resulting product cannot be used in the clinic or put in commerce until the sponsor has taken the necessary regulatory steps as outlined later. In accordance with GMP, formal change-control procedures are followed to implement the changes in a systematic manner.

MANAGING CMC CHANGES AND MAINTAINING REGULATORY COMPLIANCE

Changes to established CMC processes and procedures are routinely needed because of any one or many of the reasons in this nonexhaustive list:

1. Continuous improvement
   a. Quality
   b. Efficiency
   c. Cost
2. Adaptation of new science and technology
3. Adapting to new scientific/clinical findings
4. Adapting to supplier/vendor situations
5. Changing clinical/market needs
6. Complying with regulatory changes
7. Complying with compendial changes
8. Transfer of products/facilities to new owners
9. Expansion into new markets

Changes are more frequent during IND stages but also continue to happen after approval of the NDA. It is important to stress that effect of change in an early phase IND is usually minor compared to the same change after an approved NDA. Thus it is important that a sponsor/firm assess the nature of the change(s) and ensure that the change(s) has no significant impact on the quality/safety profile of the drug substance or drug product. Depending on the nature of the change and its potential impact on the quality of the product, the

CMC Sections of Regulatory Filings

sponsor will have to file the information to the Agency and get their acceptance or approval prior to implementation of the change and thus maintain CMC regulatory compliance. For a drug at the IND stage, significant CMC changes and new CMC information are communicated to the FDA via IND information amendments [21 CFR 312.31(a)(1)]; a summary of significant changes are also summarized in IND Annual Reports [21 CFR 312.31(a)(1)]. It should be noted that close attention should be paid to those attributes of a drug product that are requested in the "Pharmaceutical Development" section. These are the components of an application that have the greatest potential to adversely affect the identity, strength, quality, and purity of the product. For an approved product, CMC changes are submitted in multiple ways, depending on the nature of the change (21 CFR 314.70). A prior approval supplement (PAS) or a changes being effected (CBE) supplement should be filed for major and moderate changes, respectively; minor changes are reported in NDA annual reports. Only after following the appropriate regulatory process can the product resulting from the change be used in the clinic or commercial purpose. Under the GMP system of a company, a well-defined change-control process is used to make sure that all clinical and/or commercial supplies are CMC regulatory compliant.

Managing Changes during IND Stages

Generally, an IND for a new chemical entity or a new biological entity is filed with CMC processes that are not optimized. This is primarily because of time and cost constraints to develop a thoroughly optimized process. In addition, the cost of goods is not a critical factor at this stage in development. Given the industry competition, uncertainties about the viability of the drug, and uncertainties about the final dose/dosage form at the initial stages, companies file an IND with minimal CMC processes, making sure that the quality of the product does not affect the safety of the study patients. The regulations emphasize a graded nature of CMC information needed as the drug development progresses under an IND. 21 CFR 312.23(a)(7)(i) requires that an IND for each phase of investigation include sufficient CMC information to ensure the proper identity, strength, quality, purity, or potency of the drug substance and drug product.[8] Note that the regulations say "sufficient" information, as appropriate to the phase of investigation.

The phase 1 CMC regulatory review focuses on ensuring the identity, strength, quality, purity, and potency of the INDs as they relate to safety. The safety of the clinical supplies are generally assured by making sure that their quality is equal or better than the supplies used for IND-enabling toxicology studies. In addition, attention should also be paid to make sure that the following safety risk factors are absent:

1. Product made with unknown or impure components
2. Product possessing structures of known or likely toxicity
3. Product not stable through clinical study duration

4. Impurity profile indicates health hazard
5. Strength or impurity profile insufficiently defined
6. Poorly characterized master or working cell bank
7. Lack of sterility assurance for injectables

Because of the fairly rudimentary nature of CMC development at the IND stage, the CMC processes for the drug substance and drug product change routinely, as the drug progresses through the clinical phases of development. Some of the common/potential drug substance and drug product CMC changes are outlined in Table 8.1. Any of these changes, independently or in combination, has the potential to affect the identity, strength, quality, purity, or potency of the investigational drug as they relate to safety of the drug. Therefore, the FDA expects the sponsor to carefully assess the nature of the change(s) to determine if it can affect the safety of the product directly or indirectly.

Change assessment for the drug substance involves comparison of the quality by analyzing the before and after batches for purity, impurity profile, residual solvents, solid-state properties, and stability. For drug products, change assessment depends on the nature of the dosage form; usually studies include analyzing the before and after batches for dissolution/disintegration (for solid oral dosages), impurity profile, dose uniformity, pH/particulates/reconstitution time/sterility (for injectables), preservative effectiveness, functionality testing (for drug–device combination products, such as MDIs), leachables/extractables, and stability. The following factors should be kept in context, as the change assessment is carried out for the drug substance and product:

1. Clinical development stage of the drug (technical and scientific understanding of the drug substance/product and the manufacturing processes continue to increase as the development continues; commensurately, the level of complexity of change assessment will also continue to increase).
2. Where in the process is change being made? A change in the earlier step in multistep drug substance synthesis process is likely to have a lesser impact on the final drug substance, and hence the drug product than a change in the final step of the synthesis.
3. Availability of sensitive analytical methods to detect the changes pre- and postchange. Having the appropriate and highly sensitive method will allow the change assessment with higher level of confidence; for example, monitoring low levels of a highly toxic impurity.
4. Which quality criteria are affected by the change? For example, the significance of a slight increase in the levels of a highly toxic impurity is very high compared with a slight decrease in the purity of the drug substance.

An ideal case would be that the new CMC process continues to produce drug substance and/or drug product of comparable quality with no impact on safety. Changes with a significant potential to affect the safety of the product are communicated to the FDA by IND information amendments. Comparability data

TABLE 8.1
Common/Potential Drug Substance and Drug Product CMC Changes

Drug Substance

CMC Parameter	Potential Change
Physicochemical/solid state form	Salt form
	Crystal morphology or polymorphism
	Particle size
Manufacture	Site
	Scale
	Equipment
Process	Synthetic route
	Reagents/solvents
	Conditions (temperature/volume)
	Starting materials/vendor
Packaging and specifications	Container closure system
	Analytical tests
	Test methods
	Acceptance limits
Stability	NDA retest date

Drug Product

CMC Parameter	Potential Change
Dosage	Dosage form
	Strength
	Components/composition
Manufacture	Site
	Scale
	Equipment
Process	Unit operations
	Conditions (temperature/volume)
	Excipients/vendors
Packaging and specifications	Container closure system
	Different count or volume
	Analytical tests
	Test methods
	Acceptance limits
Stability	NDA shelf life

demonstrating the absence of adverse impact on the quality and safety are usually included in the submission. In the event that the drug substance and/or drug product from the new process do not meet the comparability criteria, sponsors should perform appropriate qualification and/or bridging studies to support the safety and bioavailability of the material to be used in the clinical trials.

As the investigational drug reaches phase 2 and then moves on to phase 3, the Agency expects more detailed CMC information to be submitted via IND amendments. Details about the level of CMC information required for phase 2 and phase 3 are outlined in the FDA guidance.[9]

By the end of phase 2 studies, it is more assured that the drug is likely to become a commercial product, and the likely dosage form(s), dosage strength(s), and container closure system are more clearly defined. As a result, the CMC processes start to get "locked-in" during phase 3. Usually, the anticipated CMC changes between phase 2 supplies and those to be used in phase 3 are discussed with the FDA during the EOP2 meeting.[10] Planned comparability or bridging studies are also discussed at this meeting to get a buy-in from the Agency on the scientific approaches to demonstrate equivalency of the pre- and postchange products.

The phase 3 clinical supplies and the CMC processes to manufacture them are expected to be representative of the commercial product and processes. These supplies are also used for ICH stability studies to define the retest date for the drug substance and the expiry date for the drug product. By this time, the commercial manufacturing operations also start to gear up and technology transfer activities are carried out gradually. Ideally, no major changes are expected after this stage except for unforeseen circumstances. Certain kinds of major changes in CMC processes at phase 3 stage could necessitate clinical studies to demonstrate equivalency (e.g., bioequivalency studies for modified release dosage) and also initiate new ICH stability studies. Such events could potentially cause significant delays to the completion of phase 3 clinical studies, and hence the NDA filing and approval timeline. Therefore, advanced and careful planning is recommended before finalization and initiation of phase 3 studies and subsequent major CMC changes. In the event that such a major CMC change is unavoidable, a follow-up meeting or a teleconference with the Agency is recommended to have a mutual agreement on the path forward. It should be noted that CMC changes made in the later stages of development, particularly in phase 3 clinical studies, may have to be treated similarly to postapproval changes to an NDA for demonstration of comparability.

Finally, by the completion of phase 3 studies, sponsors meet with the FDA at a pre-NDA meeting to discuss the format and content of the submission. In addition, sponsors might also provide an outline on how they intend to present in the NDA the details of resolving outstanding CMC issues discussed during the EOP2 meeting. Full details of product development information, commercial processes, and quality criteria are included in the NDA/CTD (in accordance with the ICH M4Q guidance), which get reviewed and approved by the FDA.

Biotechnology-derived products are especially sensitive to what may be perceived to be small changes in the drug substance manufacturing process. Much of the regulation and required data requirements are based on actual experience obtained in the review by the Agency on these classes of products. The following changes to a product, production process, quality controls, equipment, or facilities have been found to have caused detrimental effects on products even where validation or other studies have been performed and would require regulatory review prior to implementation because of the potential effect on identity, strength, quality, and purity of the resultant product.

- Process changes including, but not limited to, the following:
    Extension of culture growth time, leading to significant increase in number of cell doublings beyond validated parameters
    New or revised recovery procedures
    New or revised purification process, including a change in a column
    A change in the chemistry or formulation of solutions used in processing
    A change in the sequence of processing steps or addition, deletion, or substitution of a process step
    Reprocessing of a product without a previously approved reprocessing protocol
- Scale-up requiring a larger fermenter, bioreactor, and/or purification equipment (applies to production up to the final purified bulk)
- New lot of, new source for, or different in-house reference standard or reference panel (panel member), resulting in modification of reference specifications or an alternative test method
- Change of the site(s) at which manufacturing, other than testing, is performed, addition of a new location, or contracting of a manufacturing step in the approved application, to be performed at a separate facility
- Conversion of production and related area from single to multiple product manufacturing area
- Changes in the location (room, building, etc.) of steps in the production process, which could affect contamination or cross-contamination precautions

Dependent on the product and its use, these drug substance (and drug product) manufacturing changes may require *in vivo* studies to demonstrate the absence of adverse effects such as immunogenicity. The FDA guidance issued in 1996 and ICH guidance Q5E provide additional details of demonstration of comparability of biotechnological products resulting from changes in their manufacturing process.[11,12] In complex or not-so-obvious cases, it is prudent for the sponsor to discuss the comparability protocol with the FDA prior to execution to avoid any gaps in regulatory expectations.

MANAGING CHANGES DURING POSTAPPROVAL STAGES

CMC processes and procedures approved in an NDA/BLA are bound to change postapproval for any one or combination of the reasons outlined in "CMC Regulatory Compliance" section of this chapter. The common/potential drug substance and drug product changes outlined in the previous section are applicable to the postapproval stages as well.

Any of the mentioned changes, independently or in combination, has the potential to adversely affect the identity, strength, quality, purity, or potency of the drug as they may relate to the safety or effectiveness of the drug. The holder of an approved application under Section 505 of the Food, Drug, and Cosmetic Act must assess the effects of the change before distributing a drug product made with a manufacturing change [Section 314.70(a)(2)]. The NDA holder must establish if the change is a major, moderate, or a minor one on the basis of its potential to have an adverse effect on the identity, strength, quality, purity, or potency of a drug product as these factors may relate to the safety or effectiveness of the drug product.

As outlined in 314.70(b), 314.70(c), 314.70(d), and the FDA guidances on changes to an approved NDA, and the FDA draft guidance on changes reportable in an annual report,[13] a major change is a change that has a substantial potential, a moderate change is a change that has a moderate potential, and a minor change is a change that has a minimal potential to have an adverse effect on the identity, strength, quality, purity, or potency of a drug product, as these factors may relate to its safety or effectiveness. A major change requires the submission of a PAS and subsequent review and approval by the FDA prior to distribution of the drug product made using the change. A moderate change requires the submission of a CBE supplement; a CBE could be classified as CBE 30, whereby the NDA holder has to wait 30 days from the date of submission for the distribution of the drug product made using the change. If the FDA informs the applicant within 30 days of receipt of the supplement that information is missing, distribution must be delayed until the supplement has been amended to provide the missing information. Alternatively, a CBE could be classified as CBE 0, whereby for certain moderate changes, the product distribution can occur when the FDA receives the supplement. Minor changes are described in the NDA annual report. Based on the risk-based approach, the FDA published a draft guidance in 2010, which lists the CMC postapproval manufacturing changes that have been determined to be generally of low risk to product quality, and hence qualify to be submitted in the Annual Report.[13]

The type of postapproval CMC change and the reporting category are summarized in Table 8.2. An assessment of the effects of a change on the identity, strength, quality, purity, and potency of the drug product should include a determination that the drug substance intermediates, drug substance, in-process materials, and/or drug product affected by the change conform to the approved specifications. Change assessment is typically done by comparing the analytical test results of several pre- and postchange batches of the intermediate/drug substance or drug product, as appropriate and determining if the test results are equivalent and the pre- and postchange products are comparable. In addition, the FDA recommends that the NDA holder perform additional testing (chemical, physical, microbiological, biological, bioavailability, and/or stability profiles), when appropriate, to make more precise change assessment and demonstrate comparability. The FDA's scale-up and postapproval changes (SUPACs) guidance also provide very valuable information to sponsors regarding the change assessments.[14]

TABLE 8.2
Postapproval CMC Changes and the Reporting Categories

Type of Change	Extent of Potential Adverse Effect on Product Quality[a]	Reporting Category
Major	Substantial	PAS
Moderate	Moderate	CBE (CBE 0 or CBE 30)
Minor	Minimal	AR

[a] Product quality—identity, strength, quality, purity, or potency.
AR, annual report; CBE, changes being effected; PAS, prior approval supplement.

Details of reporting categories for the following major CMC parameter changes are outlined in the changes to an approved NDA or Abbreviated NDA (ANDA) guidance.

- Components and composition
- Manufacturing sites
- Manufacturing process
- Specifications
- Container closure system
- Labeling
- Miscellaneous changes
- Multiple related changes

These reporting categories are consistent with the SUPAC guidance; any difference in recommended reporting categories in previously published guidances is superseded by the changes to an approved NDA or ANDA guidance. It should be noted that reporting category of the change does not in any way change the nature of requirements or amount of data required to justify the proposed change. The required studies must be performed, and the resulting data must be supplied to the FDA. The reporting category simply changes the timing and method of providing that information to the FDA for their review. In complex or not-so-obvious cases, it is prudent for the sponsor to discuss postapproval change assessment and filing strategy with the FDA prior to execution to avoid any gaps in regulatory expectations.

If an assessment indicates that a change has adversely affected the quality of the drug product, the FDA recommends that the change be submitted in a PAS regardless of the recommended reporting category for the change.

As mentioned in the previous section "Managing Changes during IND Stages," biotechnology-derived products are especially sensitive to what may be perceived to be small changes in the drug substance manufacturing process. Dependent on the product and its use, these drug substance (and drug product) manufacturing changes may require *in vivo* studies to demonstrate the absence of adverse effects such as immunogenicity. The FDA guidance issued in 1996 and ICH guidance Q5E provide additional details of demonstration of comparability of biotechnological products resulting from changes in their manufacturing process. In complex or not-so-obvious cases, it is prudent for the sponsor to discuss comparability protocol, postapproval change assessment, and filing strategy with the FDA prior to execution to avoid any gaps in regulatory expectations.

CMC-SPECIFIC MEETINGS WITH THE FDA

Maintaining and managing effective ongoing communications with the FDA is a critical component of regulatory professional's job. In accordance with FDA Modernization Act (FDAMA), the FDA will hold formal meetings with the sponsor to reach agreements on development plans to ensure appropriate safety and efficacy of the product and successful and timely introduction of the drug in the market. (Note: Formal Meetings between the FDA and Sponsors or Applicants, available at http://www.fda.gov/downloads/Drugs/.../Guidances/ucm153222.pdf) As indicated below, there are three types of meetings that can be scheduled with the FDA:

- Type A meeting is a critical path meeting that is necessary immediately to resolve any critical issue for drug development program to proceed (e.g., dispute resolution and clinical hold).
- Type B meetings include pre-IND, EOP1, EOP2, and pre-NDA/BLA, where the sponsor discusses the development plans with the Agency at various critical stages.
- Type C meeting is any meeting other than a Type A or Type B meeting between FDA and a sponsor regarding the development and review of a drug product.

Type B meetings are well-defined meetings for all new drugs and will be discussed in detail here. The format of Type B meetings at the IND stage is usually multidisciplinary (e.g., clinical, pharmacology, pharmacokinetics, chemistry, microbiology, statistics and others, as needed). However, when the CMC issues that are too extensive or critical, a separate CMC-specific meeting can be held (and encouraged) with the CMC review division in addition to, or as an alternative to, the multidisciplinary format.[10] Because of significant time and resource requirement needed on the part of the sponsor and the FDA, the sponsor should be very diligent in assessing the need for a CMC-specific meeting and use this channel of communication and dialogue to facilitate unhindered drug development. While these meetings are usually held as a face-to-face meeting, they may also be handled via a teleconference or by written correspondence depending on the nature of issue and availability of meeting time.

Pre-IND Meeting

The purpose of pre-IND CMC meetings is to discuss safety issues related to the proper quality (i.e., identification, strength, purity/impurity, or potency) of the investigational drug and to identify potential clinical hold issues. CMC-focused pre-IND meetings are not common for conventional small molecules drugs, but such a meeting may be warranted for complex drugs, such as biotechnological drugs, biological drugs, natural products, complex dosage forms, and drug–device combinations. The meeting discussions will include unique CMC challenges that might have potential impact on the safety of the investigational drug and how the sponsor plans to address such issue and ensure patient safety during clinical trials.

EOP2 Meeting

The purpose of the EOP2 CMC meeting is to review the outcome of phase 1 and phase 2 drug development program, discuss phase 3 plans and protocols, and early identification of potential safety/scientific issues/problems and the appropriate path forward to resolve these issues. Thus, EOP2 CMC meeting will ensure that adequate CMC data are generated during phase 3 to support a successful marketing application. The Agency recommends that EOP2 CMC meeting takes place immediately prior to or after the meeting on clinical issues.

EOP2 CMC meeting should focus on the CMC-specific questions on the planned phase 3 studies. Examples of the CMC issues addressed include physicochemical and biological properties, adequacy of physicochemical characterization studies, starting material designation, role of third-party providers (e.g., DMF holders and contract manufacturers), qualification of impurities, removal or inactivation of adventitious agents, approach to specifications, formulations planned for commercial product, container/closure system, approach to sterilization process validation if applicable, phase 3/registration stability protocols, major CMC changes anticipated from phase 2 through the proposed NDA or BLA and their impact, and appropriateness of planned comparability and/or bridging studies, if applicable and others as relevant. For rDNA protein biotechnology drugs, additional topics such as appropriateness of bioassay method, adequacy of cell bank characterization, removal of product- and process-related impurities, and bioactivity of product-related substances and impurities may be discussed. For conventional biologics, additional CMC topics such as facility design, potency assay, and process validation considerations could be addressed.

Pre-NDA Meeting

The pre-NDA or pre-BLA CMC meeting is a critical interaction that will ensure the submission of a well-organized and complete NDA or BLA, and it focuses on addressing the specific questions related to filing and format issues and also a discussion to identify potential problems that can cause a refuse-to-file situation or hinder the review process. Examples of CMC topics include the format of the proposed NDA/BLA, confirmation that all outstanding CMC issues (discussed at the EOP2 meeting or subsequently) will be adequately addressed in the NDA/BLA, assurance of coordination of all third-party (e.g., DMF holders and contract manufacturers) activities, discussion of the relationship between the drug product used in phase 3 versus the intended commercial product and assurance that any comparability or bridging studies have been completed, assurance of submission of adequate stability data as agreed upon at the EOP2 meeting, inspection readiness of all GMP facilities at the time of NDA/BLA submission, and any other potential problems/regulatory issues that should be brought to the attention of the Agency.

Other Meetings (Type C)

In addition to the well-defined Type C meeting, additional CMC-specific meetings (Type C) may be scheduled in the event that new issues arise during the development program. For example, a Type C meeting can be used to address any major changes in plans during phase 3 studies from those previously discussed in the EOP2 meeting or resolve potential problems and/or refuse-to-file issues.

FDA INITIATIVE ON "PHARMACEUTICAL cGMP FOR THE 21ST CENTURY—A RISK-BASED APPROACH" AND ITS IMPACT ON CMC DRUG DEVELOPMENT AND CHANGE MANAGEMENT

The FDA introduced the pharmaceutical cGMPs for the twenty-first century initiative in August 2002 (cGMP initiative; http://www.fda.gov/Drugs/Development ApprovalProcess/Manufacturing/QuestionsandAnswersonCurrentGood ManufacturingPracticescGMPforDrugs/ucm137175.htm) to enhance and modernize the regulation of pharmaceutical manufacturing and product quality. A final report on "Pharmaceutical cGMPs for the 21st Century—A Risk-Based Approach" was published in September 2004.[15] The initiative resulted in modernization of the FDA's approach with a new framework of science-based regulation of pharmaceutical quality that encompasses quality systems and risk management approaches. This new framework is intended to encourage companies to innovate and adopt state of the science and technology in pharmaceutical manufacturing by facilitating industry application of modern quality management techniques and encouraging implementation of risk-based approaches that focus the attention on critical areas. In addition, it is intended to ensure that regulatory review, compliance, and inspection policies are based on state-of the-art pharmaceutical science. In order to encourage the industry to adopt technological innovation and continuous improvement at the global level, the FDA and other regulatory agencies (the European Union and Japan) collaboratively developed and issued guidance documents that harmonize the science-based approaches (ICH Q8 and Q11 draft). They also issued additional guidance documents to enable the risk-based approach (ICH Q9)[16] under an effective pharmaceutical quality system (Q10).[17]

A key concept of "quality by design" (QbD) was introduced for pharmaceutical development in ICH Q8 along with many new terminologies listed below:

- *QbD*. A systematic approach to development that begins with predefined objectives and emphasizes product and process understanding and process control, based on sound science and quality risk management

- *Quality target product profile (QTPP).* A prospective summary of the quality characteristics of a drug product that ideally will be achieved to ensure the desired quality, taking into account safety and efficacy of the drug product
- *Critical quality attribute (CQA).* A physical, chemical, biological, or microbiological property or characteristic that should be within an appropriate limit, range, or distribution to ensure the desired product quality
- *Critical process parameter (CPP).* A process parameter whose variability has an impact on a CQA, and therefore should be monitored or controlled to ensure the process produces the desired quality
- *Design space.* The multidimensional combination and interaction of input variables (e.g., material attributes) and process parameters that have been demonstrated to provide assurance of quality. Working within the design space is not considered as a change. Movement out of the design space is considered to be a change and would normally initiate a regulatory postapproval change process. Design space is proposed by the applicant and is subject to regulatory assessment and approval.
- *Control strategy.* A planned set of controls, derived from current product and process understanding that ensures process performance and product quality. The controls can include parameters and attributes related to drug substance and drug product materials and components, facility and equipment operating conditions, in-process controls, finished product specifications, and the associated methods and frequency of monitoring and control.
- *Lifecycle.* All phases in the life of a product from the initial development through marketing until the product's discontinuation

QbD approach embodies the philosophy of "building the quality in" whereby the drug substance, drug product, and the respective manufacturing processes and controls are designed and developed through systematic understanding and controlling of the critical variables affecting the product quality. The resulting manufacturing process is expected to produce the product of a predefined quality target product profile (QTPP) consistently. The key elements of QbD approach are

- Defining the QTPP of the desired drug product.
- Determining the CQAs necessary for safety and efficacy of the product.
- Performing risk assessment to understand the impact of variability in material attributes and/or process parameters on product quality.
- Developing a design space consisting of critical material attributes and critical process parameter where the process performs satisfactorily, based on the understanding of the link between drug attributes, raw material attributes, and process parameters to CQAs and how they interact and affect the overall product quality.

- Designing and implementing a control strategy to ensure overall control of the CQAs using a combination of appropriate elements such as control of input materials, product specifications, in-process controls, in-process or real-time tests, and monitoring programs based on enhanced product and process understanding and quality risk management.
- Managing product life cycle, including continual improvement.

Pharmaceutical development using the QbD approach offers significant postapproval regulatory flexibility in terms of change management because process changes and improvements within an approved design space in not considered as a major change.

Under the science and risk-based approach, the FDA has also created a mechanism for the sponsors to manage postapproval CMC changes via "Comparability Protocol," and a draft guidance was issued on this topic in 2003.[18]

A comparability protocol is a well-defined, detailed, written plan for assessing the effect of specific CMC changes on the identity, strength, quality, purity, and potency of a specific drug product, as these factors relate to the safety and effectiveness of the drug substance or drug product. A comparability protocol describes the changes that are covered under the protocol and specifies the tests and studies that will be performed, including the analytical procedures that will be used, and acceptance criteria that will be achieved to demonstrate that specified CMC changes do not adversely affect the product. This is an optional submission for companies and can be submitted as part of the original NDA/CTD or as a separate PAS. Upon approval of the comparability protocol, a sponsor can implement the change under a reduced-reporting category, as agreed in the comparability protocol submission/acceptance. Therefore, prudent use of comparability protocols is expected to reduce postapproval CMC regulatory burden to a great extent.

CONCLUSION

CMC information for the drug substance and drug product of an investigational and commercial drug is provided in the quality module of a dossier submitted in CTD format. It is inevitable that the information will continue to change because of various improvements and necessities throughout development and postapproval. It is the sponsor's responsibility to perform adequate assessment of the changes to demonstrate that the changes have adverse affect on the identity, strength, quality, purity, or potency of the drug as they may relate to the safety or effectiveness of the drug. Communicating the changes and the change assessments to the FDA by appropriate regulatory process is critical for maintaining CMC regulatory conformance and complying with the Food, Drug, and Cosmetic Act. Constant communication and coordinated team work among the research and development (R&D), manufacturing, quality (QA/QC), toxicology, clinical, commercial, and regulatory team members of the project is extremely important to proactively plan for the changes and implement them in a timely manner to avoid any interruptions in product supply and potential delays in drug development and marketing, as well as maintaining regulatory compliance.

REFERENCES

1. ICH M4: Organization of the CTD, available at http://www.fda.gov/downloads/Drugs/GuidanceComplianceRegulatoryInformation/Guidances/UCM073257.pdf
2. ICH M4Q: The CTD–Quality, available at http://www.fda.gov/downloads/Drugs/GuidanceComplianceRegulatoryInformation/Guidances/UCM073280.pdf

    ICH M4: The CTD–Quality questions and answers/location issues, available at http://www.fda.gov/downloads/Drugs/GuidanceComplianceRegulatoryInformation/Guidances/UCM073285.pdf
3. ICH Q6A Specifications: Test procedures and acceptance criteria for new drug substances and new drug products: Chemical substances, available at http://www.fda.gov/Drugs/GuidanceComplianceRegulatoryInformation/Guidances/ucm134966.htm

    ICH Q6B Specifications: Test procedures and acceptance criteria for biotechnological/biological products, available at http://www.fda.gov/downloads/Drugs/GuidanceComplianceRegulatoryInformation/Guidances/UCM073488.pdf
4. ICH Q3A(R2) impurities in new drug substances, available at http://www.fda.gov/downloads/Drugs/GuidanceComplianceRegulatoryInformation/Guidances/UCM073385.pdf

    ICH Q3C residual solvents, available at http://www.fda.gov/downloads/Drugs/GuidanceComplianceRegulatoryInformation/Guidances/UCM073394.pdf
5. ICH Q1A(R2) stability testing of new drug substances and products, available at http://www.fda.gov/downloads/Drugs/GuidanceComplianceRegulatoryInformation/Guidances/UCM073369.pdf

    ICH Q1B photostability testing of new drug substances and products, available at http://www.fda.gov/downloads/Drugs/GuidanceComplianceRegulatoryInformation/Guidances/UCM073373.pdf

    ICH Q1C stability testing for new dosage forms, available at http://www.fda.gov/downloads/Drugs/GuidanceComplianceRegulatoryInformation/Guidances/UCM073374.pdf

    ICH Q1D bracketing and matrixing designs for stability testing of new drug substances and products, available at http://www.fda.gov/downloads/Drugs/GuidanceComplianceRegulatoryInformation/Guidances/UCM073379.pdf

    ICH Q1E evaluation of stability data, available at http://www.fda.gov/downloads/Drugs/GuidanceComplianceRegulatoryInformation/Guidances/UCM073380.pdfpdf.

    ICH Q5C stability testing of biotechnological/biological products, available at http://www.fda.gov/downloads/Drugs/GuidanceComplianceRegulatoryInformation/Guidances/UCM073466.pdf
6. ICH Q8 (R2) pharmaceutical development, available at http://www.fda.gov/downloads/Drugs/GuidanceComplianceRegulatoryInformation/Guidances/UCM073507.pdf

    Q8, Q9, and Q10: Questions and answers (R4), available at http://www.fda.gov/downloads/Drugs/GuidanceComplianceRegulatoryInformation/Guidances/UCM210822.pdf

    Q8, Q9, and Q10: Points to consider, available at http://www.fda.gov/Drugs/GuidanceComplianceRegulatoryInformation/Guidances/ucm313087.htm
7. ICH Q3B(R2) impurities in new drug products, available at http://www.fda.gov/downloads/Drugs/GuidanceComplianceRegulatoryInformation/Guidances/UCM073389.pdf

8. INDs for phase 1 studies of drugs and biotech products (November. 1995), available at http://www.fda.gov/downloads/Drugs/GuidanceComplianceRegulatoryInformation/Guidances/UCM071597.pdf
9. INDs for phase 2 and phase 3 studies: Chemistry, manufacturing, and controls information (May 2003). See http://www.fda.gov/downloads/Drugs/GuidanceComplianceRegulatoryInformation/Guidances/UCM070567.pdf
10. IND meetings for human drugs and biologics: Chemistry, manufacturing, and controls information, available at http://www.fda.gov/downloads/Drugs/GuidanceComplianceRegulatoryInformation/Guidances/UCM070568.pdf
11. Demonstration of comparability of human biological products, including therapeutic biotechnology-derived products, available at http://www.fda.gov/Drugs/GuidanceComplianceRegulatoryInformation/Guidances/ucm122879.htm
12. ICH Q5E comparability of biotechnological/biological products subject to changes in their manufacturing process, available at http://www.fda.gov/downloads/Drugs/GuidanceComplianceRegulatoryInformation/Guidances/UCM073476.pdf
13. Changes to an approved NDA or ANDA (April 2004), available at http://www.fda.gov/downloads/Drugs/GuidanceComplianceRegulatoryInformation/Guidances/UCM122871.pdf

    Changes to an approved NDA or ANDA questions and answers (January 2001), available at http://www.fda.gov/downloads/Drugs/GuidanceComplianceRegulatoryInformation/Guidances/UCM122871.pdf

    CMC postapproval manufacturing changes reportable in annual reports (Draft, June 2010), available at http://www.fda.gov/downloads/Drugs/GuidanceComplianceRegulatoryInformation/Guidances/UCM217043.pdf
14. SUPAC IR: Immediate release solid oral dosage forms, available at http://www.fda.gov/downloads/Drugs/GuidanceComplianceRegulatoryInformation/Guidances/UCM070636.pdf

    SUPAC IR questions and answers, available at http://www.fda.gov/Drugs/GuidanceComplianceRegulatoryInformation/Guidances/ucm124826.htm

    SUPAC MR: Modified release solid oral dosage forms, available at http://www.fda.gov/downloads/Drugs/GuidanceComplianceRegulatoryInformation/Guidances/UCM070640.pdf

    SUPAC IR/MR manufacturing equipment addendum, available at http://www.fda.gov/downloads/Drugs/GuidanceComplianceRegulatoryInformation/Guidances/UCM070637.pdf

    SUPAC SS: Nonsterile semisolid dosage forms, available at http://www.fda.gov/downloads/Drugs/GuidanceComplianceRegulatoryInformation/Guidances/UCM070930.pdf

    SUPAC SS: Manufacturing equipment addendum, available at http://www.fda.gov/downloads/Drugs/GuidanceComplianceRegulatoryInformation/Guidances/UCM070928.pdf
15. Pharmaceutical cGMPs for the 21st Century—A Risk-Based Approach Final Report—Fall 2004, available at http://www.fda.gov/downloads/Drugs/DevelopmentApprovalProcess/Manufacturing/QuestionsandAnswersonCurrentGoodManufacturingPracticescGMPforDrugs/UCM176374.pdf
16. ICH Q9, Quality risk management, available at http://www.fda.gov/downloads/Drugs/GuidanceComplianceRegulatoryInformation/Guidances/UCM073511.pdf
17. ICH Q10, Pharmaceutical quality system, available at http://www.fda.gov/downloads/Drugs/GuidanceComplianceRegulatoryInformation/Guidances/UCM073517.pdf

18. Comparability protocols—Chemistry, manufacturing, and controls information (DRAFT), available at http://www.fda.gov/downloads/Drugs/GuidanceCompliance RegulatoryInformation/Guidances/UCM070545.pdf

   Comparability protocols protein drug products and biological products—Chemistry, manufacturing, and controls information (DRAFT), available at http://www.fda.gov/downloads/Drugs/GuidanceComplianceRegulatoryInformation/Guidances/UCM070262.pdf

9 Overview of the GxPs for the Regulatory Professional

Bob Buckley, Robert Blanks, Kimberly J. White, and Tonya White-Salters

CONTENTS

Introduction to the GxPs	236
Introduction to the GLPs	237
History of the GLPs	238
GLPs Today?	239
US GLP Regulations	239
GLP Guidance Documents	239
Organization and Personnel	242
Facilities and Equipment	244
Testing Facility Operations	244
Test and Control Article	244
Protocols, Records, and Reports	245
Compliance with GLPs	246
Introduction to Current GMPs	248
History of the cGMPs	248
cGMPs Today	250
GMP Regulations and Guidance	252
Compliance with cGMPs	254
GMP Inspections and Consequences	255
cGMP References	256
Introduction to the GCPs	256
History of the GCPs	257
GCPs Today	258
US GCP Regulations	258
GCP Guidance Documents	262
Who Is a Sponsor and How Do They Meet Their Obligations?	263
IRB Responsibilities	263
Investigator Responsibilities	264

Sponsor Responsibilities and Oversight of Clinical Trials 269
Compliance with GCPs 275
Sponsor Audits 277
Summary 278
Internet Resources 278
GxP References 278
GLP References 279
GMP References 279
International GMP References 280
GCP References 280
References 281

INTRODUCTION TO THE GxPs

Good... practices (GxPs) is an acronym commonly used in many regulated industries such as the drug, device, or biologic industry to describe a collection of quality guidelines that are based on regulations, guidance, and industry standards to ensure that a product or process will produce results that are consistently fit for the intended purpose. In the pharmaceutical industry, GxP is employed in several aspects of drug development and commercialization, including nonclinical research, manufacturing, and clinical research. The big three GxPs, for which we will attempt to provide a high-level and practical overview in this chapter, include good laboratory practices (GLPs), good manufacturing practices (GMPs), and good clinical practices (GCPs).

The GxPs sole purpose is to ensure that medication available to the public is both safe and effective. Unfortunately, the previous statement does not always hold true, and the evolution of the GxPs is grossly, but fairly, generalized as a series of reactions to scientific misconduct, medical tragedies, human rights violations, and the resulting media attention each received. These events have directly, if not immediately, resulted in a number of ethical doctrines and regulations while many more have evolved over time. Not surprisingly, the concepts described in all three GxPs are similar. These concepts include the need for independent oversight, written procedures, change control, and good documentation practices.

GxP requirements in the United States are administered by the Food and Drug Administration (FDA) through a series of regulations and guidances whose definitions are described later. In general, the regulations and guidances do not specify how a company meets the requirements defined in these documents but allows each company to determine their own path to compliance. However, there are expectations within the industry that have become the accepted norm or commonly practiced procedures in achieving compliance often referred to generically as "industry standards."

Overview of the GxPs for the Regulatory Professional 237

Regulation	This is an enforceable instruction that is codified and is meant to provide the minimum standard that must be met to comply with the law.
Guidance	Guidance document is a recommended approach to comply with regulations for a given process. An FDA guidance document typically represents the Agency's current thinking on a topic; however, alternative approaches may be taken to comply with regulations. Guidances are not enforceable by the FDA, but if an alternative approach is taken, it should be discussed with the Agency.
Industry standards	This is a generic term used to describe the accepted norms and commonly practiced procedures to achieve a given task within the industry.

In this chapter, we will briefly review the origins of the GxPs to provide a brief historical basis for today's standards, and we will focus on what we feel are a few of the "key" regulations and guidelines that one should be familiar with as a Regulatory Affairs professional; however, the reader must be cautioned that this is merely the tip of the iceberg. We will also attempt to provide practical tips and examples of how to implement GxP in the real world and supply the reader with a list of valuable Websites to expand their GxP knowledge (exposing more of the iceberg) and to keep handy as a GxP reference guide.

INTRODUCTION TO THE GLPs

GLPs are the regulations that govern the nonclinical laboratory studies that support investigational new drug (IND) applications, investigational device exemptions (IDEs), and ultimately marketing applications. Nonclinical studies, as per the FDA definition, includes *in vivo* and *in vitro* experiments in which test articles (including drugs, devices, and biologics) are studied prospectively in test systems under laboratory conditions to determine their safety.[1] The results from these non-human experiments provide predictive evidence that the dose selected to move into human trials should be safe, and GLPs are intended to provide assurance that the data are credible. GLPs define the quality processes and working environment under which studies are planned, performed, monitored, recorded, archived, and reported.[2] It is important to note that GLP is not synonymous with good science, as GLPs provide the procedural controls, not the detail to design scientifically sound studies. A study conducted according to GLPs gives the sponsor the ability (in theory anyway) to submit data to a number of regulatory authorities worldwide, as the data were derived in accordance with globally recognized standards.

Some scientists, with limited experience working in an FDA-regulated environment, oftentimes will claim that they "use" GLPs and may become rather offended when you challenge them to the contrary. While they may use good laboratory techniques, the fact that the GLPs are confined to nonclinical studies is not always well understood and the phraseology associated with GLPs is commonly confused.

You will recognize some common themes within this overview of GLPs, which we will also discuss in the GMP and GCP overviews in this chapter—the importance of independent oversight by a quality unit, written procedures, proper change control practices, traceability, and good documentation practices. To have an understanding of the GLPs today, it is important to recognize some of the reasons why they became necessary and from where they evolved.

History of the GLPs

As with GMPs and GCPs, the origination of GLPs is rooted in a series of missteps by industry. In the 1970s, suspicion arose as to the validity of nonclinical safety data submitted in a couple of new drug applications (NDAs) by a major pharmaceutical company, Searle (Omaha, NE). Further inspections of several nonclinical studies and test facilities, of both Searle and other companies, indicated the lack of oversight during nonclinical studies. There was evidence of inadequate control of these studies, including replacing dead animals with new ones without the proper documentation, deleting necropsy observations because the pathologist received no specimens of lesions, or substituting hematology results from a control group not associated with the study. Congress held a series of hearings (Kennedy Hearings, 1975) to address these issues. Subsequently, the FDA promulgated a series of proposed GLP regulations in 1976, which were finalized in 1978 under 21 Code of Federal Regulations (CFR) 58[3] and came into effect in 1979. In the preamble to the final rule, the FDA stated that the GLP regulations "is based on the investigatory findings by the agency that some studies in support of the safety of regulated products have not been conducted in accord with acceptable practice, and that accordingly data from such studies have not always been of quality and integrity to assure product safety."[3]

The Environmental Protection Agency (EPA) soon followed with their own proposed GLP regulations, equivalent to the FDA's, in 1979 and 1980 because of similar problems with the integrity of submitted nonclinical safety data for chemicals. In fact, one of the same testing facilities (IBT) that the FDA had found as having inadequate controls was responsible for a majority of the nonclinical safety studies submitted in support of new pesticides. These regulations were finalized in 1983 (40 CFR 160[4] and 40 CFR 792[4]). In 1981, the Organization for Economic Cooperation and Development (OECD) published their version of the GLPs[5] to promote the mutual acceptance of nonclinical safety data for chemicals and, thereby, eliminate a nontariff trade barrier for the import or export of chemicals.

The FDA revised the GLP regulations in 1987[6] to allow for greater flexibility in the operation of nonclinical laboratories while still protecting the public safety. Most of these changes were to clarify the wordings of the original GLP regulations that test facilities were interpreting too strictly. Other changes included altering the definition of control articles to exclude feed and water given to control animals, allowing the quality assurance unit (QAU) to determine the phase of the study to inspect instead of having to inspect all phases, and eliminating the need to put the proposed start and completion dates in the protocol.

Overview of the GxPs for the Regulatory Professional 239

GLPs Today?

US GLP Regulations

The US regulations that cover the GLPs of interest for this textbook (drugs/devices/biologics) are within Title 21 of the US CFR. Title 21 of the CFR applies to products regulated by the FDA and 21 CFR Part 58 is titled "Good Laboratory Practice for Nonclinical Laboratory Studies." FDA GLP regulations apply to safety studies of food and color additives, animal food additives, human and animal drugs, medical devices for human use, biologic products, and electronic products.[7] The generic term "good laboratory practice" also applies to regulations governed by the US EPA, regulating studies conducted on pesticide products,[8] and studies relating to health effects, environmental effects, and chemical fate testing[9]; however, we will not be discussing EPA GLPs here (Table 9.1).

GLPs are applicable to those nonclinical studies that are intended to support INDs/IDEs and marketing applications (e.g., NDA, BLA, and PMA or premarket approval). Nonclinical studies conducted by sponsors in preparing for their IND/IDE supporting studies (such as a dose range-finding study) do not need to be GLP compliant, and these exploratory studies will be less expensive to conduct non-GLP. Once the exploratory studies have yielded enough data to provide a solid scientific hypothesis, the regulatory pathway (nonclinical studies required) to advance the product into human clinical trials via an IND/IDE are described in general terms in the following guidance documents:

FDA Guidance for Industry: M3 (R2) Nonclinical Safety Studies for the Conduct of Human Clinical Trials and Marketing Authorization for Pharmaceuticals, available at http://www.fda.gov/downloads/Drugs/GuidanceComplianceRegulatoryInformation/Guidances/UCM073246.pdf

FDA Guidance for Industry ICH S6(R1): Preclinical Safety Evaluation of Biotechnology-Derived Pharmaceuticals, available at http://www.fda.gov/Drugs/GuidanceComplianceRegulatoryInformation/Guidances/ucm304390.htm

FDA Guidance for Industry ICH S7A: Safety Pharmacology Studies for Human Pharmaceuticals, available at http://www.fda.gov/downloads/drugs/guidancecomplianceregulatoryinformation/guidances/ucm074959.pdf

FDA Guidance for Industry: Estimating the Maximum Safe Starting Dose in Initial Clinical Trials for Therapeutics in Healthy Volunteers, available at http://www.fda.gov/downloads/Drugs/Guidances/UCM078932.pdf

GLP Guidance Documents

Guidance documents related to GLPs are issued by the FDA and ICH, two organizations discussed in almost every chapter of this book but also by the OECD. The OECD was originally an organization established in post-World War II Europe to help reconstruct the economic stability of the region; however, today it is an international organization (including the United States) which issues guidance documents on a number of very broad topics including economic, environmental, and social issues. While the OECD guidance documents are viewed as guidance

TABLE 9.1
Complete List of 21 CFR 58 GLP Requirements

Subpart A: General Provisions
- 58.1 Scope
- 58.3 Definitions
- 58.10 Applicability to studies performed under grants and contracts
- 58.15 Inspection of a testing facility

Subpart B: Organization and Personnel
- 58.29 Personnel
- 58.31 Testing facility management
- 58.33 Study director
- 58.35 Quality Assurance Unit

Subpart C: Facilities
- 58.41 General
- 58.43 Animal care facilities
- 58.45 Animal supply facilities
- 58.47 Facilities for handling test and control articles
- 58.49 Laboratory operations areas
- 58.51 Specimen and data storage facilities

Subpart D: Equipment
- 58.61 Equipment design
- 58.63 Maintenance and calibration of equipment

Subpart E: Testing Facility Operation
- 58.81 Standard operating procedures
- 58.83 Reagents and solutions
- 58.90 Animal care

Subpart F: Test and Control Articles
- 58.105 Test and control article characterization
- 58.107 Test and control article handling
- 58.113 Mixture of articles with carriers

Subpart G: Protocol for and Conduct of a Nonclinical Laboratory Study
- 58.120 Protocol
- 58.130 Conduct of a nonclinical laboratory study

Subpart H and I (Reserved)

Subpart J: Records and Reports
- 58.185 Reporting of nonclinical laboratory study results
- 58.190 Storage and retrieval of records and data
- 58.195 Retention of records

Subpart K: Disqualification of Testing Facilities
- 58.200 Purpose
- 58.202 Grounds for disqualification
- 58.204 Notice of and opportunity for hearing on proposed disqualification

(Continued)

TABLE 9.1
(Continued) Complete List of 21 CFR 58 GLP Requirements
- 58.206 Final order on disqualification
- 58.210 Actions upon disqualification
- 58.213 Public disclosure of information regarding disqualification
- 58.215 Alternative or additional actions to disqualification
- 58.217 Suspension or termination of a testing facility by a sponsor
- 58.219 Reinstatement of a disqualified testing facility

in the United States, the OECD Principles of GLP have been adopted as legislation by some European community member states and is enforceable in these countries. The OECD has issued 15 guidance documents titled *The Series on Principles of Good Laboratory Practice and Compliance Monitoring.*

No 1: OECD Principles on Good Laboratory Practice (as revised in 1997)
No 2: Revised Guides for Compliance Monitoring Procedures for Good Laboratory Practice (1995)
No 3: Revised Guidance for the Conduct of Laboratory Inspections and Study Audits (1995)
No 4: Quality Assurance and GLP (as revised in 1999)
No 5: Compliance of Laboratory Suppliers with GLP Principles (as revised in 1999)
No 6: The Application of the GLP Principles to Field Studies (as revised in 1999)
No 7: The Application of the GLP Principles to Short-Term Studies (as revised in 1999)
No 8: The Role and Responsibilities of the Study Director in GLP Studies (as revised in 1999)
No 9: Guidance for the Preparation of GLP Inspection Reports (1995)
No 10: The Application of the Principles of GLP to Computerised Systems (1995)
No 11: The Role and Responsibilities of the Sponsor in the Application of the principles of GLP (1998)
No 12: Requesting and Carrying Out Inspections and Study Audits in Another Country (2000)
No 13: The Application of the OECD Principles of GLP to the Organisation and Management of Multi-Site Studies (2002)
No 14: The Application of the Principles of GLP to In Vitro Studies (2004)
No 15: Establishment and Control of Archives that Operate in Compliance with the Principles of GLP (2007)[10]

For those who are interested in learning more about the difference in GLP regulations, a comparison chart of the FDA, EPA, and OECD GLPs is posted on the FDA Website: http://www.fda.gov/ora/compliance_ref/bimo/comparison_chart/.

One key OECD guidance document is "The Application of the OECD Principles of GLP to the Organisation and Management of Multi-Site Studies."[11] This document defines the management of a GLP study that is conducted at more than one site, a trend that has been growing for the past decade. A simplistic example of a multisite GLP study would be a study where one laboratory may analytically confirm the uniformity, concentration, and stability of the test article and dose the animals with the test article, but send the animal serum samples to a separate laboratory to conduct the bioanalytical analyses. Though the work conducted for this type of study setup may take place in separate laboratories within a single company (either geographically remote or organizationally distinct locations) or at two (or more) separate companies, the study is viewed as a single study. Since a multisite study is a single study, communication becomes critically important.

The GLP concepts within 21 CFR 58 Subparts A through K and the OECD guidance documents mirror some of the common themes repeated throughout this chapter, that is, the need for independent oversight by a quality unit, written procedures, proper change control, and good documentation practices. Subparts A through K are shown in full in Table 9.1, and the full list of OECD guidance documents are listed earlier. Instead of regurgitating the contents of the regulations and guidance, we will summarize some of the key highlights from them.

Organization and Personnel

Both the FDA and the OECD delineate the requirement for nonclinical laboratory facility management to create several key personnel roles to conduct a GLP study, mainly the role of the study director, principal investigator, and the QAU, all of which are defined below[12,13]:

Study director	The individual who is responsible for the overall conduct of the nonclinical laboratory study.
Principal investigator	The individual who, for a multisite study, acts on behalf of the study director and has defined responsibility for delegated phases of the study. The study director's responsibility for the overall conduct of the study cannot be delegated to the principal investigator(s); this includes approval of the study plan and its amendments, approval of the final report, and ensuring that all applicable principles of GLPs are followed.[14]
QA program/QAU	A QA program is a defined system, and the QAU is the personnel who is independent of the study conduct that execute the program, which is designed to assure test facility management of compliance with the principles of GLPs.

Overview of the GxPs for the Regulatory Professional

As with GMPs, the GLPs require that management ensure that there is adequate staffing, equipment, and facilities to conduct the nonclinical experiments in compliance with GLPs in addition to an adequate training program to ensure personnel are properly trained to perform their duties.

One key management function is the formal appointment of the study director and investigators. The study director is responsible for the scientific conduct of the study, including planning, documentation, and approval of its protocols and reports. They must have the appropriate experience and knowledge of GLP principles and regulations to allow them to successfully perform their task and coordinate with the QAU and management to ensure GLP compliance. Should management determine that a study director is not capable of fulfilling their duties, management's decision to replace a study director must be documented.[15] Within a multisite study, the study director faces an additional challenge of being responsible for all study activities; some of these may be conducted by other companies. The study director should ensure that all test sites used in a multisite study are acceptable. To do this, the study director may visit the different facilities; but in practice, this is more of the exception than the rule, and is not required. Most often what is required is that each test site names a principal investigator whom the study director will communicate with directly. Direct communication is critical and must be allowed by the sponsor,[11] though the sponsor would be best served to be copied on all communications between all sites in a multisite study.

The role of QAU is critical in establishing and maintaining GLP compliance. The QAU, as defined by GLPs, must be independent of the study conducted, which is a common element of a quality organization across the GxPs. Some of the key responsibilities of the QAU are to

- Maintain a copy of the master schedule of all GLP studies conducted at the test facility by test article.
- Maintain copies of protocols.
- Conduct inspections of studies at intervals adequate to assure integrity of the study.
- Submit written reports to management and study directors.
- Assess and document deviations from protocols and standard operating procedures (SOPs).
- Review final study reports for data accuracy and sign a statement specifying dates of inspection and when findings were reported to management and study director.
- Maintain a SOP/policy detailing the roles and responsibilities of the QAU.

In the case of a multisite study, there should be a distinction between the lead QAU and test site QAU. The lead QAU is the one that oversees the study director. A test site QAU is the unit that provides the quality oversight of the additional facilities where principal investigators conduct certain phases of the study. The lead QAU maintains all QAU responsibilities except for the oversight and inspection of the activities carried out by the principal investigator(s). Test site QAU(s) provides a statement relating the GLP compliance of the principal investigator's activities to the lead QAU prior to the completion of the study report.

Facilities and Equipment

Both the facilities and equipment used in a GLP study must be appropriately designed. For a facility, this means the building must be of adequate size and design to quarantine incoming and/or sick animals, segregate different studies and animal species, and provide appropriate workflow and work areas for all the different functional areas conducting a GLP study. Environmental controls, to prevent infestation and contamination, must be employed. Test/control article storage must also be adequate to preserve the identity, strength, purity, and stability of the test/control articles and mixtures. Equipment, or analytical instrumentation, used in a GLP study must be appropriately qualified, inspected, cleaned, maintained, calibrated, and standardized, all of which must be accompanied by supportive documentation. The design, setup, and maintenance of both GLP facilities and equipment must be adequate to minimize the risk of test/control article, test system, or sample mix-ups and/or cross-contamination. These same principles apply to facilities and equipment in a GMP environment.

Testing Facility Operations

One key aspect under the Subpart E heading of the FDA GLPs and the OECD guidelines is the requirement that SOPs shall be established for, but not limited to, the following topics (Table 9.2).[16,17]

It should be noted that Table 9.2 is a listing of required SOP topics and not SOP titles. While this is only a small list, these topics will encompass a great number of procedures. It is commonplace for contract nonclinical laboratories to maintain hundreds of SOPs; such is the life of the QAU. In conducting a GLP study under such a large number of SOPs, deviations from the SOPs may occur. In such circumstances, as per GLPs, the study director, and the principal investigator in the case of multisite studies, should acknowledge these deviations in the raw data.

Other notable requirements of Subpart E are the labeling requirements for reagent/solutions and animal care requirements. The animal care requirements, generally speaking, highlight the need for isolation of newly received animals from outside sources until their health status is determined—identification and segregation of animals, keeping the animal cages and area cleaned, and providing the animals with uncontaminated feed and water.[18]

Test and Control Article

The drug, device, or biologic under development during nonclinical evaluation is generically referred to as a test article. All test articles used in GLP studies must have a documented "chain of custody," that is, records of receipt, use, and

TABLE 9.2
FDA and OECD Required SOP Topics/Categories

FDA	OECD
Animal room preparation	Test and reference items
Animal care	Apparatus, materials, and reagents
Receipt, identification, storage, handling, mixing, and method of sampling of the test and control articles	Recordkeeping, reporting, storage, and retrieval
Test system observations	Test system (where appropriate)
Laboratory tests	Quality assurance procedures
Handling of animals found moribund or dead during study	
Necropsy of animals or postmortem examination of animals	
Collection and identification of specimens	
Histopathology	
Data handling, storage, and retrieval	
Maintenance and calibration of equipment	
Transfer, proper placement, and identification of animals	

return, including dates and amounts of material.[19] Methods of synthesis or manufacturing of the test article must be documented.[20] However, test articles during this stage of development do not need to be made under GMPs (an overview of GMPs will be provided in the next section). Test articles are required to be "characterized," meaning that the identity, strength, purity, and composition are determined and documented. The stability of the test article as well as the test article in the mixture given to animals needs to be determined either prior to study initiation or concomitantly with the ongoing study, and a retention sample of the test and control articles must be kept for all studies lasting more than four weeks.[21] Characterization and stability may be conducted by the nonclinical laboratory itself or the study sponsor in accordance with GLPs. Test article characterization is essential to ensure that the drug, device, or biologic being advanced through nonclinical development is representative of what will be used in the "first-in-man" study. Failure to do so will result in inadequate toxicology coverage for human clinical trials and potentially necessitate the need to repeat nonclinical safety studies.

Protocols, Records, and Reports

Each GLP study must be conducted according to an approved, written protocol (or study plan as referenced in the OECD). The FDA and the OECD required content of GLP protocols are listed in 21 CFR 58.120 and OECD Principles of GLP Section 8, respectively. The FDA requires protocols to be signed by the study director and the sponsor. The OECD requires the study director's approval signature on all GLP protocols and the sponsor and test facility management's approval where required by national regulation. Alternatively, deviations from or amendments to a GLP protocol require only the study director's approval signature as per both the FDA and the OECD,[22,23] although sponsor review and approval of significant deviations and amendments is often the recommended practice.

All original test facility records and documentation that are the result of original observations and activities in a study are considered the raw data. This includes both manual (laboratory notebooks, worksheets, etc.) and automated data (chromatograms, telemetry data, balance printouts, etc.). Where raw data are acquired through automated computer systems for the generation, measurement, or assessment of data intended for regulatory submission, 21 CFR 11 applies, and these systems must be validated to assure the integrity and quality of the raw data. Additional guidance on the application of GLP to computerized systems is available from OECD.[24]

As with the study protocols, the contents of study reports as per the FDA and OECD requirements are listed in 21 CFR 58.185 and OECD Principles of GLP Section 9, respectively. The FDA and OECD regulations require that the study director sign the final report. The QAU must prepare and sign a statement indicating when the QAU inspections of the study took place and when the inspection findings were reported to facility management, the study director, and to principal investigators (in the case of multisite studies).

It is important to remind Regulatory Affairs professionals that study reports do not necessarily have to be final or audited reports to submit the tabulated, integrated summarized data to the FDA for initiation of the IND. If, however, the integrated summary is based on unaudited reports, sponsors should submit an update to their integrated summary within 120 days of the FDA's receipt of the integrated summary. If the audit of the individual reports did not result in a change to the integrated summary data, an update still must be submitted stating no changes.[25,26]

COMPLIANCE WITH GLPS

The FDA's inspection program for nonclinical laboratories is their Bioresearch Monitoring Program, often referred to as BIMO. The BIMO was established in 1977 by a task force with representation from the drug, biologic, device, radiologic product, veterinary drug, and food branches of the FDA. This task force established an inspection program for nonclinical (animal) laboratories as well as clinical investigators, research sponsors, contract research organizations (CROs), bioequivalence laboratories, and Institutional Review Boards (IRBs).

The FDA uses compliance program guidance manuals (CPGMs) as procedures for its field personnel to conduct these inspections. The purpose of each CPGM is to assure the quality and integrity of safety data submitted to the FDA, with the ultimate goal of protecting human research subjects. The FDA has issued the following CPGMs enforced by the BIMO, and they are available on the Web at http://www.fda.gov/oc/gcp/compliance.html (Table 9.3).

TABLE 9.3
CPGM and the Regulations They Enforce

CPGM	FDA Regulation BIMO Enforces
CPGM for good laboratory practice for nonclinical laboratories	Part 58—GLP for nonclinical laboratory studies
CPGM for clinical investigators	Part 50—protection of human subjects
CPGM for sponsors, monitors, and contract research organizations	Part 312—INDA Part 812—IDE Part 511—new animal drugs for investigational use
CPGM for IRBs	Part 56—IRBs
CPGM for *in vivo* bioequivalence compliance program	Part 320—bioavailability and bioequivalence requirements Parts 50, 56, and 312 also enforced by this CPGM

CPGM, Compliance Program Guidance Manual; FDA, Food and Drug Administration; IDEs, investigational device exemptions; INDA, investigational new drug application; IRB, Institutional Review Board.

The objective of BIMO inspections of nonclinical laboratories is

- To verify the quality and integrity of data submitted in a research or marketing application.
- To inspect (approximately every two years) nonclinical laboratories conducting safety studies that are intended to support applications for research or marketing of regulated products.
- To audit safety studies and determine the degree of compliance with GLP regulations.[27]

The FDA's BIMO inspects commercial nonclinical laboratories without prior notification in most cases. There are two classifications of inspections: surveillance inspections and directed inspections.

Surveillance inspections	These are periodic, routine determinations of a laboratory's compliance with GLP regulations. These inspections include a facility inspection and audits of ongoing and/or recently completed studies.

Directed inspections	These are assigned to achieve a specific purpose, such as
	  Verifying the reliability, integrity, and compliance of critical safety studies being reviewed in support of pending applications.
	  Investigating issues involving potentially unreliable safety data and/or violative conditions brought to the FDA's attention.
	  Reinspecting laboratories previously classified as official action indicated, OAI (usually within six months after the firm responds to a warning letter).
	  Verifying the results from third-party audits or sponsor audits submitted to the FDA for consideration in determining whether to accept or reject questionable or suspect studies.[28]

Both the biennial surveillance inspection program and any directed inspections of nonclinical laboratories are data-driven processes and are often conducted by a team of BIMO investigators, depending on the nature of the facility being inspected or the reason (for cause) for the directed inspection. The FDA field personnel conducting facility audits are referred to as investigators. We also use the term "investigator" in multisite GLP activities and the conduct of GCP studies at clinical trial sites, so the terminology can sometimes be confusing. The FDA investigators will interview key personnel who typically will include a QAU representative, the study director(s) for those studies being reviewed, and potentially facility management representatives, the archivist, and technicians within the laboratories and/or animal care facilities. Tours of the facility will be taken to determine the conditions, layout, and workflow of the employees, test systems, test/control articles, and in-process samples. SOPs will be reviewed and raw data will be reviewed to confirm that the laboratory is in compliance with its own SOPs, and that the SOPs adequately ensure compliance with GLPs. The inspection will typically include the evaluation of at least one completed GLP study. The audit of the completed study will include a comparison of the protocol and amendments, raw data, other records, and specimens against the final report to ensure that the protocol requirements were met and that the final report accurately reflects the conduct and findings of the GLP study. A typical audit may last for three to four business days, but can vary greatly.

As with all FDA GxP inspections, the inspectional and administrative follow-up procedures are similar across the GxPs, and an explanation of the FDA process is included in "GMP Inspections and Consequences" section.

INTRODUCTION TO CURRENT GMPs

Current GMPs (cGMPs) are the regulations that govern the manufacture of human and veterinary drugs, biologics, and medical devices to endure the identity, strength purity, and quality of the finished product. The cGMPs are based on the fundamental principles of quality assurance: (1) quality, safety, and efficacy must be designed and built into the product; (2) quality cannot be inspected or tested into the product; and (3) each step of the manufacturing process must be controlled to maximize the likelihood that the product will be acceptable.[29] Sponsors, even if part or all of the manufacturing activities are outsourced, are responsible for ensuring that their products comply with cGMPs throughout the product development life cycle from development to commercialization. The FDA enforces these regulations primarily through inspections, and failure to comply with the regulations will result in various regulatory actions.

HISTORY OF THE cGMPs

It is important for the Regulatory Affairs professional to understand that the evolution of cGMP regulations have emerged mostly as a result of fraud and missteps by medical product manufacturers.

The regulatory basis for cGMPs is the Food, Drug, and Cosmetic (FD&C) Act, which was first published in 1938 after a public health crisis involving elixir of sulfanilamide. This sulfa drug was dissolved in diethylene glycol, an analogue of antifreeze, and as a result over a hundred people died, many of them children. Subsequently, the Act was quickly passed by Congress and required that all drugs be labeled with directions for safe use and mandated the preapproval of all new drugs. The initial use of the term, "good manufacturing practices," is recorded in the Act, as it sets tolerances for poisonous substances that cannot be avoided by the observance of GMP. The law also formally authorized facility inspections and gave the FDA authority to enforce compliance with the Act.

However, the general use of the term "current good manufacturing practice" did not become prevalent until the Kefauver–Harris Amendments to the FD&C Act in 1962. Under these amendments, a drug is adulterated if:

> The methods used in, or the facilities or controls used for, its manufacture, processing, packaging, or holding do not conform to or are not operated or administered in conformity with cGMP to assure that such drug meets the requirements of this Act as to safety and has the identity and strength, and meets the quality and purity characteristics, which it purports or is represented to possess.[30]

The amendments require that all drugs be shown to be safe and effective before being marketed and gave the FDA greater oversight of clinical trials and access to manufacturers' production records. The amendments were precipitated by a health crisis that occurred outside the United States, involving thalidomide. Thalidomide, a sedative, was given to thousands of pregnant European women and resulted in over 8000 children with birth defects. The FDA refused to approve the drug in the United States despite pressure from the applicant. It should be noted that the FDA reviewer in charge received the highest governmental civilian award, the Civilian Medal of Honor, for not approving the thalidomide application.

Modern-day drug cGMP regulations are a result of a regulatory revisions published in 1978.[31] Additional amendments were proposed in 1996,[32] but have since been withdrawn by the Agency in anticipation of a more comprehensive cGMP initiative (see "Pharmaceutical cGMPs for the 21st Century—A Risk-Based Approach," FDA September 2004). The cGMP regulations as they apply to drugs and biologics can be found in the 21 CFR Parts 210 and 211 (21 CFR 210 and 211).

The cGMP regulations for medical devices were promulgated in 1978 as a result of the Medical Device Amendments to the FD&C Act in 1976.[33] The Medical Device Amendments were adopted in response to another health crisis caused by the Dalkon Shield intrauterine device. The Dalkon Shield caused numerous injuries to women and resulted in one of the largest class action lawsuits ever against the manufacturer. The subsequent passage of the Medical Device Amendments requires different levels of the FDA premarketing oversight, depending on the classification of the device. In the ensuing years after 1978, the

FDA interpretation of the cGMPs evolved to more closely resemble the medical device requirements set by the International Organization of Standardization (ISO). In 1996, the FDA published a final rule revising the cGMP requirements for medical devices, incorporating them into a quality system regulation (QSR).[34] The regulations are codified in 21 CFR 820.

Although not discussed in detail in this chapter, additional cGMP regulations exist to cover more nonstandard medical products. These include the cGMPs for blood and blood products that can be found in the 21 CFR 606 and cGMPs for positron emission tomography (PET) Drugs Products 21 CFR 212 (proposed), which is yet to be codified into law.

cGMPs Today

In recent regulatory publications by the FDA and ICH, the cGMPs are converging toward a single set of regulations for all medical products that are applicable throughout the ICH community. Concepts found in medical device cGMPs, as well as international standards (e.g., ISO, EU GMPS, and PIC/S), are being integrated into drug cGMPs. There is a greater emphasis on understanding process performance, use of modern analytical techniques, and continual process improvement. cGMPs are being applied as appropriate throughout the product's life cycle from pharmaceutical development to product discontinuation. Management responsibilities are being identified to assure compliance to cGMP regulations. The hope for this evolution in cGMPs is for sponsors to have greater regulatory latitude and subsequently reduce the number of regulatory changes submitted to the health authorities while continuing to manufacture products that are consistently safe and effective.

A major initiative by the FDA in August 2002 titled "Pharmaceutical cGMPs for the 21st Century—A Risk-Based Approach" launched a series of guidelines to promote a more modern, comprehensive approach to cGMPs for drug products. The initiative's five major objectives are

1. To encourage the early adoption of new technologic advances by the pharmaceutical industry.
2. To facilitate industry application of modern quality management techniques, including implementation of quality systems approaches, to all aspects of pharmaceutical production and quality assurance.
3. To encourage implementation of risk-based approach that focuses on both the critical areas of the industry and Agency.
4. To ensure that regulatory review, compliance, and inspection policies are based on state-of-the-art pharmaceutical science.
5. To enhance the consistency and coordination of the FDA's drug quality regulatory programs, in part, by further integrating enhanced quality systems approached into the Agency's business process and regulatory policies concerning review and inspection activities.[35]

Overview of the GxPs for the Regulatory Professional 251

Two major guidelines published by the FDA to address the needs described in the "Pharmaceutical cGMPs for the 21st Century" report were "PAT—A Framework for Innovative Pharmaceutical Development, Manufacturing and Quality Assurance" and "Quality Systems Approach to Pharmaceutical cGMP Regulations." The FDA's guidance "Quality Systems Approach to Pharmaceutical cGMP Regulations" outlines concepts where process analytical technology (PAT) can be used to increase manufacturing efficiency while still working within the framework of cGMPs. This guideline emphasizes the need for the sponsor to build in or design a quality process on the basis of the desired attributes of the product, a concept originally found in the medical device cGMPs. In addition to quality by design, the guideline emphasizes the need for extensive process development and understanding. Process understanding requires the identification of critical sources of variability, the ability to manage variability, and the knowledge that the management of variability will result in a quality product. Using the knowledge obtained during process development and analytical tools during the processing allows for more regulatory flexibility in defining the parameters that produce a quality product. The FDA guidance "Quality Systems Approach to Pharmaceutical cGMP Regulations" provides a systematic approach to meeting 210 and 211 CFR while incorporating more modern, universally recognized concepts of quality. The quality system approach, another concept originally found in the medical device cGMPs, identifies four critical elements associated with a successful quality program: (1) senior management support, (2) sufficient resources, (3) manufacturing operations partnership, and (4) continual self-evaluation. The guidance also describes the FDA's six systems approach to inspections, dividing manufacturing into five systems: (1) production, (2) facilities and equipment, (3) laboratory controls, (4) materials, and (5) packaging and labeling. The sixth system, quality, encompasses the other five.

Concurrent with the recent cGMP publications by the FDA, there has been the publication of several ICH cGMP applicable guidelines: "Q8 Pharmaceutical Development," "Q9 Quality Risk Management," and "Q10 Pharmaceutical Quality System." These guidelines espouse the same concepts put forth by the FDA in "Pharmaceutical cGMPs for the 21st Century," such as quality by design, risk management, and quality systems. The ICH guideline "Q8 Pharmaceutical Development" promotes the idea of prospectively designing quality on the basis of the product's indication, method of administration, and physiologic properties. The concept of product life cycle management and improvement is introduced with a focus on the early development of a drug product. "Q9 Quality Risk Management" integrates risk management—a common notion that is found throughout the medical device regulations—into pharmaceutical development and provides the tools to achieve the goals of ICH guideline "Q8 Pharmaceutical Development." Risk identification, analysis, and evaluation allow manufacturers to manage variability in their processes with greater regulatory flexibility. "Q10 Quality Systems" is the international equivalent of the FDA's guidance "Quality Systems Approach to Pharmaceutical cGMP Regulations" and provides a modern, quality framework for meeting cGMPs, as appropriate, throughout the different stages of product development from inception to discontinuation. Based on ISO standards, this guideline's requirements for a successful quality system include management support plus continual process and quality system improvement.

The basic premise for cGMPs is that "quality should be built-in to the product, and testing alone cannot be relied on to ensure product quality."[36] Through quality, one can achieve the desired identity, strength, purity, and other quality characteristics of the final product and, therefore, be assured that the final product meets the required levels of safety and effectiveness. In general, the cGMPs require

- The establishment of a quality system and an independent group to oversee the quality system.
- A system for monitoring process performance and product quality to ensure that a state of control is maintained.
- The documentation of process performance and product quality through written records.
- A change management system to assure that all changes are properly evaluated and documented.
- A corrective action and preventative action system to address items that may affect process performance and product quality.

GMP Regulations and Guidance

The specific cGMP regulations for drugs and medical devices, as summarized in Table 9.1, address all areas that impact process performance and product quality: personnel, components, procedures, equipment, and facilities. Personnel must be qualified and trained to perform their function. Materials used in the process must meet specified quality attributes and controlled in a manner to prevent mix-ups. Procedures must be established and followed for the manufacture, testing, cleaning, and validation activities associated with the product. Equipment must be properly identified, cleaned, and maintained to prevent cross-contamination. Facilities must be suitable for their intended purpose with proper lighting, air handling, plumbing, and sanitation. Although design controls, which require that the desired product performance characteristics be established prior to production, are found only in the medical device cGMPs, this concept is now being promulgated for pharmaceutical cGMPs in the FDA's guidance "Quality Systems Approach to Pharmaceutical cGMP Regulations" and ICH's guidance "Q10 Quality Systems" (Table 9.4).

The cGMP expectations for active pharmaceutical ingredients (APIs) are equivalent to those outlined in the drug product cGMP regulations (21 CFR 210 and 211). ICH's guidance for Industry "Q7A Good Manufacturing Practice Guidance for Active Pharmaceutical Ingredients" further clarifies the application of cGMPs to APIs (both chemical and biologic) and intermediates. This guidance recognizes the differences between API and drug product production and, most importantly, defines the point at which API production should be under cGMP control.

TABLE 9.4
Subparts of 21 CFR 211, 21 CFR 600, and 21 CFR 820

	Subpart Topic		
Subpart	21 CFR 211	21 CFR 600	21 CFR 820
A	General provisions	General provisions	General provisions
B	Organization and personnel	Establishment standards	Quality system requirements
C	Buildings and facilities	Establishment inspection	Design controls
D	Equipment	Reporting of adverse events	Document controls
E	Control of components and drug product containers and closures		Purchasing controls
F	Production and process control		Identification and traceability
G	Packaging and labeling control		Production and process controls
H	Holding and distribution		Acceptance activities
I	Laboratory controls		Nonconforming product
J	Records and reports		Corrective and preventive action
K	Returned and salvaged drug products		Labeling and packaging control

Other guidelines exist that apply cGMPs to a specific area or topic [e.g., "Sterile Drug Products Produced by Aseptic Processing—Good Manufacturing Practice" or "Application of the Device Good Manufacturing Practice (GMP) Regulation to the Manufacture of Sterile Devices"], and still others will continue to be published by the FDA and ICH.

The cGMPs apply throughout the product's life cycle, but the stringency of cGMPs increases from initial clinical trials to commercialization. Early stage development products are given greater flexibility in their approach to cGMP compliance than commercial products because process knowledge is limited at this stage. Therefore, the level of controls needed to achieve investigational product quality differs from that of a commercial product. For example, pharmaceutical GMPs require that the production process be validated, which is impractical for an investigational product where sometimes only one or two batches have been manufactured. This concept is expressed in several FDA guidelines, including the "INDS-Approaches to Complying with cGMP During Phase 1" (since withdrawn, but still representative of the Agency's thinking), "1991 Guideline on the Preparation of Investigational New Drug Products (Human and Animal)," "Content and Format of investigational new drug applications (INDs) for Phase 1 Studies of Drugs, Including Well-Characterized Therapeutic, Biotechnology Product," and "INDs for Phase 2 and Phase 3 Studies Chemistry, Manufacturing, and Controls Information." The ICH guidance "Q7A Good Manufacturing Practice Guidance for Active Pharmaceutical Ingredients" contains a specific section regarding the application of cGMPs to investigational APIs. On a cautionary note, reasoned judgment should apply in establishing the level of cGMPs for investigational products as the Agency can put on clinical hold or terminate an IND if there is "evidence of inadequate quality control procedures that would compromise the safety of an investigational product."[37]

Given the number of companies today that outsource the manufacture, packaging, and labeling activities for their medical products, how GMPs apply to these companies deserves special mention. The ICH guidance for industry "Q10 Quality Systems" specifically states that the pharmaceutical quality system extends to the oversight and review of outsourced activities, and this idea is reiterated in the FDA guidance for industry "Quality Systems Approach to Pharmaceutical cGMP Regulations." Statutory pharmaceutical cGMP requirements, under 21 CFR 211, require that the quality control unit "accepts/rejects products manufactured, packed or held under contract by another company."[38] Therefore, even companies that do not manufacture their own products must have an appropriate quality system. Oversight is established through written agreements, called quality agreements, which define quality roles and responsibilities between the contract giver and contract acceptor. In addition, quality agreements describe change management expectations, audit activities, and communication mechanisms.

Compliance with cGMPs

The FDA and other regulatory authorities assure compliance with cGMPs through their routine and preapproval inspection program. In general, routine inspections are scheduled to occur biennially, while preapproval inspections occur prior to the approval of an NDA or PMA. In reality, due to limited FDA resources, inspections may not happen in these time frames and is more dependent upon the compliance history of the firm. It is important to note that the FDA may inspect a firm at anytime.

There are several FDA internal inspection-related documents available, all of which contain various compliance programs and instructions for agency personnel to follow during inspections. These documents include the FDA CPGM[39] and the FDA Inspection Reference.[40] Both the CPGM and Inspection Reference provide important information regarding the Agency's expectations during the inspections of different systems and/or product types. Some examples follow

- Drug manufacturing inspections
- Sterile drug process inspections
- Guide to inspection of computer systems in drug processing
- High purity water systems
- Inspection of medical device manufacturers
- Medical device PMA and postmarket inspections
- Inspection of biologic drug products

The biennial inspection program strictly audits the firm for compliance with GMPs, while an FDA preapproval inspection also includes ensuring that the information, especially the chemistry, manufacturing, and controls (CMC) information, submitted in the regulatory application is in agreement with the company's data on site (e.g., stability, manufacturing processes, and test methods). Any inaccuracies found can be considered a cGMP violation and can lead to the approval of the application being withheld.

GMP Inspections and Consequences

An inspection is initiated by the FDA's presentation on site of the Form FDA 482 (Notice of Inspection) and, if there are observations of noncompliance with cGMPs upon completion of the inspection, concluded with the presentation of the Form FDA 483 (Inspectional Observations). From the 483 observations, an establishment inspection report (EIR) is generated by the FDA for internal review and classification. The inspection is classified as no action indicated (NAI), no substantive GMP noncompliance, voluntary action indicated (VAI), substantive cGMP noncompliance but no further regulatory action required, and OAI, meaning further administrative and/or judicial regulatory actions are required.[41]

The consequences of noncompliance are multifold. Administrative actions include the issuance of a warning letter that states that the company is in violation of laws or regulations (i.e., specific 21 CFR 211, 21 CFR 600, or 21 CFR 820 violations) and failure to correct such violations may result in further FDA action without warning. The company has 15 days to respond to the letter. Other administration actions include application action (e.g., withdraw approval of IND, IDE, NDA, BLA, or PMA) and the FDA-initiated product recalls. The FDA's Application Integrity Policy (AIP)[42] and Compliance Policy Guide (CPG) 7150.09, "Fraud, Untrue Statements of Material Facts, Bribery, and Illegal Gratuities" allows for the Agency to defer substantive review of applications if there is evidence of fraudulent data and requires the company to undergo a series of corrective actions to reestablish the integrity of the data.[43] If warranted, the violations may be referred to the FDA's Office of Criminal Investigations for potential judicial actions, including injunction, seizure, and prosecution. A felony conviction on the FDA-related charges results in debarment.

A firm must respond to any inspectional observations in a thorough and timely manner, either by outlining the corrective action to be taken and timing of these corrective actions or by disputing the findings with supportive data. Adequately addressing an FDA inspector's concern during an inspection can prevent an observation from being noted on the 483 Form. Adequately addressing any observations on the 483 Form can prevent the issuance of a warning letter or, if a warning letter is issued, prevent further administrative or judicial action. Even for a finding of VAI, a firm should respond to the observations as the FDA can issue an untitled letter or regulatory meeting to notify the firm that the findings are expected to be corrected. It is important to keep the Agency informed of the firm's progress in instituting the corrective actions, especially if there are numerous serious observations that require a long-term corrective action plan or if the timing or corrective action plan itself changes. The FDA will conduct a follow-up inspection to ensure that all of the items in the warning letter have been addressed appropriately.

The failure to respond to the substantive 483 observations can be a costly mistake, and the responsibility ultimately falls upon the head of the company. In the landmark decision, *United States v. Park*, the president of a 36,000-employee company was found responsible for storing food in a warehouse under unsanitary condition. The court found that "persons responsible for exercising supervisory authority have a duty not only to seek out and remedy violation but to prevent them, thus imposing upon that person a duty to take affirmative action."[44]

The FDA's latest enforcement tool to ensure cGMP compliance is the policy of disgorgement, a sanction based on the premise that the firm is not entitled to profits gained by illegal means, which allows the FDA to impose huge fines. In 2001, Schering Plough paid a US$500 million fine as part of a consent decree for cGMP violations in their New Jersey and Puerto Rico manufacturing facilities. In addition to the fine, the approval of Clarinex® was delayed for about a year and ultimately the president and chief operating officer resigned.[45]

cGMP References

There are multiple cGMP references available to the Regulatory Affairs professional. By accessing the information available in the regulations, *Federal Register*, guidelines, and policies, the Regulatory Professional can keep abreast of the Agency's thoughts in this area. The *Federal Register* is the daily official publication of the US government and where all proposed and final federal rulemaking (including the FDA regulations) can be accessed by the public. The FDA-related *Federal Register* items can be accessed at www.accessdata.fda.gov/scripts/oc/ohrms/index.cfm, or one can subscribe via e-mail to a daily list of the *Federal Register* table of contents at http://listserv.access.gpo.gov. Many other useful e-mail subscriptions are available from the FDA and can be accessed at https://service.govdelivery.com/service/user.html?code=USFDA, including updates on the latest regulatory guidelines. Another important site is electronic freedom of information (FOI) room that contains useful information on cGMPs, including recent warning letters and 483 observations. This information is available for drugs, biologics, and devices and for the Office of Regulatory Affairs (ORAs). The FDA's ORA (http://www.fda.gov/aboutfda/centersoffices/officeofglobalregulatoryoperationsandpolicy/ora/default.htm) contains valuable information on compliance and inspectional activities. Additional references to access information regarding cGMPs available on the Internet are listed in the reference section of this chapter. This listing of Internet resources also contains a few useful international GMP references because of the harmonization cGMPs and the global nature of most companies today.

INTRODUCTION TO THE GCPs

The regulations, guidance, and industry standards that make up the GCPs are intended to provide assurance that the safety and well-being of human subjects

Overview of the GxPs for the Regulatory Professional

participating in research has been protected *and* that the research yields quality scientific data. A trial conducted in full adherence to GCPs gives the sponsor the ability (in theory anyway) to submit data to a number of regulatory authorities worldwide as the data were derived in accordance with a globally recognized standard, if not in compliance with each and every local regulation.

It is commonly accepted that the GCPs are far less descriptive than both the GLPs and GMPs, resulting in significant "gray area" open to interpretation. While a sponsor enjoys the benefit of choosing their best practice for GCP compliance, *ad hoc* interpretation without supporting written procedures can breed inconsistency and inefficiency. You will recognize some common themes within this overview of GCPs, which were also discussed in the GLP and GMP overviews in this chapter, the importance of independent quality unit oversight, written procedures, proper change control, and good documentation practices. To have an understanding of the GCPs today, it is important to recognize some of the reasons why they became necessary and from where they evolved.

HISTORY OF THE GCPs

The human subject has always been and, at least for the time being, is still the gold standard for human physiology experimentation. Today in the industry, we use the more palatable words "investigation," "research," or "trial" when describing experimentation in man; however, the evolution of the GCPs and our collective awareness to human subject rights and protection is surprisingly a relatively recent development.

Certainly, a regulator or historian would argue that the following selection of events leading to the creation of the GCPs is an oversimplification. Guilty as charged. But for the purposes of this chapter, which is intended to be more practical than theoretical, we chose to highlight just a couple of post–World War II events which laid the foundation for the GCPs.

1947	The Nuremberg Code is written following the Nuremberg Trials, in which Nazi doctors were tried (and some sentenced to death) for the bizarre human experimentation they conducted in the name of science during World War II. The first of the 10 principles of the Code states that "the voluntary consent of the human subject is absolutely essential"; however, the Code falls short, in that it is interpreted to apply only to nontherapeutic human research, and is not applied uniformly for all human research subjects.
1962	US Congress passes the Kefauver–Harris Amendments to the FD&C Act. In addition to requiring the FDA to evaluate new drugs for efficacy, the amendments establish the requirement for obtaining the informed consent of human research subjects.
1964	The World Medical Association meets in Helsinki, Finland, and adopts a document setting forth the ethical principles for medical research involving human subjects. The Declaration of Helsinki, as it came to be known, makes some of the principles set forth in the Nuremberg Code applicable to clinical (therapeutic) research, and thus applicable to drug development studies. The Declaration of Helsinki has been amended several times since its inception, most recently clarified with a release in Tokyo, 2004.[46]

(Continued)

1966	*The New England Journal of Medicine* publishes a landmark article by Dr. Henry K. Beecher titled "Ethics and Clinical Research." In his article, Dr. Beecher describes 22 research studies published in major medical journals, which he believed were examples of "unethical or questionably ethical studies."
1972	*The New York Times* publishes an expose on the Tuskegee syphilis study conducted by the US Public Health Service (PHS). The study began in 1932, documenting the natural progression of syphilis in African American sharecroppers in Macon County, Alabama. When these men enrolled into the study, there was no effective treatment for the disease; however, a decade into the study, penicillin was shown to be a safe and effective treatment for syphilis. The men in the study went decades without receiving penicillin for their syphilis though they were led to believe they were receiving treatment. The study continued in this fashion until the *Times* article exposed their mistreatment.
1974	The National Commission for the Protection of Human Subjects of Biomedical and Behavioral Research is formed with the primary goal to establish the basic ethical principles and policies to conduct human subject research in the United States. The *Times*' uncovering of the Tuskegee study and Dr. Beecher's article six years earlier are often cited as precursors for the Commission's formation.
1979	The National Commission for the Protection of Human Subjects of Biomedical and Behavioral Research publishes the Belmont Report, identifying three basic ethical principles of human subject research: respect for persons, beneficence, and justice. Very simply stated:
	Respect for persons = Acknowledge the subject's autonomy and protect those subjects whose autonomy is diminished.
	Beneficence = Minimize potential harm to the subject and maximize their potential.
	Benefit Justice = Distribute the benefits and burdens of research fairly. Avoid exploiting a subject population who would not benefit from the research for the sake of convenience.
1980	Federal regulations governing the Protection of Human Subjects (21 CFR 50) are published in the *Federal Register*.

The Protection of Human Subjects regulations changed the way clinical research was conducted in the United States during the 1980s, and paved the way for additional regulations governing human subject research in the 1980s and 1990s.

GCPs Today

US GCP Regulations

The US regulations that cover the GCPs are contained within Titles 21 and 45 of the US CFR. Unlike GLP and GMP, there is no part of the US CFR titled "Good Clinical Practice"; so you can stop looking.

Title 21 of the CFR applies to products regulated by the FDA. The CFR regulations under Title 21 that apply equally across drug, device, and biologic trials are cited in Table 9.5.

The CFR regulations under Title 21 that apply to a specific type of product (drug, biologic, or device) research regulated by the FDA are cited in Table 9.6.

Title 45 of the CFR applies to Public Welfare. The CFR regulations under Title 45 apply to research conducted by the Department of Health and Human Services

TABLE 9.5
Title 21 Parts/Subparts That Apply to Drugs, Devices, and Biologic Clinical Trials

Title/Part	21 CFR 11	21 CFR 50	21 CFR 54	21 CFR 56
	Electronic records; electronic signatures	Protection of human subjects	Financial disclosure by clinical investigators	INDA
Applies to	Drug/device/biologic	Drug/device/biologic	Drug/device/biologic	Drug/device/biologic
Subpart		**Subpart topic**		
A	General provisions	General provisions	General provisions	
B	Electronic records	Informed consent of human subjects	Organization and personnel	
C	Electronic signatures	Reserved	IRB functions and operations	
D		Additional safeguards for children in clinical investigations	Records and reports	
E			Administrative actions for noncompliance	

INDA, investigational new drug application; IRB, Institutional Review Board.

(HHS) or conducted or funded in whole or in part by any of the governmental agencies that have adopted these standards, and are contained in 45 CFR Subtitle A—Department of HHS; Part 46 Protection of Human Subjects, which is similar to those FDA regulations governing Protection of Human Subjects and IRBs at 21 CFR Parts 50 and 56, respectively.

Title 45 CFR Part 46 is often called the "Common Rule," referring to its common adoption by 17 US governmental agencies. It should be noted that, however, when research involving products regulated by the FDA is funded, supported, or conducted by the FDA and/or HHS, both the HHS and the FDA regulations apply.[47] There are several differences between the FDA regulations and the HHS regulations for the protection of human subjects. A chart comparing the differences in these regulations is posted on the FDA Website at http://www.fda.gov/scienceresearch/specialtopics/runningclinicaltrials/educationalmaterials/ucm112910.htm. Another regulation related to GCPs and regulated by HHS is the Privacy Rule under 45 CFR Part 160 and 164, often referred to as HIPAA, an acronym for the Health Insurance Portability and Accountability Act of 1996. HIPAA was enacted to provide efficiencies in the transfer of health-related electronic data and to provide protection for the confidentiality and security of health data identifiable to an individual patient, "protected health information." HIPAA is not regulated by the FDA, and as such, will not be discussed in depth here; however, it is important to note that the investigator, as the health care provider, is considered a "covered entity" under HIPAA, but the sponsor is not. Although sponsors are not regulated under HIPAA, it is good

TABLE 9.6
Title 21 Parts/Subparts That Apply to a Specific Type of Product, Either Drugs, Devices, or Biologic Clinical Trials

Title/Part	21 CFR 312	21 CFR 314	21 CFR 601	21 CFR 803	21 CFR 812	21 CFR 814
Applies to	INDA	Applications for FDA approval to market a new drug	Licensing	Medical device reporting	IDE	Premarket
Subpart	Drug/biologic	Drug	Biologic	Device	Device	Device
			Subpart topic			
A	General provisions	General provisions	General provisions	General provisions	General provisions	General PMA
B	INDA	Applications	Establishment licensing	Generally applicable requirements for individual adverse event reports	Application and administrative action	PMA
C	Administrative actions	Abbreviated applications	Product licensing	User-friendly reporting requirements	Responsibilities of sponsors	FDA action on a PMA
D	Responsibilities of sponsors and investigators	FDA action on applications and abbreviated application	Diagnostic radiopharmaceuticals	Importer reporting requirements	IRB review and approval	Administrative review reserved

E	Drugs intended to treat life-threatening and severely debilitating illnesses	Hearing procedures for new drugs	Accelerated approval of biologic products for serious or life-threatening illnesses	Manufacturer reporting requirements	Responsibilities of investigators	Postapproval requirements
F	Miscellaneous	Reserved	Confidentiality of information		Records and reports	Reserved
G	Drugs for investigational use in laboratory research animals or *in vitro* tests	Miscellaneous provisions				Reserved
H		Accelerated approval of new drugs for serious or life-threatening illnesses				Humanitarian use devices
I		Approval of new drugs when human efficacy studies are not ethical or feasible				Humanitarian use devices

IDE, investigational device exemptions; INDA, investigational new drug application; IRB, Institutional Review Board; PMA, premarket approval application.

business practice for sponsors to ensure that investigator-informed consent forms (ICFs) comply with HIPAA requirements as per 45 CFR 164.508, and the language allows the sponsor future access to the study data obtained under that consent. For further information regarding HIPAA and its impact on clinical research, visit the HHS Website dedicated to the topic at http://privacyruleandresearch.nih.gov/.

As noted earlier, the regulations governing GCPs are not overly detailed, and in many cases are open for broad interpretation. However, this is where guidance documents (aptly titled, don't you think?) can provide useful guidance.

GCP Guidance Documents

There are a number of guidance documents related to GCPs, and a listing of the FDA and ICH guidance documents can be accessed via the FDA Website (http://www.fda.gov/opacom/morechoices/industry/guidedc.htm). The most comprehensive "how-to" GCP guidance document was created by the ICH in 1996 (ICH E6). This guidance was subsequently published by the FDA in the *Federal Register* on May 7, 1997, and is applicable to drug and biologic trials, but not applicable to device trials; though we would say that the principles of GCPs listed in the document also apply to device trials, and device trial sponsors would be well served to make use of many of the recommendations included in ICH E6. The ICH E6 guideline "is intended to define 'Good Clinical Practice' and to provide a unified standard for designing, conducting, recording, and reporting trials that involve the participation of human subjects."[48] As with all guidance documents published by FDA, the ICH E6 guideline represents the FDA's "current thinking" on GCP.

The principles of ICH E6 are paraphrased below.

- Clinical trials should be conducted ethically, consistent with the Declaration of Helsinki (which we already discussed in this section) and applicable regulatory requirements.
- Rights, safety, and well-being of subjects are paramount.
- Benefits of study must outweigh risks.
- Study to adhere to protocol that has been reviewed and approved by an ethics committee (EC) (IRB).
- Study must be scientifically sound.
- Investigator(s) must be qualified.
- Informed consent must be obtained freely.
- Records must be maintained to allow for accurate reporting, interpretation, and verification.
- Confidentiality of records must be assured to respect the privacy and confidentiality of study subjects.
- Clinical trial supplies must meet GMPs.
- Systems and procedures should be implemented to assure the quality of the trial.

Overview of the GxPs for the Regulatory Professional 263

The ICH E6 guideline defines the responsibilities of IRBs, investigators, and sponsors, all of which will be discussed in this section as well. ICH E6 also defines the minimum information that should be included in a clinical protocol, an investigator's brochure (IB), and includes a list of required "essential" documents to be maintained during a clinical trial. A copy of the ICH E6 GCP guideline is a must for every regulatory, quality, or clinical professional conducting clinical trials on regulated investigational drugs, devices, or biologics. This document can be found online at http://www.fda.gov/downloads/Drugs/Guidances/ucm073122.pdf

WHO IS A SPONSOR AND HOW DO THEY MEET THEIR OBLIGATIONS?

The sponsor of a clinical trial may be an individual, a drug/device/biologic company, or a CRO that has been paid to take on specific (or all the) obligations of the sponsor. The primary responsibility of a study sponsor is to ensure that trials are being conducted and quality data are generated, documented, and recorded in compliance with the IRB-approved study protocol, GCPs, and applicable regulatory requirements. As part of the IND description in Chapter 2, we have discussed the process for submitting a protocol that is ethically and scientifically sound. We know that these protocol attributes are vetted by the IRB that has reviewed and approved the protocol in addition to the ICF, so we will begin discussing the practical application of GCPs with a review of IRB responsibilities.

IRB RESPONSIBILITIES

The IRB [or research ethics board (REB)/EC/independent EC (IEC)] is regulated by the FDA under 21 CFR 56, IRBs for drug, device, and biologic trials. ICH guidance on investigator responsibilities is included in ICH E6 GCP Section 3. An IRB's primary responsibility is to provide independent oversight of a clinical trial to safeguard the rights, safety, and well-being of human subjects. IRBs generally are categorized as "local" IRBs, which are institutionally based entities responsible for the rights, safety, and well-being of research subjects at their own institution (e.g., hospital), or "central" IRBs, which are "for-profit" IRBs that hold the same responsibilities for patient safety.

The central IRB is not affiliated with, or may not even be in close geographic proximity to, the research.

The membership requirements of IRBs are defined in the FDA regulation 21 CFR 56.107 and also in the ICH E6 guideline Section 3.2.1. IRBs are responsible for the initial and continuing review and approval of protocols and amendments, ICFs, recruitment advertising, IBs, available safety information, subject payments, and any other written information to be given to the subject. By regulation and guidance, IRBs must maintain written procedures to ensure that these responsibilities are met and must maintain adequate documentation of their review and approval activities. Review and approval of the research and related documents must be carried out during a convened meeting of the IRB in which a quorum is present, with the exception of those research activities allowed an "expedited review." An expedited review is a review of research conducted by an IRB chairperson, or by a single member or members of the IRB designated by the chairperson, whereby he/she can approve research (expedited review cannot disapprove research) that meets certain criteria listed in the *Federal Register*,[49] or approve of a minor change to previously approved research during the existing previously authorized approval period, one year or less, without a full quorum IRB meeting. The list of research activities allowed expedited review in the *Federal Register* can best be summarized as those not requiring an IND/IDE or those that involve no greater than minimal risk, with the most invasive procedure being twice weekly finger-stick blood draws. Expedited reviews are regulated by the FDA and HHS regulation in 21 CFR 56.110 and 45 CFR 46.110, respectively.

The minimum standard for continuing review of research by an IRB requires annual reapproval; however, IRBs commonly require more frequent updates regarding the trial in the form of periodic written progress reports from the investigator. IRBs, generally speaking, are also becoming more proactive in conducting periodic investigator site inspections.

In theory, the obligation to obtain IRB approval sits with the investigator rather than the sponsor; however, in practice, and especially when a central IRB is involved, the sponsor may communicate directly with the IRB. Sponsors may submit protocols, ICFs, recruitment advertising, and any other documents required by the IRB for approval (IB, additional information given to study subjects, etc.) on behalf of all investigator sites using the central IRB. The role that the sponsor plays in communicating with central IRBs is changing, and the FDA issued guidance in March 2006 titled "Using a Centralized IRB Review Process in Multicenter Clinical Trials". Sponsor communication directly with IRBs can be very beneficial to clearly understand if a review must be conducted by the full board or via expedited review, as this may impact clinical trial timelines. A couple of regulatory exceptions to the standard investigator-IRB reporting obligation is with devices where it is a sponsor's responsibility to evaluate adverse device effects and report the results directly to the IRB[50] and when an informed consent waiver has been invoked under 21 CFR 50.24.[51]

The FDA-issued Information Sheet, Sponsor–Investigator–IRB Relationship (http://www.fda.gov/oc/ohrt/irbs/toc4.html), is a good resource for further understanding the interrelationship and interaction of these entities. This guidance also applies to devices as well as drugs and biologics.

INVESTIGATOR RESPONSIBILITIES

An investigator's responsibilities are regulated by the FDA under 21 CFR 312 60–69 for drug and biologic trials and under 21 CFR 812, Subpart E for device trials. ICH guidance on investigator responsibilities is included in ICH E6 GCP Section 4. For studies conducted under a US IND/IDE, the sponsor submits either a Form FDA 1572 for drug and biologic trials or an investigator agreement for device trials, which is signed by the principal investigator. Sponsors should ensure that the investigator understands that they are committing to compliance via a signed document, in essence, a contract. While the specific responsibilities for investigators are similar, but not identical for drugs/biologics trials and device trials, the investigator commitments discussed later are paraphrased from Section 9 of the Form FDA 1572; however, the FDA has stated that "the general responsibilities are essentially the same."[52]

At the time of signing the Form 1572, the investigator has committed in writing to do the following:

Conduct the study in accordance with the protocol. The investigator is to follow the protocol as written, unless justified, to protect the safety, rights, or well-being of the subject. Nonemergency changes to the protocol may be made via a protocol amendment with prior sponsor and IRB approval. Minor logistical or administrative changes in the protocol [change of monitor(s), change of telephone number(s)] may not require a protocol amendment,[53] and may be documented through the use of an erratum or errata page. ICH E6 (Section 4.5.3) states "the investigator, or person designated by the investigator, should document and explain any deviation from the approved protocol." The limitation here is that neither ICH nor FDA provides a definition of a protocol deviation. Sponsors use terms such as "deviation" and "violation," sometimes interchangeably, and some classify them as critical, major, or minor. There is no standardization in this terminology from sponsor to sponsor, so it is critical that the investigator receives clear direction from the sponsor on terminology of protocol deviations, violations, their classification (if any), and a means to document them to comply with ICH E6. The FDA's device regulations are more descriptive than the drug or biologic regulations with regard to deviations from the protocol and the reporting requirements.[54]

Also of note, the protocol is expected to be followed as written. For example, if the protocol calls for a physician to conduct a physical exam, but state law allows nurse practitioners to conduct physical exams, the protocol must be followed.[55] Another example may be if sponsors indicate in their protocols that they are being conducted in accordance with ICH GCP, then ICH GCP must be followed.

Personally conduct or supervise the study. The investigator is wholly responsible for the care of study subjects and the conduct of the trial within his/her institution. Study tasks and investigator responsibilities may be delegated to appropriately qualified (and licensed in some cases) individuals, but the investigator is responsible for their supervision. ICH E6 requires documentation of these delegated tasks; and in May

2007, the FDA published a draft guidance titled "Protecting the Rights, Safety, and Welfare of Study Subjects—Supervisory Responsibilities of Investigators," which clarifies the FDA's expectations regarding an investigator's responsibilities (http://www.fda.gov/OHRMS/DOCKETS/98fr/07d-0173-gdl0001.pdf).

Inform patients and obtain their consent, ensuring IRB requirements are met. Informed consent is the process by which subjects are consented to participate in the study. IRBs review and approve the ICF to verify that the document contains all required elements. Title 21 CFR 50.25 describes the eight required elements of informed consent and another six additional elements to be included, if appropriate. ICH E6 has its own list of elements of informed consent in Section 4.8.10. A checklist of required elements of informed consent is included as an appendix at the end of this chapter. Obtaining informed consent from a study subject or their legally authorized representative is more than securing a signature on a document; it is a multistep process. An investigator must provide the subject with information regarding the study, the potential risks and benefits (as well as letting them know that there may be no benefit at all), the subject's role in the study, the opportunity for Q&A, and time to think about their decision and consult with family members or friends. Finally, the investigator must provide the subject with a copy of the consent form once it is signed and personally dated by the subject or their legally authorized representative. The process of obtaining consent should be appropriately documented so that it is clear that the subject was recruited and enrolled appropriately and that "informed consent was obtained prior to participation in the study."[56] The point at which the study actually begins is sometimes a gray area. The FDA recognized this and issued an information sheet on the topic titled "Screening Tests Prior to Study Enrollment." This information sheet states that

> consent must be obtained prior to initiation of any clinical procedures that are performed solely for the purpose of determining eligibility for research, including withdrawal from medication (wash-out). Procedures that are to be performed as part of the practice of medicine and which would be done whether or not study entry was contemplated, such as for diagnosis or treatment of a disease or medical condition, may be performed and the results subsequently used for determining study eligibility without first obtaining consent.[57]

There are exceptions from the requirement to obtain informed consent from a research subject before receiving an investigational product. These exceptions from informed consent are detailed in 21 CFR 50.23 and 50.24 and 45 CFR 46.116. There are also additional requirements that may require the assent of children in pediatric studies. 21 CFR 50.55 is titled "Requirements for permission by parents or guardians and for assent by children."

Report all adverse experiences. The FDA regulations require an investigator to promptly report to the sponsor any drug or biologic adverse experience (AE) that may reasonably be regarded as caused by, or probably caused by, the drug. If the AE is alarming, the investigator shall report the AE immediately. A good source for definitions of AEs and serious AEs (SAEs) and related terms for drugs and biologics is the ICH E2A guideline titled "Clinical Safety Data Management: Definitions and Standards for Expedited Reporting," March 1995. The following are the definitions of an AE and SAE taken from the ICH guidance document[58].

1. An adverse event (or adverse experience) can therefore be any unfavorable and unintended sign (e.g., including an abnormal laboratory finding), symptom, or disease temporally associated with the use of a medicinal product, whether or not considered related to the medicinal product.
2. An SAE or reaction is any untoward medical occurrence that at any dose
   a. Results in death.
   b. Is life-threatening.
      (*Note:* The term "life-threatening" in the definition of "serious" refers to an event in which the patient was at risk of death at the time of the event; it does not refer to an event which hypothetically might have caused death if it were more severe.)
   c. Requires inpatient hospitalization or prolongation of existing hospitalization.
   d. Results in persistent or significant disability/incapacity.
   e. Is a congenital anomaly/birth defect of a medicinal product, whether or not considered related to the medicinal product.

   The FDA regulations require an investigator to report unanticipated adverse device effects to the sponsor and to the IRB as soon as possible but no later than 10 working days after the investigator learns of the event.[59]

   The FDA definition of an unanticipated adverse device effect is
3. Any SAE on health or safety or any life-threatening problem or death caused by, or associated with, a device, if that effect, problem, or death was not previously identified in nature, severity, or degree of incidence in the investigational plan or application (including a supplementary plan or application), or any other unanticipated serious problem associated with a device that relates to the rights, safety, or welfare of subjects.[60]

Read and understand the IB. An IB is a compilation of the clinical and nonclinical data on the investigational product(s) that is relevant to the study of the investigational product(s) in human subjects.[61] The investigator must understand the relevant data to adequately oversee the investigation and adequately protect the safety and well-being of the subjects he/she enrolls in the trial.

Ensure all associates involved are informed of obligations. It is typically expected that an investigator will delegate tasks and investigator obligations to a study team. The study team may consist of nurses, physicians, pharmacists, unlicensed individuals, etc. Section 6 of Form FDA 1572 is where subinvestigators are listed. Subinvestigators listed on the 1572 should be limited to those individuals who play a critical role in the treatment and/or evaluation of the study subjects in the study. The investigator is responsible for ensuring that all members of the study team are appropriately trained. The FDA's draft guidance, "Protecting the Rights, Safety, and Welfare of Study Subjects—Supervisory Responsibilities of Investigators," indicates that the investigator should ensure that the study team is familiar with the protocol, understands the details of the investigational product, understands and are competent to perform tasks they have been delegated, are aware of their regulatory obligations, are informed of any pertinent changes, and are retrained during the conduct of the trial (if necessary). While this draft document does not mention that the training must be documented, the old FDA adage could be applied in this case, "if it isn't written, it didn't happen."

Maintain adequate and accurate records (device/drug use records, subject case histories, record retention) and make available for inspection. The fundamental elements of data quality are that documentation be attributable, legible, contemporaneous, original, and accurate. This is often referred to as the ALCOA principle.[62]

The FDA defines an adequate and accurate case history to include the case report form (CRF) and supporting data including, for example, ICFs, medical records, including physicians' and nurses' progress notes. 21 CFR 312.62(b) also notes that the case history for each subject shall document that informed consent was obtained prior to study participation. This is important to note that for those studies whereby the date consented and date that the subject undergoes study-related procedures is the same. Unless times of consent are on the ICF and in the investigator's progress notes to verify the subject consented prior to participation, the case history file must contain documentation that the subject's consent was obtained prior to undergoing any study-related procedures.

Document retention periods must be satisfied by the investigator. The FDA requires retention for two years following the date of marketing application approval or two years after the investigation is discontinued and the FDA notified. ICH E6 is similar, though slightly different; however, it is important to note that different countries require different retention periods for certain documentation. Sponsors typically specify a period of time for document retention in their clinical trial agreement/contract with the investigator.

The FDA regulations speak to the investigator's obligation to allow the FDA access to the study records, while ICH E6 indicates that the investigator is to allow record access to the monitor, auditor, IRB, and regulatory authority on request.[63,64]

Ensure an IRB that complies with 21 CFR Part 56 reviews and approves research, any changes to the research, and any unanticipated problems. The investigator is responsible for ensuring that an appropriately constituted IRB oversees the research, and that he/she enables the IRB to comply with its requirements by providing the IRB with all required reports and documents for review in compliance with regulations and the IRB's written procedures.

Comply with all other requirements. This all-encompassing phrase could be interpreted as meaning compliance with all applicable state and local regulations or written procedures.

Measuring an investigator's compliance often does not include their adherence to SOPs, as investigators are not required by the FDA regulations to have SOPs. Many investigators still do not have SOPs in place, a notable exception being commercial clinical research entities—doctors who have gone into the business of conducting clinical trials instead of carrying a patient load. There have been recent FDA guidances published, both final and draft, that may be changing the expectations for SOPs at investigator sites. In May 2007, the FDA published the final guidance document, "Computerized Systems Used in Clinical Investigations." This guidance document provides a list of "suggested" SOPs that need to be in place when using computerized systems to create, modify, maintain, or transmit electronic records, including when collecting source data at clinical trial sites. Also in May 2007, the FDA published a draft guidance titled "Protecting the Rights, Safety, and Welfare of Study Subjects—Supervisory Responsibilities of Investigators." This draft guidance makes several references to an investigator's responsibility to have procedures for the overall supervision and oversight of the trial. While this draft document does not specifically say "written" procedures, the old adage "if it isn't written, it didn't happen" applies. The Supervisory Responsibilities of Investigators guidance, as noted earlier, also clarifies and more explicitly explains the FDA's expectations of investigators with regard to appropriate delegation of tasks and training of study staff as well as highlighting the fact that the investigator is wholly responsible and accountable for the conduct of the study and for protection of the rights, safety, and welfare of study subjects. Time will tell, but this guidance may become a useful tool for the sponsor when establishing quality expectations with an investigator and his/her study staff.

SPONSOR RESPONSIBILITIES AND OVERSIGHT OF CLINICAL TRIALS

A sponsor's responsibilities are regulated by the FDA under 21 CFR 312.50–59 for drug and biologic trials and under 21 CFR 812, Subpart C for device trials. ICH guidance on sponsor responsibilities is included in ICH E6 GCP Section 5.

Although it appears that 13 out of a list of 13 principles of ICH GCP,[65] the very first obligation of a sponsor listed in the document is to implement quality systems and SOPs to ensure that trials are conducted and data are generated, documented, and reported in compliance with the protocol, GCP, and other applicable regulatory requirements.[66] The FDA device regulations require written monitoring procedures as part of the investigational plan.[67] The FDA drug and biologic regulations do not specifically require that sponsors have written procedures; however, guidance documents, which reflect the FDA's current thinking, do require SOPs. Additionally, implementing written SOPs is industry standard, definitely expected, and just good business practice. The number and type of SOPs a sponsor institutes may vary widely depending on how many activities are outsourced versus conducted in house. Below is a suggested list of SOPs to achieve GCP compliance.

- Investigator site selection
- Regulatory document collection, review, and submission
- Financial disclosure
- Investigator site initiation
- Investigational product distribution and tracking
- Clinical monitoring of investigator site
- Investigator site close-out
- AE reporting
- Quality assurance audits
- Required documents for study master file and document retention
- Vendor (CRO) qualification and oversight
- Protocol deviations
- Amending a protocol
- FDA inspection at sponsor facility

To conduct the clinical trial, the study sponsor must ensure that investigators are qualified by education and experience and are trained on the conduct of the protocol. It is often a misconception in the popular press that investigators are qualified by the FDA, when in fact this is a sponsor responsibility, mandated by regulation. Sponsors typically collect curricula vitae and applicable licenses from investigators to check to see if they are qualified by training and experience, that the research is related to their field of practice, and that they are appropriately licensed to conduct those procedures required by the protocol.

Although actions on the part of the sponsor regarding 21 CFR 54 "Financial Disclosure by Clinical Investigators" is not required until filing a marketing application, the process of collecting financial disclosure information from investigators begins at the time of study start-up for covered clinical trials [i.e., those that the applicant or the FDA relies on to establish that the product is effective (including studies that show equivalence to an effective product) or any study in which a single investigator makes a significant contribution to the demonstration of safety[68]]. Sponsors are required to disclose to the FDA any proprietary or equity interests held by the investigators, as this could potentially bias the study results by submitting a Form FDA 3455. If no financial interests and arrangements that fall under the 21 CFR 54 definitions exist, then the sponsor must certify the absence of these potential biases to the Agency by submitting a Form FDA 3454. Since the marketing application may be submitted a number of years after completion of the study, this information is typically collected at the beginning of the study by having each investigator directly involved in the research complete and sign a questionnaire. Most sponsors have an SOP covering financial disclosure (see discussion earlier). In addition to the regulation, the FDA has published a guidance document on the topic, Financial Disclosure by Clinical Investigators.[68]

To gain assurance that the study is being conducted according to set standards, the sponsor must monitor the progress of human research on an ongoing basis. Monitoring is defined in ICH E6 as the act of overseeing the progress of a clinical trial and of ensuring that it is conducted, recorded, and reported in accordance with the protocol, SOPs, GCPs, and the applicable regulatory requirements. The monitoring of a clinical trial may employ varying levels of oversight (e.g., frequency of study visits, depth, and detail of document review) depending on the size, duration, and complexity of the clinical trial design, plus the safety risk to study subjects. The most common method of clinical trial monitoring is through on-site visits made to the clinical trial site before the study begins and on a periodic basis until the study has been completed. Sponsors also use audits as another means to ensure compliance to GCPs.

The FDA originally issued a guideline for the monitoring of clinical investigations in 1988. In this document, the FDA references a "preinvestigation" visit, which in practice is often carried out as a two-step process commonly referred to as investigator qualification (or site selection) and initiation.

An investigator qualification assessment may be conducted during an on-site visit to the investigator's site or by phone. From the Agency's perspective, the purpose of the qualification assessment is to obtain information to assess the investigator's appropriateness to conduct the clinical trial, that is, experienced staff, adequate facilities, time, and resources to assure patient safety. The sponsor will also want to ensure that the investigator has access to appropriate subjects for recruitment as well as gauging the investigator's interest in conducting the trial.

The initiation covers more protocol-specific and GCP training. The initiation is typically conducted in one of two ways: an investigator's meeting, where all investigators participating in the study are trained on the protocol and GCPs by the sponsor or an on-site initiation visit where a sponsor representative, or clinical research associate (CRA), or team of sponsor representatives, visits the clinical trial site and trains the investigator and his/her staff on the protocol and GCPs. Documentation of the investigator's training on the protocol, through attendance at an investigator's meeting or an on-site initiation visit should be maintained in the investigator site's study files as well as the sponsor's files before the site's enrollment of study subjects.[69]

- The investigator site's facilities continue to be acceptable for study purposes.
- The investigator is following the study protocol/investigational plan.
- Any changes to the protocol have been reported to the sponsor and approved by the IRB.
- The investigator is maintaining accurate, complete, and current records for each study subject.
- The investigation is making accurate, complete, and timely reports to the sponsor and IRB.
- The investigator is carrying out the activities he/she agreed to and has not delegated responsibilities to other previously unspecified staff.

Once a trial has begun, it is the sponsor's responsibility to monitor the conduct of the study at the investigator's site. Monitoring frequency is dependent on the size of the study, complexity of the protocol, safety risk to the study subjects, and the sponsor's philosophy on GCP compliance. A CRA will periodically visit an investigator's site during the active phase of the study when subjects are being seen and patient data are being collected. Visit frequency is study dependent and may vary greatly. Periodic (or interim) monitoring, as outlined in the FDA Monitoring Guideline, is required to assure that:

> During an interim monitoring visit, the CRA is responsible for ensuring that all required documentation is maintained on site, that the protocol is being followed, the investigational product is accounted for, and that the rights, safety, and well-being of the study subjects are being protected. By conducting personnel interviews reviewing supporting documentation verifying source data during interim monitoring visits, the CRA can assure compliance with the protocol and GCPs.

The ICH GCP guideline, Section 8, "Essential Documents for the Conduct of a Clinical Trial," provides a quick and easy reference for required documents that need to be maintained at the study site. CRAs can use Section 8 of the ICH GCP guideline as a reference, or a study-specific checklist to ensure that the site is maintaining all required documents. To ensure that the investigator is following the study protocol, the CRA should review study subject medical records, study charts, and all appropriate documentation to ensure the subjects were being treated as dictated by the approved protocol.

The CRA should review investigational product dispensing/accountability logs and conduct a physical count of all investigational products on site to ensure that the investigator is appropriately dispensing and accounting for all investigational products. All investigational products must be stored in a manner that limits its distribution to those qualified and delegated by the investigator to do so. There must be adequate documentation to verify the chain of custody, that is, shipping records that account for every unit of investigational product received and is maintained at appropriate storage conditions plus an accurate inventory accounting for all investigational product received, dispensed, recollected from study subjects, and returned to the sponsor or destroyed.

The CRA should ensure that the rights, safety, and well-being of study subjects were protected by the investigator and his/her staff. This is done initially through review of the ICF before study start-up to ensure that it contains all the required elements, and on an ongoing basis via a review of patient records to ensure they were appropriately consented before study participation, and that they are receiving quality care.

The FDA Monitoring Guideline discusses the sponsor's responsibility to "compare a representative number of subject records and other supporting documents with the investigator's reports...." To fulfill this obligation, CRAs verify the accuracy of study data entered by the investigator's staff into the CRF against source documentation, which is commonly referred to as source document/data verification (SDV). "Source documentation" is the term used to describe where a study subject's information is first recorded. In some cases, the CRF may be considered source data; however, per ICH E6, the identification of any data to be recorded directly on the CRF and considered source documentation should be prospectively defined in the protocol.[70] As with monitoring visit frequency, study sponsors conduct SDV using different formulas to determine a representative sample. Some sponsors may choose to conduct 100% SDV. Others may choose a plan whereby key safety and efficacy data, or a percentage of data, are only verified.

> 1. Information recorded in the investigator's reports is complete, accurate, and legible.
> 2. There are no omissions of specific data, such as concomitant medications or AEs.
> 3. Any missed study visits are noted in the reports.
> 4. Subjects who were dropped from or failed to complete the study are noted in the report with the reason adequately explained.
> 5. Informed consent was executed and adequately documented in accordance with federal regulations.

SDV provides assurance that the data recorded in the subject's records is completely and accurately transcribed to the CRFs. The CRF data eventually become the basis for marketing authorization submissions to the FDA and other regulatory authorities. SDV is described in the FDA's monitoring guideline as a means to provide assurance that[71]:

> Another key item to look for during SDV is to ensure the presence of an appropriate "audit trail." An audit trail is required in both paper and electronic documentation systems in GCPs. An "audit trail" is a documentation that allows the reconstruction of the course of events. This allows someone reviewing the documentation to determine

what data was changed, the original entry that was changed, why it was changed, by whom and when it was changed, and in cases where the need for the correction is not readily obvious, a brief explanation of why the change was necessary. The requirement for such documentation is referenced in ICH E6 Sections 4.9.3 and 5.18.4(n).

The description of a computerized system used in a clinical trial to create, modify, maintain, archive, retrieve, or transmit clinical data required to be maintained, or submitted to the FDA, can apply to many different types of computer applications used by IRBs, investigators, and sponsors. Sponsors have been dependent on computerized systems to store and manipulate study data for years; however, investigators were somewhat behind the times from a technology standpoint, but that is now rapidly changing. Although investigators have progressed with computerized recordkeeping, the application of 21 CFR 11 regulations may not be clear to all. The FDA's guidance for its field investigators states that "records in electronic form that are that created, modified, maintained, archived, retrieved, or transmitted under any records requirement set forth in agency regulations must comply with 21 CFR 11."[72] Guidance issued subsequently on the topic indicates that while this is true, the Agency intends to interpret the scope of 21 CFR 11 narrowly and will exercise enforcement discretion with regard to some of the Part 11 requirements.[73]

All three entities, the IRB, investigator, and sponsor, bear the obligation to oversee, conduct, and monitor a clinical trial involving human subjects in compliance with GCP. The sponsor is additionally obligated by regulation to obtain the investigator's compliance. If the sponsor is unable to obtain compliance from the investigator, the sponsor is required to terminate the investigator's participation in the study as per 21 CFR 312.56(b) and 21 CFR 812.46(a), for drugs, biologics, and devices, respectively. Additionally, the drug regulations require the sponsor to notify the FDA of the investigator's termination for noncompliance. However, the device regulations are silent on this topic.

Other key sponsor responsibilities covered by the GCPs are safety monitoring and clinical data management. The Pharmaceutical Research and Manufacturers of America (PhRMA) defines ongoing safety monitoring as a process whereby

> All safety issues are tracked and monitored in order to understand the safety profile of the product under study. Significant new safety information will be shared promptly with the clinical investigators and any Data and Safety Monitoring Board or Committee (DSMB),[74] and reported to regulatory authorities in accordance with applicable law.[75]

An investigator's responsibilities for safety reporting were briefly discussed in the previous section, and a sponsor's obligations for reporting safety data to regulatory authorities are discussed in the chapters covering INDs and medical device regulations. Further information regarding a sponsor's obligations for reporting of drug and biologic safety data can be found in ICH E2A guideline, "Clinical Safety Data Management: Definitions and Standards for Expedited Reporting," March 1995. For devices, additional detail can be found in the guidance for "Medical Device Reporting for Manufacturers," March 1997.[76]

Overview of the GxPs for the Regulatory Professional

The discipline of Clinical Data Management includes paper and electronic case report form (CRF) design, clinical trials database design and programming, data acquisition and entry into the clinical trials database, data review, validation, coding and database finalization. Independent of how individual companies perform these tasks within their company, each company is obligated to ensure that the individuals performing these tasks follow Good Clinical Practices.[77]

The records and reports received and manipulated by the clinical data management function are regulated by the FDA and addressed in ICH E6 guidance, and the electronic systems employed to handle the data are also governed by regulations and guidance. However, few guidances have been issued from the FDA or ICH to provide these groups with direction with regard to clinical data management processes. The Society for Clinical Data Management, a professional organization not affiliated with a regulatory body, has issued a document titled "Good Clinical Data Management Practices," which is commercially available to members and nonmembers of the organization.[78]

COMPLIANCE WITH GCPS

The objective of a BIMO inspection at a clinical trial site is to assess, through audit procedures, if the clinical records adequately and accurately substantiate data submitted to the FDA to demonstrate safety and efficacy in support of an FDA-regulated product marketing application to determine that the rights and well-being of human subjects were adequately protected during the course of the research and to verify compliance with applicable FDA regulations and guidelines.[79] The objective of a BIMO inspection of a sponsor and/or CRO is to assess how sponsors assure the validity of data submitted to them by clinical investigators, and verify their adherence to applicable regulations.[80]

There are three classifications of BIMO inspections of a clinical investigator: study-oriented inspections, investigator-oriented inspections, and bioequivalence study inspections.

The study-oriented inspection is conducted by the FDA field office personnel and is usually assigned by the FDA headquarters on the basis of a pending sponsor application to market a new drug, device, or biologic, rather than as per an FDA-defined schedule. When the FDA reviewers are considering a marketing application or supplement for approval, they will choose clinical trial sites for inspection. The selection of a clinical trial site(s) for a study-oriented inspection is usually based on the amount of data contributed by the clinical trial site (the highest enrolling sites will most commonly be considered for inspection).

Once a site has been selected, the FDA field office will contact the investigator to arrange an inspection date. In general, the FDA will try to schedule the inspection within 10 business days of contact. On arriving at the clinical site, the FDA field investigator will present the investigator with a Form FDA 482 "Notice of Inspection" along with their credentials.

The FDA investigators are trained to conduct the inspection using the CPGM for clinical investigators, which outlines the minimal scope of the inspection.[81] The investigator will first obtain facts about the study conduct through interviews with the investigator, study coordinator, or responsible party at the clinical site to understand[82]:

- Who did what
- The degree of delegation of authority
- Where specific aspects of the study were performed
- How and where data were recorded
- How test article accountability was maintained
- How the CRA communicated with the clinical investigator
- How the CRA evaluated the study's progress

The FDA investigator will audit the study data, comparing what was submitted to the Agency with all supporting documentation. The FDA investigator will request a clinical trial subject's medical records, which may come from a doctor's office, hospital, nursing home, laboratory records, outpatient clinic records, or other sources.

An investigator-oriented inspection may be conducted when a single investigator's data may prove crucial to a product's approval, if the investigator has participated in many studies or if the investigator has conducted a study outside of his specialty. An investigator may also be targeted for a "for-cause" inspection if a study sponsor, patient, or any anonymous "whistle-blower" contacts the FDA with a complaint about the investigator's conduct. An investigator-oriented inspection may also be conducted to investigate any unusual findings or trends noted in the data submitted to the Agency. The conduct of an investigator-oriented inspection is much the same as a study-oriented inspection with the exception that the FDA investigator may dig deeper into the data audit and may audit data from more than one study.

NAI: A notice that no significant deviations from the regulations were observed. This letter does not require any response from the clinical investigator.

VAI: A letter that identifies deviations from regulations and good investigational practice. This letter may or may not require a response from the clinical investigator. If a response is requested, the letter will describe what is necessary and give a contact person for questions.

OAI: A letter that identifies serious deviations from regulations requiring prompt correction by the clinical investigator. Receipt of an OAI letter may lead to other regulatory actions by the FDA, such as the issuance of a warning letter, requiring the sponsor to throw out the investigator's data, trigger a sponsor inspection, or other regulatory actions up to and including disqualifying the investigator from clinical research, injunction, and criminal prosecution.

The bioequivalence study inspection may be conducted on the basis of a pending NDA or Abbreviated NDA (ANDA) for which a bioequivalence study is critical to product approval. Bioequivalence studies often support the approval of generic versions of innovator drug products or the approval of new formulations of marketed drugs. Bioequivalence studies have both a clinical component and an analytical component, thus bioequivalence study inspections differ from study and investigator-oriented inspections in that there is often participation from an FDA chemist who can assess the validity of the analytical methods used to indicate bioequivalence.[83]

The vast majority of all BIMO inspections are study oriented. An FDA investigator will generally take two to four days on site to conduct a study-oriented inspection. At the conclusion of the inspection, an exit interview will be held with the clinical investigator, where all findings will be discussed and clarified. If deviations from applicable regulations have been noted during the inspection, the FDA investigator will issue a Form FDA 483 "Inspectional Observations," to the clinical investigator. Note that deviations from guidance documents are not considered inspectional observations and should not be included on a Form FDA 483; although deviations from the FDA guidance may be included in the FDA investigator's written report submitted to the FDA headquarters for evaluation, the EIR.[84]

After the FDA headquarters evaluates the field investigator's EIR, the FDA headquarters issues a letter to the clinical investigator categorizing the field investigator's findings. The letter can be one of the following three types as described in the FDA Information Sheets.[85]

BIMO inspections of sponsors and/or CROs occur less frequently than investigator inspections. These inspections are generally unannounced, meaning they greet the company receptionist flashing their credentials and a Form FDA 482 Notice of Inspection, and the frantic telephone calling begins from there.

EIRs are now routinely supplied by the FDA after the report has been evaluated by the FDA headquarters. Redacted copies of EIRs are available through FOI and should be requested by whoever was audited, and the sponsor. Accessing the EIR can provide additional insight to an FDA investigator's inspection strategy and expectations and can prove to be a useful learning tool to design future trials to be conducted in a manner that fulfills current FDA expectations.

Sponsor Audits

Auditing is defined in ICH E6 as a systematic and independent examination of trial-related activities and documents to determine whether the evaluated trial-related activities were conducted, and the data were recorded, analyzed, and accurately reported according to the protocol, sponsor's SOPs, GCP, and the applicable regulatory compliance.[86] Study sponsors are required by regulation to monitor the conduct of a clinical trial; however, auditing is not specifically mentioned. Although not required by the FDA regulation, clinical site and vendor (CRO) audits have become an industry standard and are recommended by ICH GCP guidelines. The auditor should be an independent reviewer who is removed from the actual day-to-day conduct of the study, so that they can provide an unbiased opinion on the setup and conduct of the study.

Audits are often conducted according to the same principles that the FDA BIMO follows. The decision to audit a study is usually based on what phase of study is being conducted, whether or not the data are intended to support a regulatory application, the complexity of the study, and the level of risk to the study subjects. The number of investigator sites to be audited for the trial is determined either by a preexisting sponsor policy, for example, 10% of phase II study sites are audited, 20% of phase III study sites, or on the basis of other factors such as trial duration. The selection of investigator sites to be audited is generally based on enrollment (high enrollers are more likely to be audited), problems discovered by CRAs, AE reporting (abnormally high or low AE rates), the presence of an investigator's financial interest in the sponsor company, or previous experience with the investigator.

SUMMARY

A successful Regulatory Affairs professional recognizes the importance of ensuring their company's compliance with regulations while at the same time working to meet submission timelines and business objectives. One must keep abreast of an ever-changing GxP environment by constantly monitoring the evolution of new regulations and guidance documents. The Regulatory professional is the conduit and gatekeeper in the data flow process from GLP, GMP, and GCP activities to the FDA and other global health authorities, and while you are not expected to be an expert in all stages of drug development, having an understanding of the GxPs, how they are enforced, and the consequences of noncompliance is essential.

INTERNET RESOURCES

GxP References

Application Integrity Policy: Regarding the integrity of data and information in applications submitted for FDA review and approval. Available at http://ww.fda.gov/ora/compliance_ref/aip_page.html

Bioresearch Monitoring Program: BIMO main page providing links to regulations and CPGMs as well as to lists of inspections and lists of firms/individuals who have been disbarred/disqualified/have made assurances. Available at http://www.fda.gov/ora/compliance_ref/bimo/default.htm

Compliance Program Guidance Manuals (CPGM): Compliance programs and program circulars (program plans and instructions) directed to field personnel for project implementation. Available at http://www.fda.gov/ora/cpgm/default.htm

Electronic Records; Electronic Signatures, 21 CFR Part 11: Background information and updates on the rule that allows the use of electronic records and electronic signatures for any record that is required to be kept and maintained by other FDA regulations. Available at http://www.fda.gov/ora/compliance_ref/part11/Default.htm

FDA's Warning Letters and Responses Search Engine. Available at http://www.fda.gov/foi/warning.htm

FDA Industry Portal: Information for the FDA regulated industry. Available at http://www.fda.gov/oc/industry/.

Freedom of Information: ORA documents frequently requested by the public through the Freedom of Information Act. Available at http://www.fda.gov/foi/foia2.htm

From Test Tube to Patient Protecting America's Health Through Human Drugs: A Special Report from the *FDA Consumer Magazine* and the FDA Center for Drug Evaluation and Research providing the general public with an overview of the FDA's oversight during of the drug development life cycle. Available at http://www.fda.gov/fdac/special/testtubetopatient/default.htm

Guidance Documents: A link to various guidance documents that outline the Agency's and/or other regulatory authorities thinking on a variety of topics. There are links to CDER, CBER, CDRH, CFSAN, CVM, and ORA. Available at http://www.fda.gov/opacom/morechoices/industry/guidedc.htm

Public Health Service (PHS) Administrative Actions Listings: Individuals who have had administrative actions imposed against them. The list is maintained by the PHS Office of Research Integrity (ORI). Available at http://silk.nih.gov/public/cbz1bje.@www.orilist.html

Public Use Forms and How to Obtain Them: Access page to current versions of FDA forms. Available at http://www.fda.gov/opacom/morechoices/fdaforms/fdaforms.html

Revisions, Drafts, and Updates to ORA compliance references: A chronologic listing of updates to the FDA Office of Regulatory Affairs compliance references. Available at http://www.fda.gov/ICECI/ComplianceManuals/RevisionstoComplianceManuals/ucm289198.htm

GLP References

21 CFR 58: Good Laboratory Practice for Nonclinical Laboratory Studies. Available at http://www.accessdata.fda.gov/scripts/cdrh/cfdocs/cfcfr/cfrsearch.cfm?cfrpart=58&showfr=1

Bioresearch monitoring Good laboratory Practices: GLP references and Guidance. Available at http://www.fda.gov/ICECI/EnforcementActions/BioresearchMonitoring/default.htm

Bioresearch Monitoring Good Laboratory Practice compliance program. Available at http://www.fda.gov/ICECI/EnforcementActions/BioresearchMonitoring/ucm133789.htm

FDA Guidance: Good Laboratory Practices, Questions and Answers. Available at http://www.fda.gov/ICECI/EnforcementActions/BioresearchMonitoring/NonclinicalLaboratoriesInspectedunderGoodLaboratoryPractices/default.htm

FDA Preamble to the GLPs. Available at http://ovpr.uga.edu/qau/resources/glps/fda/preamble

OECD Series on Principles of Good Laboratory Practice and Compliance Monitoring: Link to all 15 guidance/advisory documents and position papers regarding GLPs. Available at http://www.oecd.org/chemicalsafety/testing/oecdseriesonprinciplesofgoodlaboratorypracticeglpandcompliancemonitoring.htm

GMP References

Center for Devices and Radiologic Health: Medical device GMP reference information. Available at http://www.fda.gov/medicaldevices/deviceregulationandguidance/postmarketrequirements/qualitysystemsregulations/default.htm

Compliance Policy Guides (CPG): Contains FDA compliance policy and regulatory action guidance for the FDA staff. Available at http://www.fda.gov/ora/compliance_ref/cpg/default.htm

Field Management Directives: The primary vehicle for distributing procedural information/policy on the management of Office of Regulatory Affairs (ORAs) field activities. Available at http://www.fda.gov/ICECI/Inspections/FieldManagementDirectives/ucm056246.htm

Guides to Inspections of...: Guidance documents written to assist the FDA personnel in applying the FDA's regulations, policies, and procedures during specific types of inspection or for specific manufacturing processes. Available at http://www.fda.gov/ora/inspect_ref/igs/iglist.html

Inspection Technical Guides: Guidance documents that provide the FDA personnel with technical background in a specific piece of equipment or a specific manufacturing or laboratory procedure, or a specific inspectional technique, etc. Available at http://www.fda.gov/ICECI/Inspections/InspectionGuides/InspectionTechnicalGuides/default.htm

Investigations Operations Manual: Primary procedure manual for the FDA personnel performing inspections and special investigations. Available at http://www.fda.gov/ora/inspect_ref/iom/default.htm

Medical Device QSIT Inspection Guide. Available at http://www.fda.gov/ora/inspect_ref/igs/qsit/default.htm

Questions and Answers on Current Good Manufacturing Practices. Available at http://www.fda.gov/Drugs/GuidanceComplianceRegulatoryInformation/Guidances/ucm124740.htm

Regulatory Procedures Manual (RPM): Contains FDA regulatory procedures for use by the FDA personnel. A reference document for enforcement procedures, practices, and policy guidance. Available at http://www.fda.gov/ora/compliance_ref/rpm/default.htm

INTERNATIONAL GMP REFERENCES

European Union GMPS. Available at http://ec.europa.eu/enterprise/pharmaceuticals/pharmacos/gmp_doc.htm

Pharmaceutical Inspection Convention and Pharmaceutical Inspection Cooperation Scheme (PIC/S): International group dedicated to developing harmonized GMP standards and guidance documents. Available at http://www.picscheme.org/index.php

World Health Organization (WHO) GMPs: WHO is the directing and coordinating authority for health within the United Nations system. Available at http://www.who.int/medicines/areas/quality_safety/quality_assurance/production/en/index.html

GCP REFERENCES

Clinical Research Training: A course developed by the National Institutes of Health to train its own investigators. It may be accessed by others to enhance their knowledge of clinical research. Available at http://crt.nihtraining.com/login.php

Declaration of Helsinki and Belmont Report: OHRP Web page providing a link to these documents. Available at http://www.hhs.gov/ohrp/international/.

E6 Good Clinical Practice: ICH Consolidated Guideline. Available at http://http://www.fda.gov/downloads/Drugs/Guidances/ucm073122.pdf

FDA Office of Good Clinical Practice: Homepage with links to all GCP regs/guidance/hot topics, etc. Available at http://www.fda.gov/oc/gcp/.

GCP Questions E-Mail Messages: An FDA Web page providing links to email Q&A. Available at http://www.fda.gov/ScienceResearch/SpecialTopics/RunningClinicalTrials/RepliestoInquiriestoFDAonGoodClinicalPractice/default.htm

GCP Regulatory Activities: Provides links to lists of all disqualified investigators, warning letters, listing of all investigators who have been inspected by the FDA. Available at http://www.fda.gov/ICECI/EnforcementActions/ucm321308.htm

Office for Human Research Protections (OHRP): Under Department of Health and Human Services, references on protecting the rights of human research subjects. Available at http://www.hhs.gov/ohrp/.

Online training on human subject protection: Provided by the Office for Human Research Protections. Topics include HHS regulation and institutional responsibilities, investigator responsibilities and informed consent, and human research protections program. Available at http://ohrp-ed.od.nih.gov/CBTs/Assurance/login.asp?Submitted=True&UserType = Login

Pharmaceutical Research and Manufacturers of America: Principles on Conduct of Clinical Trials and Communication of Clinical Trial Results. Available at http://www.phrma.org/about/principles-and-guidelines

Proposed Regulations and Draft Guidances on Good Clinical Practice and Clinical Trials. Available at http://www.fda.gov/oc/gcp/draft.html

REFERENCES

1. Code of Federal Regulations, Title 21, Section 58.3(d).
2. Organization for Economic Cooperation and Development Principles on Good Laboratory Practice, ENV/MC/CHEM(98)17, 1997.
3. 43 Federal Register, 59986, December 22, 1978.
4. 48 Federal Register, 53946, November 29, 1983.
5. "Decision Concerning the Mutual Acceptance of Data in the Assessment of Chemicals," 1981 [C(81) 30 (Final)].
6. 52 Federal Register, 33768, September 4, 1987.
7. Code of Federal Regulations, Title 21, Section 58.1(a).
8. Code of Federal Regulations, Title 21, Section 160.
9. Code of Federal Regulations, Title 21, Section 792.
10. http://www.oecd.org/general/official-unclassied-documents.htm
11. The Application of the OECD Principles of GLP to the Organization and Management of Multi-Site Studies, ENV/JM/MONO(2002)9, 2002.
12. Code of Federal Regulations, Title 21, Section 58.3.
13. OECD Principles on Good Laboratory Practice, ENV/MC/CHEM(98)17, 1997, Section 2, Definitions of Terms.
14. It should be noted that the term "principal investigator" does not appear in 21 CFR 58 but is an OECD term that has been adopted as multisite studies are now commonplace.
15. Pharmaceutical Technology Europe Magazine, Applying Good Laboratory Practice Regulations, 6/1/03 by Hana Danan. Available at http://connection.ebscohost.com/c/articles/10076015/applying-good-laboratory-practice-regulations
16. Code of Federal Regulations, Title 21, Section 58.81(b).
17. OECD Principles on Good Laboratory Practice, ENV/MC/CHEM(98)17, 1997, Section 7, Standard Operating Procedures.
18. Code of Federal Regulations, Title 21, Section 58.90.
19. Code of Federal Regulations, Title 21, Section 58.107(d).
20. Code of Federal Regulations, Title 21, Section 58.105(a).
21. Code of Federal Regulations, Title 21, Section 58.105(d).
22. Code of Federal Regulations, Title 21, Section 58.120(b).

23. OECD Principles on Good Laboratory Practice, ENV/MC/CHEM(98)17, 1997, Section 8.1, Study Plan.
24. The Application of the Principles of GLP to Computerized Systems, OCDE/GD(95)115, 1995.
25. FDA Guidance for Industry. Content and Format of Investigational New Drug Applications (INDAs) for Phase I Studies of Drugs, Including Well-Characterized, Therapeutic, Biotechnology-Derived Products. November 1995.
26. FDA Guidance for Industry. Q&A Content and Format of Investigational New Drug Applications (INDAs) for Phase I Studies of Drugs, Including Well-Characterized, Therapeutic, Biotechnology-Derived Products. October 2000.
27. Compliance Program Guidance Manual. Bioresearch Monitoring: Good Laboratory Practice (Nonclinical laboratories). February 21, 2001. Available at http://www.fda.gov/ICECI/EnforcementActions/BioresearchMonitoring/ucm133789.htm
28. Compliance Program Guidance Manual. Bioresearch Monitoring: Good Laboratory Practice (Nonclinical Laboratories). February 21, 2001. Part II, Section C, Types of Inspections.
29. Juran, J. M. and Gryna, F. M., (1988). *Quality Control Handbook*. 4th edn. New York: McGraw-Hill.
30. United States Code, Title 21, Section 351(a)(2)(B) [FDCA § 501 (a) (2) (b)].
31. 43 Federal Register, 45104, September 29, 1978.
32. 61 Federal Register, 20104, May 3, 1996.
33. United States Code, Title 21, Section 351(a)(2)(B) [FDCA § 515 (a) (2) (b)].
34. 61 Federal Register, 52448, October 7, 1996.
35. Pharmaceutical cGMPS for the 21st Century—A Risk-Based Approach Final Report, September 2002.
36. Food and Drug Administration Quality Systems Approach to Pharmaceutical cGMP Regulations.
37. Food and Drug Administration Guidance for Industry "INDS-Approaches to Complying with cGMP during Phase 1," p. 4.
38. Code of Federal Regulations, Title 21, Part 211.22 (a).
39. www.fda.gov/ora/cpgm/default.htm
40. www.fda.gov/ora/inspect_ref/default.htm
41. Office of Regulatory Affairs Field Management Directive No. 86.
42. FR FDA 09/10/91 Notice 56 Fr 46191—Fraud, Untrue Statements of Material Facts, Bribery, and Illegal Gratuities; Final Policy.
43. http://www.fda.gov/ICECI/ComplianceManuals/CompliancePolicyGuidanceManual/ucm073837.htm
44. 421 US 658 (1975).
45. Food & Drug Letter, June 7, 2002.
46. The Declaration of Helsinki is included in the US Federal Regulations 21CFR 312.120(c)(4) as a minimum standard for the FDA's acceptance of clinical trial data gathered from foreign studies not conducted under an IND.
47. http://www.fda.gov/oc/ohrt/irbs/faqs.html
48. 62 FR 25692 (5/7/97) International Conference on Harmonization; Good Clinical Practice: Consolidated Guideline; Availability.
49. Federal Register: November 9, 1998 (Volume 63, Number 216), Protection of Human Subjects: Categories of Research That May Be Reviewed by the Institutional Review Board (IRB) Through an Expedited Review Procedure. Available at: http://www.fda.gov/oc/ohrt/irbs/expeditedreview.html
50. Code of Federal Regulations, Title 21, Section 812.150(b)(1).
51. http://www.fda.gov/oc/ohrt/irbs/faqs.html

52. http://www.fda.gov/OHRMS/DOCKETS/98fr/07d-0173-gdl0001.pdf
53. International Conference on Harmonization E6 4.5.2.
54. Code of Federal Regulations, Title 21, Section 812.150(a)(4).
55. Food and Drug Administration Draft Guidance Document "Protecting the Rights, Safety, and Welfare of Study Subjects—Supervisory Responsibilities of Investigators," May 2007.
56. Code of Federal Regulations, Title 21, Section 312.62(b).
57. http://www.fda.gov/oc/ohrt/irbs/toc4.html#screening
58. International Conference on Harmonization E2A Guideline, "Clinical Safety Data Management: Definitions and Standards for Expedited Reporting," March 1995.
59. Code of Federal Regulations, Title 21, Section 812.15(1).
60. Code of Federal Regulations, Title 21, Section 812.3(s).
61. International Conference on Harmonization E6 1.36.
62. http://www.fda.gov/OHRMS/DOCKETS/98fr/04d-0440-gdl0002.pdf
63. Code of Federal Regulations, Title 21, Section 312.68.
64. International Conference on Harmonization E6 4.9.7.
65. International Conference on Harmonization E6 2.13.
66. International Conference on Harmonization E6 5.1.1.
67. Code of Federal Regulations, Title 21, Section 812.25(e).
68. FDA's Guidance "Financial Disclosure for Clinical Investigators." Available at: http://www.fda.gov/downloads/regulatoryinformation/guidances/ucm341008.pdf
69. International Conference on Harmonization E6 8.2.20.
70. International Conference on Harmonization E6 Section 6.4.9.
71. http://www.fda.gov/ICECI/EnforcementActions/BioresearchMonitoring/default.htm
72. Compliance Program Guidance Manual for FDA Staff, Bioresearch Monitoring: Clinical Investigators, October 1997. Available at http://www.fda.gov/iceci/enforcementactions/bioresearchmonitoring/ucm133562.htm
73. FDA Guidance for Industry, Computerized Systems Used in Clinical Investigations, May 2007. Available at http://www.fda.gov/OHRMS/DOCKETS/98fr/04d-0440-gdl0002.pdf
74. An independent data monitoring committee that may be established by the sponsor to assess at intervals the progress of a clinical trial, the safety data, and the critical efficacy endpoints and to recommend to the sponsor whether to continue, modify, or stop a trial. Guidance on when a DSMB is needed is issued by the FDA, Establishment and Operation of Clinical Trial Data Monitoring Committees. Available at http://www.fda.gov/downloads/RegulatoryInformation/Guidances/ucm127073.pdf
75. PhRMA "Principles on Conduct of Clinical Trials and Communication of Clinical Trial Results," revised June 2004. Available at http://www.phrma.org/principles-and-guidelines-clinical-trials
76. http://www.fda.gov/medicaldevices/deviceregulationandguidance/guidancedocuments/ucm094529.htm
77. http://www.livinglinks.net/biotech.html (definition attributed to Society for Clinical Data Management).
78. Society for Clinical Data Management. Available at http://www.scdm.org
79. Compliance Program Guidance Manual. Bioresearch Monitoring: Clinical Investigators. October 1, 1997. Available at http://www.fda.gov/iceci/enforcementactions/bioresearchmonitoring/ucm133562.htm
80. Compliance Program Guidance Manual. Bioresearch Monitoring: Sponsors, Contract Research Organizations and Monitors. February 21, 2001. Available at http://www.fda.gov/iceci/enforcementactions/bioresearchmonitoring/ucm133777.htm
81. http://www.fda.gov/ora/compliance_ref/bimo/7348_811/Default.htm

82. http://www.fda.gov/oc/ohrt/irbs/operations.html#inspections
83. http://www.fda.gov/ora/compliance_ref/bimo/7348_001/Default.htm#PART%20I%20-%20BACKGROUND
84. http://www.fda.gov/aboutfda/centersoffices/officeofglobalregulatoryoperationsandpolicy/ora/default.htm
85. FDA Information Sheets—Guidance for IRBs and Clinical Investigators—FDA Operations. Available at http://www.fda.gov/oc/ohrt/irbs/operations.html#inspections
86. International Conference on Harmonization E6 1.6.

10 FDA Regulation of the Advertising and Promotion of Prescription Drugs, Biologics, and Medical Devices

Karen L. Drake, Esq.

CONTENTS

Introduction 286
Regulation of the Advertising and Promotion of Prescription Drugs 287
Promoting Prescription Drugs—General Policies 288
    Prior Approval and Preclearance of Promotional Materials 288
        When Is Preclearance Required? 289
        The "Fair Balance" Requirement 289
    The Brief Summary Requirement for Prescription Drug Advertisements
    Directed to Physicians 290
    Advertisements Exempt from the Brief Summary Requirement 290
    Product Name and Placement 291
    Submission of Promotional Materials to OPDP 292
    Product Claims for Prescription Drugs 292
    Unapproved Products and Unapproved Uses for Approved Products 292
    Accelerated Approvals—Relationship to Advertising and Promotion 293
    Off-Label Promotion 293
        Public Request/Response 295
        Private Request/Response 295
    Comparative and Superiority Claims 295
    Pharmacoeconomic Claims 296
    Quality-of-Life Claims 296
    "New" and "Now-Available" Claims 297
    Promotion to Health-Care Professionals 297

Drug Detailing 297
　　Medical Conferences and Exhibits 298
　　Exchange of Scientific Information 298
　　Single-Sponsor Publications 299
　　Use of Spokespersons 299
　　DTC Advertising 299
　　Broadcast and Print Media 299
　　Press Releases, Video News Releases, and Materials for the
　　Financial Community 300
　　Industry Organizations 301
Regulation of the Advertising and Promotion of Biologic Products 301
　　General Policies 301
Regulation of the Advertising and Promotion of Medical Devices 302
Differences between Policies for Drugs and Devices 302
General Promotion of Devices 303
　　PE Promotion 303
　　Investigational Device Advertising 303
　　DTC Promotion of Medical Devices 303
　　Similarities between CDER Policies and CDRH Policies
　　on Advertising and Promotion 303
　　FDA Enforcement—Violation of the FD&C Act or FDA Regulations
　　for Advertising and Promotion of Prescription Drugs, Biologics, and
　　Medical Devices 304
　　　　Primary Enforcement Tools 304
　　　　Untitled Letters 304
　　　　Warning Letters 304
　　　　Remedies for Warning Letters 305
Conclusion 306
References 306

INTRODUCTION

The Food and Drug Administration (FDA) has been regulating food and drugs since 1906, when Congress enacted the Federal Food and Drugs Act. In 1938, the Federal Food, Drug, and Cosmetic (FD&C) Act was enacted and gave the FDA authority over the safety of drugs and food additives and established the enforcement processes now followed by the FDA.

In 1962, Congress enacted the 1962 Drug Amendments or the Kefauver–Harris Amendments named after the two congressional representatives who sponsored the bill. The law required that all drugs be shown to be safe and effective, and it broadened the FDA's authority over other aspects of drug manufacturing and marketing. Section 502(n) of the FD&C Act gives the FDA specific authority over the advertising of prescription drugs, and the provision is also applied to biologic products, medical devices, and prescription animal drugs [1].

Since 1962, the FDA has issued a number of regulations under Section 502(n) of the FD&C Act, which can be found in Title 21 of the Code of Federal Regulations (CFR). These regulations specifically address prescription drug advertisements, what needs to be included in such advertisements, the definition of types of advertising (e.g., reminder advertisements and price advertisements), and the requirements of presenting product information relative to its safety and efficacy [2].

The FDA regulations for advertising and promotion have had very few revisions since their enactment. The regulations were written at a time when medical journal advertising and industry sales representatives calling on physicians were the primary ways in which drug products were promoted to physicians. In today's world, the FDA policies for the regulation of the advertising and promotion have been created on a case-by-case basis or through the FDA guidance documents, rather than a formal rule-making process.

In essence, the FDA's view is that any product-related material issued by a pharmaceutical or biotechnology company is subject to the FDA regulation, and the industry has generally accepted this view. Companies are reluctant to challenge the broad authority of the FDA's jurisdiction. To ensure a successful risk assessment of their marketing materials and activities relative to the regulations, companies remain current on the FDA's regulatory thinking by monitoring the Agency's activity in this area, reviewing regulatory correspondence (e.g., warning letters[*] issued to companies when in violation of the regulations), and guidance documents.

The following sections will describe the regulation of prescription drugs, biologics, and medical devices.

REGULATION OF THE ADVERTISING AND PROMOTION OF PRESCRIPTION DRUGS

Within the Center for Drug Evaluation and Research (CDER) is the Office of Prescription Drug Promotion (OPDP) formerly known as the Division of Drug Marketing, Advertising, and Communications. OPDP has the responsibility for regulating prescription drug advertising and promotion. The regulations that set

---

[*] Regulatory communications from the FDA to a company are usually in the form of a notice of violation (NOV) letter, sometimes referred to as an "untitled" letter, or a warning letter. NOV letters are usually sent first and may involve minor health or safety issues. Warning letters involve serious health issues or occur when similar NOVs have been submitted to the same company.

forth the rules applicable to prescription drug advertising and promotion also include promotional labeling.*

The regulations also address specific requirements for the content of labeling; for example, the label must have the name and place of business of the manufacturer, packer, or distributor [3]. More general rules for labeling include the prohibition against making labeling claims that are false or misleading regarding another company's product [4]. While the labeling regulations do not contain provisions relative to advertising, the general position of the FDA is that the advertising regulations apply to promotional labeling.

PROMOTING PRESCRIPTION DRUGS—GENERAL POLICIES

Prior Approval and Preclearance of Promotional Materials

Except in certain circumstances, the FDA cannot require preclearance of the advertising and promotional materials [5]. This includes launch materials† and direct-to-consumer (DTC) advertising.‡ If such materials are submitted for an OPDP review, the submission is done on a voluntary basis. However, most companies do submit their launch and DTC materials (at first use) for prior approval or preclearance for two reasons: (1) there is considerable time, money, and effort spent on a company's part to create these materials; therefore, if the materials are submitted to OPDP at first use, rather than obtaining OPDP's prior approval, the company runs the risk of OPDP requiring them to pull the materials if they are found to be in violation of the regulations. (2) By obtaining OPDP's prior approval, the company will have a very good sense of the acceptance of their promotional claims and how OPDP views them relative to the regulations, before creating more materials within a certain marketing campaign.

OPDP will review advertising and promotional materials at a company's request and will try to accommodate requests for rapid approval of time-sensitive materials [6]. Comments from OPDP to the company are always in writing, and OPDP may change its decision about an advertisement after approving it. These situations are rare; however, if they do happen, the Agency notifies the company of the change in opinion with a change-of-opinion letter and provides a reasonable time for correction before taking any regulatory action.

---

* Brochures, booklets, mailing pieces, detailing pieces, file cards, bulletins, calendars, price lists, catalogs, house organs, letters, motion picture films, film strips, lantern slides, sound recordings, exhibits, literature, and reprints and similar pieces of printed, audio, or visual matter descriptive of a drug and references published for use by medical practitioners, pharmacists, or nurses, containing drug information supplied by the manufacturer, packer, or distributor and which are disseminated by or on behalf of its manufacturer, packer, or distributor [Code of Federal Regulations Title 21 Section 202.1(1)(1)].

† Launch materials are generally defined as initial marketing materials created at the time when a drug is new to the market or has been approved for a new indication.

‡ DTC materials are advertising and promotional materials intended to be seen or used by a consumer and mention directly or indirectly a specific product.

When Is Preclearance Required?

If a company has committed serious or repeated violations of the advertising and promotion regulations, the FDA can require preclearance of the company's advertising and promotional materials. The preclearance requirement remains for six months to two years.

The regulations also require preclearance of all promotional materials for drugs approved under the FDA's accelerated approval process.[*]

The "Fair Balance" Requirement

Fair balance in advertising and promotional materials is regulated at 21 CFR Section 202.1(e) [6].[†] It is one of the most important requirements for advertising and promotional materials and is one of the most frequent requirements violated by companies. As such, it is very often the subject of regulatory correspondence citing violations of the regulations [7].

21 CFR Section 202.1(c)(5)(ii) states "fair balance must be achieved between information relating to side effects and contraindications and information relating to the effectiveness of the drug." The efficacy and safety claims must be presented in "balance" with the risks of the drug. Risk information must have a prominence and readability reasonably comparable to the presentation of effectiveness claims. For example, efficacy claims on a piece cannot be in 14-point, bold, black font, and the risk information appears at the bottom of the piece in 8-point light-color font. Fair balance applies to the content as well as the format of the material. OPDP looks at typography, layout, contrast, headlines, paragraphing, white space, and any other techniques apt to achieve emphasis [8].

The fair balance requirement does not appear in the FD&C Act or in the regulations governing labeling; it only appears in the prescription drug advertising regulations. Certain ways in which an advertisement may not meet the fair balance requirement include the following:

1. Failure to provide balanced emphasis of side effects and contraindications
2. Failure to be clear where the risk information appears when multiple pages are involved
3. Failure to refer readers to the risk information in a multiple-page advertisement, if located on a different page [9]

The decision regarding whether or not advertising materials meet the fair balance regulations is considered subjective. Many companies and industry organizations long requested a better definition of fair balance, and in May 2009 OPDP issued a draft guidance on the subject (Guidance for Industry: Presenting Risk Information in Prescription Drugs and Medical Devices Promotion).

---

[*] Certain drugs for life-threatening conditions can qualify for accelerated review, and the process requires all promotional materials to be precleared prior to dissemination.
[†] Every promotional piece must meet the fair balance requirement.

In the guidance, the FDA describes the factors they take into consideration when determining if promotional material properly describes the benefits and the risks of a prescription drug or medical devices. When considering such factors, the FDA relies on the actual presentation of the risk information, the likelihood the reader will understand the risk(s), and the overall safety data of a drug or device to determine if the quantity of the risks correlate to the known adverse events of such drug or device.

The Brief Summary Requirement for Prescription Drug Advertisements Directed to Physicians

The brief summary requirement relates to advertisements (e.g., an advertisement in a medical journal) and in essence comprises certain major sections from the product's package insert. All advertisements must be accompanied by a "true statement of information relating to side effects, contraindications, and effectiveness" [10]. Side effects, warnings, precautions, and contraindications are the four categories of risk information required for the brief summary and are taken directly from the product labeling.

The brief summary of the product labeling is usually printed on the adjacent page from the advertisement. Including the brief summary in an advertisement does not relieve the company from providing fair balance on the page or pages that contain product benefit information. The instruction to "see the full prescribing information" does not mitigate the requirement that the benefits of a product must be fairly balanced by providing the appropriate risk information in a reasonable prominence [11].

In 1994, the FDA issued an industry-wide letter in which they stated that "wraparound" advertisements (presenting advertising on the front cover and the brief summary on the back cover) do not comply with the brief summary requirement. The brief summary must appear adjacent to the advertisement [12].

Advertisements Exempt from the Brief Summary Requirement

21 CFR Section 202.1(e)(2) specifies that certain advertisements are exempt from the brief summary requirement. The four categories of such advertisements are reminder advertisements, help-seeking advertisements, advertisements for bulk-sale drugs, and advertisements for prescription-compounding drugs.[*]

Reminder advertisements. These advertisements do not make product claims; therefore, they are exempt from providing risk information in the form of fair balance or a brief summary. Reminder advertisements are typically materials such as pens, notepads, or giveaways for physicians such as medical textbooks. The materials can only contain the proprietary or established name

---

[*] Advertisements for bulk-sale drugs and prescription-compounding drugs are beyond the scope of this discussion.

of the drug, the established name of each active ingredient in the product, and other types of information that do not represent the benefits of the product [13]. Reminder advertisements cannot be used for products that have "boxed" warnings.*

Help-seeking advertisements. These advertisements are used by a company to inform consumers of a disease state, or symptoms of a particular condition, and to encourage them to seek the advice of a health-care practitioner if the consumer has the particular symptoms. Help-seeking advertisements do not refer to the drug product used to treat the condition or symptoms. There is a draft guidance titled "Guidance for Industry: Help-Seeking and Other Disease Awareness Communications by or on Behalf of Drug and Device Firms" that the FDA issued in February 2004. In the guidance, the FDA explained the types of communications that constitute help-seeking advertisements. The guidance explicitly warned against attempts to use the help-seeking advertisement in combination with the reminder advertisements or any advertisement that contained a product claim that would cause a connection by the consumer between the disease or symptom conditions discussed in the help-seeking advertisement and the product used to treat the condition or symptoms.

PRODUCT NAME AND PLACEMENT

The advertising and promotion prescription drug regulations are very specific about the requirement for product name and placement. 21 CFR Section 201 and Section 202 set out the major requirements of the regulations as follows:

Brand (i.e., proprietary) name drug advertisements must reference the established (i.e., generic name) name of the drug.

1. The established name should be placed directly to the right or directly underneath the proprietary name.
2. There should be no intervening matter that would in any way detract from the established name.
3. The established name must be cited every time the brand name is featured (headlines, logo, taglines, etc.).
4. If the established name is only used in the running text, the proprietary name must only appear at first mention in the running text.
5. The established name must be a font size that is at least half the type size of the brand name and must have a comparable prominence to the brand name.
6. The type size, prominence, and juxtaposition requirements also apply to broadcast advertisements, audiovisual promotions, and electronic media such as the Internet.

---

* "Boxed box" warnings are imposed by the FDA to highlight a major risk(s) of a drug.

Submission of Promotional Materials to OPDP

The FDA regulations state "specimens of mailing pieces and any other labeling or advertising devised for promotion of the drug product" must be submitted to OPDP "at the time of initial dissemination"* (for promotional labeling) or "initial publication"† (for advertising) [14]. A specific form titled "Transmittal of Advertisements and Promotional Labeling for Drugs for Human Use" (Form FDA 2253) must be completed in its entirety and submitted with all advertising and promotional materials at their first use. Failure to submit materials at first use may result in regulatory action against the company.

OPDP has limited resources for reviewing materials; they receive thousands of submissions annually and do not review all materials that are submitted. They reserve the right to request the medical reviewers of each division to assist in the review of materials to address scientific and medical content.

Product Claims for Prescription Drugs

Within the regulations for advertising and promotion of prescription drugs, guidance documents, letters to industry, and public pronouncements, the FDA has expressed its view on how companies can promote their products. Tension continues to exist between the FDA and companies as to the "how" in terms of a company promoting its product and being able to do so in a way that differentiates its product from the competition.

Unapproved Products and Unapproved Uses for Approved Products

When a drug is under investigation (i.e., not yet approved for an indication), or is an approved drug being reviewed for a new use, claims of safety about the drug are expressly prohibited [15]. The primary concern of the FDA regarding preapproval promotion is that a health-care provider may form an opinion about the drug's use on the basis of claims by the company before the drug's approval, and that opinion may be incorrect, relative to the future approved use. The incorrect opinion on the part of the health-care provider could lead to incorrect use of the drug.

The FDA does have two exceptions regarding preapproval promotion:

1. "Institutional ads" in which a company states that it is conducting research in a certain therapeutic area to develop a new drug, and the proprietary or established name of the drug cannot be mentioned in the advertisement.
2. "Coming soon" advertisements, which state the name of the product, but make no representation about the new product relative to its safety, efficacy, or intended use. A drug with a potential "boxed" warning cannot be the subject of a coming soon advertisement.

---

* "Initial dissemination" is generally defined as when the material is sent to or shown to a health-care provider or health-care audience.
† "Initial publication" is generally defined as the date on which the advertisement first appears in print in one or more publications.

Once a company has chosen either an institutional advertisement or a coming soon advertisement, they cannot change to the other type [16].

ACCELERATED APPROVALS—RELATIONSHIP TO ADVERTISING AND PROMOTION

As of 1992, certain drugs that treat life-threatening conditions (e.g., AIDS or cancer) can qualify for accelerated review. Essentially, these types of drugs are given priority for review and can be approved in a much shorter time frame than the typical NDA review.

While accelerated review can make a positive difference in terms of getting a product to market faster, thereby potentially helping patients with life-threatening illnesses sooner, there are restrictions on a company in how it handles its advertising and promotional materials.

The accelerated approval regulations require that a company submit copies of its promotional materials intended for dissemination during the first 120 days after approval as part of OPDP's preapproval review process [17]. This is different from the product launch review process in which a few core promotional materials intended for use at the beginning of the launch are submitted to OPDP for review. In addition, after the first 120 days of the launch period, the regulations require that the company continue to submit materials for preapproval, and they must do so 30 days before the material is intended to be disseminated.

The process for preapproval of promotional materials continues until the FDA informs the company otherwise, and it is very common for companies to never receive this notice from the FDA. If the Agency determines that preapproval is no longer necessary for the safe and effective use of a product, they will lift the preapproval requirement [18].

OFF-LABEL PROMOTION

Much like the concept of fair balance, off-label promotion is one of the most frequent topics of regulatory action taken by the FDA against a company. Off-label promotion occurs when a company promotes its product for an indication for which the product is not approved.

The most widely known case involving the concept of off-label promotion is the Washington Legal Foundation (WLF) case that began in October 1993, in which the WLF—a DC-based public interest group—filed a citizen's petition with the FDA, challenging the Agency's policy on off-label promotion as being in direct conflict with the First Amendment and the right to free speech.

The case was in the court system for more than seven years. During that time, Congress enacted the FDA Modernization Act (FDAMA) of 1997, which included a provision (Section 401) for the dissemination of off-label information. On the basis of FDAMA and the Agency's implementing regulations at

21 CFR Part 99,* which allowed for a "safe harbor" for companies to disseminate off-label information, and after several communications between the FDA and WLF, the FDA's position regarding the dissemination of off-label information is more clearly understood. Most companies would say they do not agree with the FDA's position and feel it is their obligation and right to disseminate any and all information about their product to have informed and educated health-care practitioners and consumers.

In addition to the FDA violations, off-label promotion also crosses over into the jurisdiction of the Department of Justice and the Office of Inspector General. This occurs if a doctor prescribes a product for a treatment that the product is not indicated for, on the basis of off-label promotional claims made by a company. If the product is reimbursed, by the government or private insurer, they have essentially paid for a product that otherwise would not have been prescribed but for the company's off-label promotion. The first most notable example of this was the Parke-Davis (Pfizer, Inc.) case in which the drug Neurontin® was promoted for the off-label use as first-line monotherapy treatment for epilepsy; the actual indication is for adjunctive therapy in the treatment of epilepsy. Parke-Davis was charged with engaging in a fraudulent scheme to promote the sale of prescription drugs for off-label uses, thereby causing the submission of false claims to the government for Medicaid benefits and reimbursement. The result was that Pfizer (who then owned Parke-Davis) was liable for over US$430 million in fines and was subjected to very negative publicity. Since the Neurontin case, there have been other companies charged with similar offenses and the resulting fines have been larger than what Pfizer had to pay.†

Since the Neurontin case, there have been several cases regarding off-label promotion, and the affiliated companies have been placed under Corporate Integrity Agreements (CIAs) and paid hefty fines. To date, the largest fine paid was by Pfizer, Inc., for US$2.3 billion for several offenses, including off-label promotion.

In addition to the large fines, companies who engage in and are charged with these types of violations are often held to a CIA, which is an agreement between the violating company and the government that outlines the restrictions on the

---

* Under 21 Code of Federal Regulations Title 21 Subpart B Section 99.101, information about the safety, efficacy, or benefit of a drug for a use not described in the approved labeling may be disseminated and shall (1) be about a drug already approved; (2) be in the form of an unabridged reprint, peer reviewed by experts and scientifically sound; (3) not pose a significant health threat; (4) not be false or misleading; (5) not be derived from clinical research conducted by another manufacturer; and (6) not be letters to the editor, abstracts, phase 1 publications, and publications containing little substantive discussion (publications regarding observations of fewer than four people are not scientifically sound and not allowed to be disseminated).
Under 21 Code of Federal Regulations Title 21 Subpart C 99.201 before disseminating information under Section 99.101, a company must submit the materials to the FDA 60 days prior, any clinical trial information the company has, a bibliography of the articles that are being disseminated, and commit to a time frame in which the company plans to conduct studies to obtain approval for the unapproved indication.

† Serono, Inc., paid over US$700 million and Purdue Pharma paid US$500 million.

company's promotional activities, additional compliance measures that must be instituted, and the consequences of not following the CIA.

In December 2011, the FDA issues a draft guidance titled "Responding to Unsolicited Requests for Off-Label Information About Prescription Drugs and Medical Devices." The guidance is the FDA's attempt to provide clarity to industry regarding when it is appropriate to respond to unsolicited requests for off-label information.

In essence, the guidance divides such responses into two categories: public and private responses. In addition, the guidance is clear that only scientific/medical personnel should be answering requests for off-label information. By allowing a sales representative to answer such questions, it is deemed as off-label promotion, even if the sales representative is providing the same information that the scientific/medical person would be providing.

Public Request/Response

If the request is made in a public forum, and pertains to a company's product, the response should not include off-label information. The response should note the off-label nature of the request, and the responder should direct the requestor to the company's medical information group to answer the question.

Private Request/Response

If the request is made in a private setting (e.g., 1:1 conversation), the responder may provide the answer. The response must be truthful and not misleading, accurate, and balanced.

In March 2012, the Pharmaceutical Research and Manufacturers Association (PhRMA), an industry organization and liaison between industry and the government, requested comments from industry regarding the guidance and its implications on current trends regarding ways companies conduct business when responding to requests for off-label information. The categorization of responses into the public domain will require many companies to significantly rethink their current business practices. The comments were submitted to the FDA in April 2012, and to date, the FDA has not responded to the comments.

COMPARATIVE AND SUPERIORITY CLAIMS

Comparative claims occur when a company implies, suggests, or represents that its product, when compared with a competitor product, is comparable or superior to the competitor's product. The FDA reviews such claims with the same standards as they review efficacy and safety claims in a product's approved label.

When such comparative claims are made, either as comparable or superior, there must be substantial evidence to support the claim. Substantial evidence is generally based on two adequate and well-controlled clinical studies that compare one drug with another in head-to-head clinical trials, and the comparison must be clinically and statistically significant [19]. Unsubstantiated superiority claims made by a company about its product when compared with a competitor product are often the subject of regulatory communications from OPDP to a company [20].

Pharmacoeconomic Claims

Pharmacoeconomic (PE) claims relate to the cost-effectiveness of a company's product, and often do so in terms of a competitor's product. Most companies do not conduct formal well-controlled PE trials in which their product is compared with that of a competitor's.

Often, PE claims are the result of a scientific approach of cost-modeling. The standard of two well-controlled clinical trials to obtain PE data is not necessary, unless a company wants to tie clinical efficacy of a product to the cost-effectiveness position. In that case, the company would have to conduct trials to be able to make claims that suggest that because of a product's efficacy, it is more cost-effective than a competitor's product.

Under FDAMA 114, the FDA established an evidentiary standard for health economic claims for pharmaceutical products. Essentially, Section 114 allows for a company to make health economic claims about its products to a formulary committee, or similar entity, if the information is based on competent and reliable evidence.

Because of the emphasis placed on economics for health care, especially since the health care reform legislation has been enacted, many pharmaceutical and device companies are now focusing on length of stay and readmission rates for their products. This economic area is a primary focus and metric for many medical institutions, private insurance payers, and the government. As such, companies want their customers to believe that the company's drug is cost-effective, and will align with the payer world regarding reimbursement.

The challenge that companies face in this area is the requirement to distinguish cost-effectiveness from the clinical efficacy of a drug. If a company wants to claim that their drug is cost-effective because it is clinically efficacious, the company must conduct clinical trials to connect efficacy with cost-savings. Because of the time, cost, and effort to conduct such a trial, many companies opt to have their economic messages disseminated in such a way that they do not run afoul of FDAMA 114. The government's view is that companies cannot make the bright-line distinction between efficacy and cost-effectiveness, therefore this is a new focus area for the government to ensure that companies are not promoting their products inappropriately.

Quality-of-Life Claims

Quality-of-life (QOL) claims are those that position the company's drug in a favorable light relative to a patient's daily life activities. It is difficult to obtain such data in a clinical trial because having QOL as a primary endpoint of a study is viewed as too broad.

OPDP has informally advised companies that measurement instruments, when adequately validated, may be appropriate to adequately support QOL claims. In a draft guidance published in February 2006 [21], the FDA takes the position that patient-reported outcome (PRO) instruments may be effective endpoints in clinical trials. A PRO is a measurement of any aspect of a patient's health status that comes directly from the patient (i.e., without the interpretation of the patient's responses to a physician or anyone else).

"New" and "Now-Available" Claims

While there is no official regulation or guidance for the use of these terms, the generally accepted time frame for a company to use these terms is six months from product launch [22].

Promotion to Health-Care Professionals

The FDA regulates virtually all contact between a manufacturer of prescription drugs and a health-care provider. Field force visits by sales representatives to hospitals and doctors' offices, promotional speaking events, and medical conferences all come under the FDA's watchful eye, as does any other venue that is or could be perceived as a place or event in which promotional activity occurs.

Drug Detailing

Potentially, any materials used by or oral statements made by a sales representative is subject to the advertising and promotion regulations. Generally, conversations between a sales representative and health-care provider are difficult to monitor, making it difficult for the FDA to learn of violations. However, there are a few ways in which the FDA receives information about violative advertising and promotional activities:

- Health-care providers can contact the FDA to report improper activity by a sales representative.
- Competitors of drug products will report inappropriate activity to the FDA.
- Many FDA employees are practicing physicians and will report inappropriate behavior.

FDA's BAD AD PROGRAM TO PROMOTE TRUTH IN ADVERTISING

Implemented in 2010, the FDA's educational outreach program is designed to educate health-care providers about the role they can play in helping the Agency make sure that prescription drug advertising and promotion is truthful and not misleading.

The "Bad Ad" Program is administered by the Agency's OPDP in the CDER. It will help health-care providers recognize misleading prescription drug promotion and provide them with an easy way to report this activity to the Agency.

Most companies have internal processes and procedures to ensure, to the best of their ability, that their sales representatives are adequately trained on the "dos and don'ts" of promotional activity. The FDA's position is not "who" makes a promotional oral statement or hands out a promotional piece; rather, it is the "what" that the FDA monitors, and they do not distinguish between sales representatives or other company representatives when monitoring and regulating promotional activities. This includes medical science liasions (MSLs)—a group typically found within a medical affairs department of a company—who are typically responsible for working in the field with sales representatives to discuss the science behind a product.

Medical Conferences and Exhibits

Promotional exhibits sponsored by prescription drug companies and accompanying promotional materials found at medical meetings or conferences are subject to the advertising and promotion regulations. Several warning letters have been issued to companies exhibiting at medical conferences.*

Exchange of Scientific Information

It is typical for prescription drug manufacturers to sponsor continuing medical education (CME) events and events in which scientific information is exchanged (e.g., a poster presentation of a disease state at a medical conference). The FDA supports such exchange of information and does not regulate it; however, if the Agency perceives that a manufacturer is unduly influencing,† such activities or exchange of information, especially by using the activity as a way to disseminate off-label information that otherwise could not be lawfully disseminated, the Agency will step in and review the activity under the advertising and promotion regulations.

Most companies today have internal programs that follow the Accreditation Council for Continuing Medical Education (ACCME) guidelines regarding CME events and exchange of scientific information. The ACCME has strict accreditation requirements that CME providers must adhere to in order to hold an event and provide the educational materials for an event. Likewise, manufacturers who support CME events (primarily financial support) must keep an arm's length approach to their collaboration with CME providers and cannot in any way influence the selection of presenters for an event or the subject matter for an event [23].

---

* In a January 31, 2005, letter to GlaxoSmithKline, the FDA cited the company for not prominently displaying the risk information for the products on display. In May 2001, OPDP issued 12 untitled letters for promotional activities at the American Society for Clinical Oncology, mostly for promotion of unapproved uses and investigational drugs. Available at www.fda.gov/warningletters

† "Unduly influencing" refers to a manufacturer providing the content of a CME event, or strongly suggesting a specific speaker for a CME event who will talk about the manufacturer's product in an off-label way and not provide fair balance by discussing other treatment options.

Single-Sponsor Publications

When pharmaceutical companies sponsor publications that bear any contextual relationship to a company's drug product, the publication is subject to the regulations as promotional labeling. These types of publication require disclosure by the sponsoring company of any support provided by the company (e.g., "This publication was funded by Company X."), an accompanying package insert, and if the publication deals with multiple products, each product must be presented in a fair and objective manner [24].

Use of Spokespersons

A company may use spokespersons, including celebrities, to promote their product(s).

If the spokesperson is a celebrity, the company must disclose the affiliation between the celebrity and the company, and the discussion or presentation, including a TV or radio advertisement, cannot go beyond the product's label.

Many companies use physicians to speak on the company's behalf about their product(s). These physicians are typically under contract with a company and present promotional programs to physician audiences. The presenting physicians are subject to the same regulations as any other member of a company and must present information in an appropriate regulatory manner. Presenting physicians can be held accountable by the FDA if their actions violate the advertising and promotion regulations.[*]

DTC Advertising

DTC advertising is subject to the advertising and promotion regulations and to the laws and regulations of the Federal Trade Commission (FTC).

In addition to the defined materials under the advertising and promotion regulations, DTC advertising and promotion consists of radio, TV, and the Internet materials. DTC materials that are not sponsored financially or influenced in any way by a drug company are not subject to the FDA regulation (e.g., a pharmacy price advertisement).

Broadcast and Print Media

A primary difference between broadcast materials and printed material is the presentation of the risk information. For print advertisements, the brief summary of risks must accompany the advertisement, and it may be written in consumer-friendly language [25]. For broadcast advertisements, the brief summary does not have to accompany the advertisement; however, there must be clear information about where the consumer can access the prescribing information.[†]

---

[*] See www.fda.gov/warningletters/gleason
[†] Providing access to the full prescribing information has four components: a toll-free number, reference to a print advertisement containing the brief summary, recommending to ask a health-care professional for the information, and an Internet address directing the consumer to the package insert.

DTC advertising must contain the critical risk factors of the drug, the indication(s) and contraindications in consumer-friendly language, not be false or misleading, and provide information as to where to obtain the package insert.

The FDA does not preclude the use of DTC advertising in any therapeutic area; however, the Drug Enforcement Administration has opposed the advertising of controlled substances (e.g., opiates and narcotics) via DTC advertising [26].

Press Releases, Video News Releases, and Materials for the Financial Community

Generally, the FDA considers product-specific press releases, video news releases (VNRs), and materials for the financial community to be subject to the advertising and promotion regulations. For press releases and VNRs, it is generally a good idea to submit "major announcement" types of materials to the Agency prior to public release (e.g., approval of a new indication) so that the agency can be prepared for questions or communications they may receive regarding the major announcement.

For routine press releases or VNRs, there is no need to preclear the materials; instead, the material can be submitted at initial dissemination via the process for submission of materials to OPDP.

When a company has material financial information, it must meet the requirements of the Securities and Exchange Commission (SEC) for reporting such information. The FDA has not taken action against a company for reporting information (including product-related information) exclusively to the financial community. This is because it is not the intent of the company in this instance to promote the product; rather, it is to make investors aware of the most current financial status of the company and its product(s).

However, red flags can be raised. For example, if a company issues a press release announcing dramatic study results and the financial content is minimal, the FDA may perceive that the company, under the guise of disseminating important financial results per the SEC regulations, is really intending to get clinical news out about its product.

A precedent-setting case occurred in April 1986 when Upjohn Co., issued a press release regarding positive study results for its drug Minoxidil®, a hair growth product. At the time, Minoxidil was not approved by the FDA. Upjohn was issued a warning letter stating that they had violated the misbranding provisions of the FD&C Act [27] by promoting a drug for an unapproved use. Upjohn contended that the press release was "financial and was released to meet the SEC requirements." However, the FDA stated that the release contained highly detailed reporting of study results, and this reporting went beyond the SEC requirements and was deemed promotional labeling under the regulations [28].

In 2004, the FDA and SEC agreed to collaborate to assist one another in protecting the public and investors. The FDA provides technical and scientific support to the SEC and has established a centralized procedure for the FDA staff to use in referring to the SEC statements by pharmaceutical companies to the investment community that may be false or misleading.

Industry Organizations

While drug, biologic, and medical device companies are regulated by the FDA, certain industry organizations have established voluntary guidelines that deal with promotion to health-care professionals, CME, gift-giving to health-care providers, and promotion to consumers.

The PhRMA has adopted the "Code on Interactions with Health Care Professionals," the "Code of Pharmaceutical Marketing Practices," and the "Guiding Principles on Direct-to-Consumer Advertising."

For medical devices, AdvaMed has approved a "Code of Ethics on Interactions with Health Care Professionals," which took effect on January 1, 2004. It provides guidance in seven areas in which device sales representatives interact with health-care professionals. In addition, in 1994, the Hearing Industries Association adopted a "Code of Principles for the Advertising and Promotion of Hearing Health Products."

REGULATION OF THE ADVERTISING AND PROMOTION OF BIOLOGIC PRODUCTS

The Center for Biologics and Research (CBER) implements the regulations of the two laws governing biologic products: the FD&C Act and the Public Health Service (PHS) Act. The procedures for the review and monitoring of biologics are almost identical to CDER. In addition to the regulations in 21 CFR Section 202, biologics are also regulated under 21 CFR Section 600 and Section 601.

GENERAL POLICIES

As a practical matter, the same substantive rules that apply to prescription drug advertising and promotion also apply to biologics advertising and promotional activities and materials. CBER applies the same basic criteria for approval of advertising and promotional materials as CDER. In the CBER Procedural Guidance Document, there are four main criteria listed for the approval of the advertising and promotional materials:

1. Materials cannot be false or misleading.
2. Materials must be consistent with the approved package insert.
3. Materials must contain fair balance.
4. Materials must include proper prescribing information (e.g., brief summary).

In addition, the generic name of the product must be used in advertising, must be in type size at least half the type size of the brand name, and must be used each time the brand name is featured. And, like prescription drug reminder advertisements, biologic product reminder advertisements do not require fair balance.

Like CDER, CBER issues notice of violation (NOV) letters and warning letters for violations of the regulations.

REGULATION OF THE ADVERTISING AND PROMOTION OF MEDICAL DEVICES

The FDA and the FTC regulate the advertising and promotion of medical devices. The FDA, via the Center for Devices and Radiological Health (CDRH), regulates the advertising and promotion of "restricted"[*] devices, and the FTC regulates the advertising of all other devices.

Medical device approval is regulated under 21 CFR Section 801 and medical device investigational device exemptions are regulated under 21 CFR Section 812.

DIFFERENCES BETWEEN POLICIES FOR DRUGS AND DEVICES

CDER's policies have provided a basis for the development of CDRH guidelines; CDRH pays close attention to CDER policies and ensures that the differences between drugs and devices are clear when setting CDRH guidelines and policies.

CDRH does view many of CDER's advertising and promotion policies to be applicable to devices. However, CDRH also recognizes that devices have important characteristics that are different from prescription drugs.[†] This translates into differences between drug advertising and promotion and restricted device advertising and promotion.

1. Restricted device advertising and promotion have no regulations. They do, however, have regulations specific to labeling (i.e., type, size, prominence, etc.) under 21 CFR Section 801.
2. CME for devices is often intended as training for technicians or doctors on the use of the specific device, whereas CME for physicians is typically for a broader subject matter such as a category of drugs for a therapeutic indication.
3. Preapproval promotion rules for devices are different than those for drugs.[‡]
4. There is less public and congressional focus on the advertising and promotion of devices, unlike the advertising and promotion of drugs, which is constantly under the scrutiny of government agencies, consumer-interest groups, and the general public.

---

[*] A device is deemed by the FDA to be "restricted" if it is sold, distributed, or used only with a licensed practitioner's oral authorization or when specific conditions established by the Agency are met. Devices are decreed restricted on a case-by-case basis by either regulation, such as has occurred with hearing aids, or as a condition of premarket approval.

[†] Devices are mechanical instruments requiring a different level of education for health-care professionals to use them, and devices are subject to ongoing modifications, and devices present a different level of investment.

[‡] If a device is pending approval, the manufacturer may advertise it, provided that the manufacturer discloses the current regulatory status (e.g., pending approval) but cannot take orders for the device, claim safety or efficacy, or make comparative claims.

CDRH does not require the routine submission of most advertising and promotion materials and relies heavily on competitor complaints to monitor companies and exert regulatory action. Device companies can request preclearance of their materials; however, comments from CDRH are considered advice rather than official clearance. Labeling that is part of a premarket approval is reviewed during the market application approval process [29].

GENERAL PROMOTION OF DEVICES

PE Promotion

There are no specific regulations regarding PE promotion of medical devices, and the CDRH has informally stated that companies may promote or discuss price as long as the information is truthful and accurate.

Investigational Device Advertising

During the investigational stage of device development, CDRH prohibits any type of commercialization of the device, unless it is advertising that is seeking to recruit clinical investigators or enroll patients in a study [30].

DTC Promotion of Medical Devices

DTC marketing of medical devices has been on the increase in recent years. For example, contact lenses are considered a device, and it is common to see advertisements on television for such devices. In February 2004, CDRH issued a draft guidance titled "Consumer-Directed Broadcast Advertising of Restricted Devices," which spells out guidelines for device manufacturers who choose to market their products to the consumer. For the most part, the guidance mirrors CDER's "Guidance for Industry: Consumer-Directed Broadcast Advertisements" issued in 1999.

Similarities between CDER Policies and CDRH Policies on Advertising and Promotion

In other areas of advertising and promotion, CDER policies and CDRH policies are quite similar, with CDRH basically following CDER policies. This applies to educational and CME events, press releases and public relations materials, materials for the financial community, single-sponsor publications, device detailing by sales representatives, medical conferences and exhibits, and Internet advertising.

FDA Enforcement—Violation of the FD&C Act or FDA Regulations for Advertising and Promotion of Prescription Drugs, Biologics, and Medical Devices

Primary Enforcement Tools

The types of enforcement mechanisms most often used by the FDA are the warning letter and untitled letter (NOV). These letters require specific remedies to correct the alleged violations of the FDA regulations, the FD&C Act, or the PHS Act.

While the number of enforcements has decreased since 1999, there has been an increase in alleged violations for certain types of claims, particularly omission or minimization of risks (i.e., fair balance) [31]. In 2000, OPDP began publishing the violative materials along with the issued letter. The Agency views this available material as instructive for industry in general.

Untitled Letters

NOV letters typically deal with the least violative advertising and promotional activities. Generally, the violation does not jeopardize the public health and can be easily remedied. There is typically a requirement that the dissemination of the violative materials is immediately ceased.

When determining whether to send an NOV letter or warning letter, the FDA considers whether there are public health implications associated with the alleged violation, the regulatory history of the company regarding previous violations, and whether there is evidence that the violation is part of a larger promotional campaign.

Warning Letters

Warning letters are issued for more serious violations of the regulations or FD&C Act. They are issued from the respective center* and require a reply from the company within 15 business days. Failure to respond and comply with the letter may result in further regulatory action, including judicial action such as seizure and injunction.† These letters are sent to the chief executive officer of the company. Doing this holds the top company official accountable for the violation and the remedies.

Companies do have the right to appeal a warning letter. The appeal must be made to the appropriate center. Warning letters become public information after they are issued.‡

---

* Center for Drug Evaluation and Research, Center for Biologics and Research, or Center for Devices and Radiological Health.

† Seizure is a civil enforcement action brought in Federal District Court and if upheld, the company's product is "seized" and not allowed in interstate commerce until the case is resolved. Injunction actions must be brought in Federal court, and if the FDA is successful, they can enjoin a company from disseminating all promotional materials and possibly the related product until the case is resolved.

‡ See www.fda.gov

Remedies for Warning Letters

When seeking remedies, the FDA considers the seriousness of the violation and the regulatory history of the company. Remedies can include the following:

1. Discontinuation of dissemination of the violative materials
2. "Dear health care professional" letters
3. Corrective advertising
4. Appropriate communication to sales representatives to discontinue use of violative materials
5. Submission of a corrective action plan by the company

Discontinuation of Dissemination of the Violative Materials

In virtually every communication from the FDA to a company informing a company of a violation, there is a requirement for the company to no longer disseminate the violative material. Upon receipt of an Untitled letter or Warning letter from the FDA, one of the first steps taken by the company is to immediately alert the company's Commercial and/or Marketing teams to discontinue use of the material. Many companies require their sales force to return the violative material to avoid the chance that the material will still be in circulation.

"Dear Health Care Professional" Letters

These letters are sent to health-care professionals involved in the purchase, use, or prescribing of a company's product(s), alerting them that the company's promotional materials are false and/or misleading. Typically, the letters are required to appear in the venue in which the violative materials appeared. For example, if the violative material is a sales detail aid, the letter would be mailed directly to the associated health-care professionals. If the violative material is a journal advertisement, the letter has to appear in the journals or publications in which the violative advertisement appeared, for the time period in which the advertisement appeared.

Corrective Advertising

These advertisements are reviewed and approved by the FDA and must run in the same publication in which the violative advertisement appeared (much like a "dear health care professional" letter). The advertisement must clearly show that it is a corrective advertisement and is FDA mandated. In addition, the advertisement must be explicit about the violations that the FDA found in the advertisement and adequately explain how the corrective advertisement addresses the violations. It is important that the corrective advertisement look and feel like the violative advertisement so that they are "linked" when they appear side by side in a publication.[*]

---

[*] For example, OPDP issued a warning letter to Cubist Pharmaceuticals in August 2004, for a violative medical advertisement. The original advertisement showed a man's bicep holding a mallet with the heading "STRIKE FAST" in capital letters. The corrective advertisement ran in the same medical journal for a period of nine months and in a prominent box stated "IMPORTANT CORRECTION OF DRUG INFORMATION, CUBICIN" above the "dear health care professional" letter, correcting the violation.

Corrective advertisements can also be run for violative television advertisements made via broadcast DTC advertisements.

Appropriate Communication to Sales Representatives to Discontinue Use of Violative Materials

At the time of the receipt of a Warning or Untitled letter from the FDA, and in addition to the receiving company informing their Commercial teams to discontinue dissemination of the violative materials, many companies will also send communications to their sales teams via phone, e-mail, and text, to no longer use the materials in question, and explain why such materials are the subject of the Warning or Untitled letter.

Submission of a Corrective Action Plan by the Company

In the case of a Warning letter, which is considered more serious than an Untitled letter, the FDA will require the violative company to submit a corrective action plan. Such a plan will outline the company's plan for discontinuing use of the violative materials, informing the sales teams and other relevant individuals, any remedial education required for the relevant company personnel, and steps for ensuring that a repeat violation will not occur.

CONCLUSION

When considering the abundance of advertising and promotion regulations for prescription drugs, biologics, and medical devices, it may seem as if companies can say very little about their products without the concern of a regulatory communication from the FDA. However, it is safe to say that a company has significant ways to legally disseminate its advertising and promotional materials for its product(s) as long as the company adheres to the regulations and remains current on the FDA policies and guidance documents.

The primary concern must always be for public health, and while the regulations for advertising and promotion of a company's product(s) may seem cumbersome and unduly burdensome, it is clear that regulation of marketing practices must exist to guard the public from false and misleading information.

It is also important, however, that companies continue to be allowed to disseminate scientific information about their product(s) to keep the health-care community and patients informed of the current state of drugs, biologics, and medical devices and to be able to freely disseminate this information without fear of overregulation by government agencies.

The balance between public health and safety and the right of a company to disseminate information about its product(s) continues to be a balance that may always be viewed as difficult to achieve.

REFERENCES

1. USC, Title 21, Section 502(n).
2. Code of Federal Regulations, Title 21, Part 202.

3. Code of Federal Regulations, Title 21, Section 201.1(a).
4. Code of Federal Regulations, Title 21, Section 201.6(a).
5. USC, Title 21, Section 352(n).
6. Code of Federal Regulations, Title 21, Section 202.1(j)(4).
7. US Food and Drug Administration, www.fda.gov/foi/warning
8. Code of Federal Regulations, Title 21, Section 202.1(e)(7)(viii).
9. Code of Federal Regulations, Title 21, Section 202.1(e)(7).
10. Code of Federal Regulations, Title 21, Section 202.1(e)(1).
11. Code of Federal Regulations, Title 21, Section 202.1(e)(3).
12. FDA Industry-wide Letter: Current Issues and Procedures. FDA, Rockville, MD, April 1994.
13. Code of Federal Regulations, Title 21, Section 202.1(e)(2)(i).
14. Code of Federal Regulations, Title 21, Section 314.81(b)(3)(i).
15. Code of Federal Regulations, Title 21, Section 312.7(a).
16. FDA Industry-Wide Letter: Pre-Approval of Rx Drug Ads. FDA, Rockville, MD, August 19, 1986.
17. Code of Federal Regulations, Title 21, Section 314.550.
18. FDA Guidance for Industry: Accelerated Approval Products—Submission of Promotional Materials. FDA, Rockville, MD, March 1999.
19. Code of Federal Regulations, Title 21, Section 202.1(e)(6)(i).
20. US Food and Drug Administration, www.fda.gov/foi/warning/2012/zovirax
21. FDA Guidance for Industry. Patient-Reported Outcome Measures: Use in Medical Product Development to Support Labeling Claims. FDA, Rockville, MD, February 2006.
22. US Food and Drug Administration, www.fda.gov/cder/OPDP/faqs
23. Accreditation Council for Continuing Medical Education, www.accme.org
24. Code of Federal Regulations, Title 21, Section 202.
25. Code of Federal Regulations, Title 21, Section 202.1(e)(i).
26. FDA Guidance for Industry: Consumer-Directed Broadcast Advertisements, September 1999; FDA Draft Guidance for Industry: Using FDA-Approved Patient Labeling in Consumer-Directed Print Advertisements. FDA, Rockville, MD, March 2001.
27. USC Section 352(a).
28. www.fda.gov/foi/warning/upjohn 1986
29. 58 Fed. Reg. 42340: Formation of CBER Advertising and Promotion Labeling Branch, April 1993.
30. Code of Federal Regulations, Title 21, Section 814.20(b)(10).
31. www.fda.gov/foi/warning

11 The Practice of Regulatory Affairs

David S. Mantus

CONTENTS

Introduction 309
What is "Regulatory Affairs"? 310
Background and Training 311
    Is There a Degree That Matters? 311
    The Importance of Self-Education 312
    Attitude and Approach 313
    Regulatory as Navigator and Architect 313
    Zealotry 314
Information 315
    What Information Matters? 315
    Gathering Information 316
    Communicating Information 317
Documentation 319
    The Memo 319
    Managing Documents 320
    Practical Example: Documenting an FDA Contact 320
Submissions 321
    Who Writes These Documents, Anyway? 322
    Regulatory Review: Continuity and Connections 324
    Presenting Data in Submissions 324
    The Art of Handling Large Documents 325
Conclusions 325
Notes 325

INTRODUCTION

There are many resources that provide guidance to the regulation of drugs, biologics, and devices, and interpretations thereof, but very few speak generally to survival and success in the profession of Regulatory Affairs (RAs). The success of a regulatory professional is less dependent on the regulations than on how they are interpreted, applied, and communicated within companies and to outside constituents. This is also the key to success for regulatory strategy. The several academic centers providing graduate and certificate training in RA also tend to focus on the hardware of the matter: the laws, regulations, science, technology, and ethics of product development/marketing/regulation. What is missing? The real "fun" stuff consists of those unseen connections between all of these spheres and the balancing act of the persons who manage the connections. It is great to know all the laws and regulations by heart; but if you do, you are probably focusing time and energy on the wrong things. What really counts is an ability to interpret and connect, and to adapt this ability based on circumstances. This is what separates *regulation* professionals from *regulatory* professionals. This chapter is an attempt to discuss the practice of RA—the fundamental tools of the trade—without resorting to specific products or classes of products. The chapter is organized in a way that moves from the most general of concepts toward the most practical. To start, a definition of RA is provided; a review of education and attitude follows, communications and documentation are then discussed, and the chapter ends with an overview of submissions.

WHAT IS "REGULATORY AFFAIRS"?

Before we can discuss the practice of RA, we have to define RA. Too often we define RA by our own limited experience—what it does at our company, in our industry, etc. To broadly define it, one must consider every interaction a company can have with a regulatory authority, be that authority national, state/provincial, or local. We then must consider every internal department or individual that might need something from, or need to provide something to, a regulatory authority. Then we consider the entire life cycle of a product, from conception to marketing (and perhaps, eventual removal), and every type of product that is regulated. The RA group is at the nexus of all of these variables—the conduit between the company and the authorities, over all times, for all products. It is an awesome and incredibly fortunate position in which to be. Figure 11.1 is derived from several different slides I have used in lectures to encompass the field of RA. It is an imperfect attempt, but gives some sense of scale both across a company and the life cycle of a product.

It is important to remember the broad possibilities of experiences when dealing with colleagues from other companies, with different perspectives and sometimes narrow views of the field. There are often times when fruitful communications can only be achieved after learning each other's perspectives and explaining one's own position, as shown in the spectrum in Figure 11.1. What do I mean by this? Consider your colleagues who are deeply into clinical trial operations. Their goals for the year, their professional ambitions, and their day-to-day activities are all focused on the timely execution of clinical studies. They may be unaware of those manufacturing processes and issues that are either running in parallel or are necessary for timely delivery of clinical trial supplies. Regulatory is uniquely positioned to see the cross-functional dependencies of these issues. Are there *other* functional departments that have the opportunity

The Practice of Regulatory Affairs

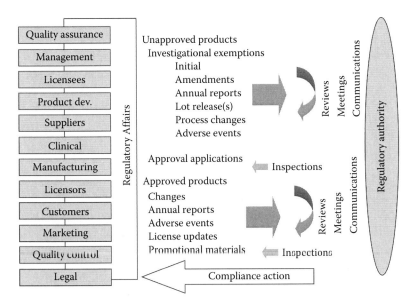

FIGURE 11.1 The spectrum of RA.

to see promotional materials for review, manufacturing batch records, informed consents, and toxicology results all in the same day? What better functional department to assist in managing the interdependencies and to bring to light the gaps in cross-functional plans? In many respects, regulatory is a shadow project management organization—aware and responsible for large, cross-functional networks, but not explicitly identified as project leaders.

BACKGROUND AND TRAINING

Is There a Degree That Matters?

What is the "right" education for an RA professional? When I first entered the field, there was no answer to this question, and this was one of the reasons I entered the field. It does not require any *one* area of technical expertise, but rather the ability to distill multiple technical fields; manage internal and external politics, and write, edit, and collate documents. I have known successful professionals with all manner of degrees (or lack thereof), and I think this diversity is one reason the profession is considered inclusive and has prospered. A trend toward specialization in recent years is worrisome—a chemistry degree is not necessary to manage chemistry, manufacture, and controls (CMC) issues, nor is a medical degree necessary to edit or write an investigator brochure (IB). One notable trend is the growth of graduate and certificate degree programs that seek to provide "basic training" in RA. To their credit, most of the programs provide diverse training across multiple disciplines, in addition to some practical training across industries/product areas.

The open nature of "required training" should encourage more people to enter the RA field. I also hope hiring managers, and managers considering existing employees' career development do not limit options due to degrees or specific training. A person whose initial training is in devices can succeed in drugs. A person without an undergraduate degree in science can develop CMC sections of submissions.[1] What matters most is an ability to question concepts and data with a critical eye and the courage to ask these questions.

THE IMPORTANCE OF SELF-EDUCATION

I believe that the importance of developing one's self cannot be overstated. A plethora of courses and workshops across a vast spectrum of technical, legal, and regulatory matters are available. Take advantage of these opportunities—what cannot be applied in the short term is liable to be useful in the long term. Such courses also provide a great opportunity for networking. Reading about topics that are less than familiar or intimidating is also recommended. While this *sounds* dull as dirt, there are plenty of authors who have been able to write fairly readable nonfiction books about normally very dry topics. Seek such books out even if they are not your typical read—you will get a good story *and* learn some things that are useful for work. Some examples include the books of John Allen Paulos.[2] These are well-written tours through the world of mathematics, with few scary formulas and a lot of "back-of-the-envelope" discussions that are useful. Another great book on statistics and decision-making is *Why Not Flip a Coin?* by H. W. Lewis—an easy read and a sometimes scary insight into how decisions are made, especially very important ones that affect millions of lives! These books on specific topics are just examples of the types of reading that you can do on your own time and that both entertain and inform. There are also books about broader topics that can help provide insights useful for regulatory work. Malcolm Gladwell's books are brilliant reviews of trends and changes in our society and the nature of decisions and expertise.[3] Mr. Gladwell has written extensively in the *New Yorker*, always using a critical and analytical eye, always making it interesting, and always educating. His article "The Art of Failure" is a terrific piece on how things fall to pieces.[4] Another book from my personal reading list is *Complications* by Atul Gawande (interestingly, a friend of Gladwell's), a memoir of a surgeon's training. This last book is mentioned not because it is a technical reference, but because it provides insight into the world of physicians—the folks who study our products, prescribe them, endorse them, and critique them.

All of these books are presented as examples of the type of reading that can provide self-training that can help with regulatory work—help not only with the technical issues that arise but also the way one needs to think: broadly, critically, with an open mind, unafraid to ask questions and be questioned.

Attitude and Approach

"No." This is the most common word associated with RA and is even more common if you consider it the origin of "not" in the contraction "don't." So, "no" easily becomes "don't do that," "don't do this," and "don't even think about that." Add "can't" into the equation and you have summed up RA for 99% of the people who work *with* regulatory. In commercial circles, regulatory is sometimes referred to as the "sales prevention group." While the perception of "no" as regulatory's favorite word is a pervasive, it is fundamentally wrong, and the fact that products get approved and marketed is evidence. The perception is based on plenty of valid experiences—almost everyone can recall a regulatory person holding up a copy of the Code of Federal Regulations (CFR) and emoting that the proposed action is "in clear violation of subparagraph 345 of paragraph B of section a, subheading iii, chapter 193."[5] The author can reluctantly confess to having said such a thing (or similar) on more than one occasion. The problem is how frequently such a position is taken, and whether any other options exist in terms of opinion and contribution to a project. As we will see, what a regulatory professional needs to say is "no, but," as in "no, we can't get to our goal in that way, but here's a viable path that works."

Regulatory as Navigator and Architect

One of the most useful analogies for product development (although it can apply to any team moving toward a goal, even if that goal is abstract, such as compliance) is that of a voyage at sea.[6] Think of management (or the board of directors, or investors) as those financing the voyage—providing the ship and supplies with a specific global objective in mind, for example, getting to point X by date Y. The crew of the ship consists of the various functional groups—the folks who really do the work. I would like to say that regulatory is the captain, but in the drug, biologic, and device industry this is not the case. We work in the regulated health-care industry, so medical issues (safety and efficacy) are paramount. So imagine clinical as the captain of the vessel, chartered with the goal stated earlier. The question remains, how to get from our point of origin (O) to point X by date Y? The navigator is usually given the job of determining the specific route to follow, and this is a very good analogy for the job of RA. In any navigation problem, there are choices—the slow, deliberate, safe route per the chart; the dangerous route over uncharted reefs and rocks; and a middle course, where risks are balanced and timing may be everything (e.g., timing the tides). Even the largest, most resource-rich company should not take the safest, most expensive, and time-consuming path. Every path taken should be a balance of resource, risks, and timing. The "damn the torpedoes" approach is stereotypical of small companies, but it is usually taken in ignorance rather than using a risk-assessment approach. This navigation analogy works well in considering the function of RA in drug development—laying a strategic and tactical path to the goal. Anyone can plot the safest of courses. It takes skill

FIGURE 11.2 Bridge design as an analogy for drug development. The beautiful Zakim Bunker Hill Bridge—it spans the river, satisfies many audiences (e.g., aesthetes, politicians, and commuters), and was completed with a timeline and budget.

and experience to plot a course that gets us there in one piece with speed and well-utilized resources.

Another analogy that fits stems from a quote I once heard from a building engineer on a large bridge project. I paraphrase, but he said, "Anyone can build a bridge strong enough to carry a load. It takes skill to build a bridge *just* strong enough to carry a load." The implication is that the goal is a bridge that is affordable and can be built in time is aesthetically pleasing, and yet carries the load required (Figure 11.2). Approval of a drug or device is like that bridge—anyone can get a good drug approved with infinite time and money. What takes skill is to get it approved in a timely manner with reasonable resources and risk. This is simply a statement of true regulatory strategy—a plan that fits budgets, time lines, and still meets the goals of the business.

Zealotry

Far too often projects, products, even company cultures become "religions" within an organization. A healthy positive attitude is replaced by a blind belief that success, even perfection, will be obtained. Again, while this seems like an extreme observation, many failures in drug development are rooted in the inability to see the obvious, heed prudent advice, and/or recall that we all must obey the rules. Zealotry is a good descriptor for this approach, which is borne from good intentions based on a strong desire for success and/or a belief in a particular technology/science. In regulated product development, it can be a fatal attitude. I am not advocating cynicism and despair, simply a healthy dose of skepticism and a reliance on sound data and equally sound advice. Regulatory must often bear the burden of keeping proper perspective. This sometimes makes us easy targets for accusations of "negativism," but in the end a balanced approach is in the interest of the product, the company, and our careers. How do you maintain the balance? Remember that your product is one of many, your company is one of many, and all of us believe we are on *that* project, working at *that* company, which just *has* to succeed! Look over your shoulders, and you see plenty of failed companies and products with very good teams in charge who believed the same things. Saying we are *not* smarter than others sounds like heresy, but usually we are not smarter. We can be faster at learning, faster at making changes, more responsive; but in general, we have the same brains.

One of the most important times to maintain a balanced (and nonzealous) perspective is in communications with regulatory authorities, such as the Food and Drug Administration (FDA). Another chapter deals with the details of face-to-face meetings, so I will only touch upon the company's *attitude* in these interactions. The FDA has seen plenty of companies claiming to have the best technology, most dedicated clinicians, and the most brilliant management teams. One or two of them have even gotten products approved. But these approvals came based on data, not because the FDA "liked" the company or was in awe of their science. Stick to data, logic, and realistic approaches. It is important to work to understand your FDA counterpart's perspective—what pressures are they facing in terms of other application reviews, congressional oversight, and public opinion? As in any negotiation, understanding the motivations of your partner in negotiation is key to success. In fact, the FDA will usually appreciate a more thoughtful approach, and a more humble attitude will improve the probability of working partnership with the Agency.

INFORMATION

Information is often described as the currency of the twenty-first century, and for RA this has been the case since the earliest days of the profession. Regulatory is the interface between the company/sponsor and the outside world (in terms of regulators/regulatory authorities). As a conduit or a funnel, the regulatory department is a focal point of information, both incoming and outgoing. In order to practice regulatory and succeed, both in objective public measures (e.g., approvals) and internal ones (e.g., recognition and reward), recognizing the power of information and learning to manage information are critical.

WHAT INFORMATION MATTERS?

Other chapters in this book have shown that there are considerable written resources (e.g., books and regulations) to help guide a regulatory professional. However, there are inherent limitations to this information. Published guidance, public presentations, and weekly industry newsletters cannot convey mood, body language, and subtexts. They are also available to anyone with access, and many are publicly available. While there will be certain "yes or no" answers in these materials, the questions they answer do not require a regulatory person to interpret. Most questions and decisions depend on subtle judgments from regulators, and predicting these judgments, perhaps even influencing these judgments requires a mastery of information gathering and management. So the most valuable information is logically the information that is hardest to get—gleaned from informal conversations, e-mails, etc. In addition, information is taken from unlikely or difficult-to-find sources. So how does one gather this?

Gathering Information

There should be no need to go over published sources of information, both commercial and governmental.[7] So what are other sources? Any opportunity to see, hear, or talk with a regulator, a more experienced drug development expert, a colleague, or a sworn enemy is an opportunity to gather information. Never be afraid to ask a question, never be afraid to approach a new person who might have information you need, and always be willing to listen. Table 11.1 provides some basic guidelines for information gathering.

TABLE 11.1
Dos and Don'ts of Information Gathering

Dos	Don'ts
Prepare questions ahead of time.	*Be overly aggressive.*
Research who you might meet at a conference, dinner, etc. Think about what you might learn!	As any good reporter will tell you, people prefer to talk to people who make them comfortable in an exchange that appears two-way.
Make small talk.	*Expect too much.*
There is nothing wrong with breaking the ice, finding out more about a person than what you need to know.	The information they hold is powerful, and they are not going to tell you that you are approved in the hallway of the Minneapolis Convention Center.
Look again where others have.	*Assume a source has been checked.*
Rereading or re-researching sources is OK. You may bring new perspective or a new eye for detail to the matter.	"I assumed someone already checked …" is a very common statement. Never get caught in it.
Look where no one else would.	*Consider the gathering complete.*
Think of novel sources—this may be academic, former colleagues, old textbooks, or non-FDA government agencies. You have to think of all the ways the information might be important to someone.	You should always be on the lookout for new information. Just because the formal process of searching for data ended does not mean you close your eyes.

So, what do I mean by novel sources or approaches? A simple anecdote relates to a project I worked on, involving an older chemical entity for which no prior approval appeared to exist. This assumption was based on input from consultants, and even implied in responses to inquiries from the FDA. The Internet, the FDA, and the freedom of information (FOI) had no data on this entity. The assumption was that it was therefore new to the regulatory arena. Then a trip to the library (yes, they still exist and are still useful!) and a review of a more than 30-year-old *Physician's Desk Reference* (PDR) helped find the drug—branded and on the market prior to the modern era of regulatory approvals. Included were dosage form information, data on pharmacokinetics, etc. This led to a wealth of valuable information to guide the development process and to better inform research on the intellectual property of the compound.

A second anecdote relates to informal conversations with regulators. At a drug development conference recently, a box lunch was provided and served in a large ballroom. Such situations usually lead to people distributing to maximize their distance from new people and populating in clusters of familiar faces. I happened to notice the director of an FDA division that our company would probably begin working with in the next 9–12 months. He was in line for a poorly prepared sandwich. What followed was an informal chat over a meal on general topics related to the state of drug development (not enough truly novel chemical entities), improving communications with industry (more frequent chats and meetings), and how quickly kids grow up nowadays. The company got face time with the FDA, established how follow-up communications with the division work, and the potential for a collaboration started. My new friend/colleague at the Agency learned about my company, one new industry person's view of drug development, and perhaps a collaborator on a future conference session.

If all this gathering sounds simple, that is because it is. But look around and see how few people execute it.

COMMUNICATING INFORMATION

What one does with information related to regulatory is as important as the information itself. Who do you tell? Who don't you tell? How do you tell it? The easiest information to share and communicate is *noncritical information*. These are findings and data from public presentations and widely available sources that simply need to be put into a logical and relevant form and shared within the organization. The main issue with such information is getting to the right audience without boring them into forgetting that they are getting useful data. Most companies subscribe to news updates or have internal regulatory information updates via e-mail. However, these updates often have a hard time grabbing attention and actually being used as a resource. One suggestion is to make them playful and user-friendly, using popular Web pages as guides.

What about data you have found from unique sources? Something "dredged up" from an obscure FOI request based on a hunch from a former colleague you met at a conference? I would never suggest hiding these data, but there is no reason to explain openly how they were obtained. Why not keep your regulatory information-gathering secrets secret? The information and your tricks to get it are part of your armamentarium of regulatory tools.

The most difficult information to communicate is critical information. This could mean anything vital to the success or failure of a project, specific and important feedback from the FDA, subtle insight that weighs heavily on the future of the company, etc. While it would be simple to just shoot off an e-mail to the entire company, it is neither in the company's interest or your interest to take that approach. The first thing to do is document the information carefully, so that you fully understand it and its implications. Then think of those individuals who are that combination of "need to know" and "know who else needs to know." At small start-ups, this might be the CEO or the president. At larger companies, the head of clinical, a project manager, or a similar middle- to senior-level manager fits the bill. Using these first points of contacts allows the information to pass through appropriate channels. It also allows for the dissemination of the information in the proper context.

One of the most difficult challenges is passing along negative information—bad news. There is a visceral desire to quickly get such information off one's hands, so oftentimes this happens carelessly and winds up feeding "rumor mills" and moving outward without appropriate management. Table 11.2

TABLE 11.2
Hints for Passing Negative Information

1. Be painfully accurate. Make sure your information is data-rich. If conversations were involved, quote comments verbatim. Avoid adding your own opinions to the information, supply the facts.
2. Think about and research (if necessary) the implications of the "bad news" in terms of resources (costs) and time. You may or may not know the full implication of the information, but if you know a new study costing US$500,000 and taking one year is the outcome, you might as well share it. It may also be that your first contact—the person you need to tell the information to—is not fully aware of such impacts.
3. Consider an informal, first contact. This should be someone you trust implicitly. Practice your conversation, getting all of the nerves and emotions out. Make sure you are sticking to the first hint!
4. *Never* e-mail this stuff. You may not be fully aware of the ramifications of the information, both legally and in terms of internal politics. E-mail puts the information in written (and therefore documented and available upon legal discovery) form before you have fully researched all the possible meanings and, perhaps, interpretations.
5. Do your best to suggest alternate paths for success. Just saying "FDA says no" does not help the organization. Look for ways goals can still be met. Even if all you can do is to determine what other resources might be available to help extricate the project or company from the situation, suggest it. How much better will it sound to say, "FDA says no, but I would suggest calling so-and-so at company Z, she has been in this situation before."

provides some hints on handling such information. Do not take this as a cynical approach to regulatory; it is realism: just as regulatory is often the recipient of positive approval news, regulatory is the first point of contact when the FDA has to provide negative feedback.

DOCUMENTATION

One of the first things one learns in regulatory and compliance is "if it is not documented, it was not done." Not following this basic principle leads to a large number of compliance failures and can also lead to the downfall of critical development projects. Projects in drug, device, and biologics development can take upward of 10 years to complete and cost tremendous amounts of money.[8] The time involved can be upward of five times longer than the average stay in a regulatory job, depending on location and industry.[9] This means projects need to outlast the people who work on them, and the only way they can do this is to have solid documentation to support them: document progress, document decisions, document information (see earlier), document failures, and document successes. This need to document is important at large companies, where complex dynamics may move a project through the hands of multiple teams, and at small companies, where key decisions may be questioned by advisory boards, investors, potential investors, and potential partners. If you have a well-thought-out defense or opinion on a key issue related to the success or failure of the company or its projects, why not write it down so others can look at it, you can share it, and it outlasts you?

The Memo

When I started in industry at Procter & Gamble, one of the first trainings they provided was in writing a memo. At first, I thought it laughable, but since then I believe in the power of the memo. It need not be long, it need not be in one specific format, but it should contain the following elements:

- Your name and initials and/or signature
- The recipient's name
- The date
- A subject line
- Text and references (if necessary)

What power is contained in such a document! Who said what to whom and when! It can document decisions that may have taken years to come to, summarize volumes of data, and correct mistakes. This last "action" is critical to understand. We produce smoothly written standard operation procedures (SOPs), master batch production records (MBPRs), clinical protocols, and policies. It is a very common misperception that in order to "comply," a company must follow the very letter of all of these standard procedures. The reality is that few, if any, actions take place perfectly in line with written procedures. More often than not some level of deviation occurs. The key to deviating and complying is to document the deviation. Use a memo! Explain what happened and why those individuals who understand the process and the deviation do not think it is a big deal. It sounds so simple—but read a few warning letters at the FDA Website to get a sense of how infrequently it is done.

Managing Documents

Much has been written about document management. I seek only to remind the reader that we have to control the writing, dissemination, filing, and archiving of documents in order for them to be useful. By all means, I strongly suggest doing so in the most efficient means possible. Clearly, if resources were no concern, this would be a fully electronic document control and management system. I will confess that I am a poor manager of documents. Therefore, I delegate and depend on others to maintain files. The concept of filing is not beyond me; I am merely poorly disciplined at starting and maintaining filing systems. Table 11.3 provides some useful hints for document management, whether the system is a fully electronic archive or an asbestos-lined fireproof cabinet.

Practical Example: Documenting an FDA Contact

The level of detail and the approach to documenting a contact with a regulatory authority is an ideal example of "good documentation practices." It represents one of the most important functions of RA and should reflect the professionalism and

---

TABLE 11.3
Document Management Ideas

1. Redundancy is OK. It is acceptable and even useful to maintain files in duplicate. For example, maintain an IND-specific file, where each submission to the FDA is included, along with the FDA feedback and supporting documents. At the same time, a chronological file of all the FDA contacts can be kept, which includes FDA feedback on an IND submission. In a pinch this redundancy can save you.
2. Use any and all means to keep it simple. Use color codes, use multiple cabinets, and label file folders elaborately. The system has to be able to outlast any one person, without an extensive training required for someone else to use the system.
3. Log files. That is to say, keep a table of contents or an index of what is in a file. This helps immensely in tracking redundancy (no. 1 given earlier) and in keeping a system simple (no. 2 given earlier).

expertise of the person making the record. A generic example follows, and I have tried to add advice and ideas for each section.

Record of Contact with Regulatory Agency

Product Identifier: Product Code or Name

Originator: Your name!	Date of contact: Date
	Time: Don't laugh! Multiple calls in one day can get confusing four years later.
IND number: XX,XXX	Initiated by: Company Other
NDA number:	Type of contact: e-mail/phone/face-to-face
Other file number:	
Contact name and title:	Agency: Other
Get this right, and get every detail.	Center:
If specific titles don't come up, look them up! Be sure to know where the person stands in terms of decision making. Know the organizational chart of the division/group!	Division:
	Phone: Get actual phone numbers—not general department numbers!
	Fax:
	E-mail:

Subject: Why did you talk?
Summary:
Describe in as impersonal a way as possible what transpired. This is not a novel or an attempt at fascinating dialogue. Stick to data. Recording verbatim comments can be incredibly powerful. The specific words people use says a lot about attitude, and this can then be relayed without editorial or subjective filtering by the reporter.
Action(s):
A clear list of actions deriving from this contact needs to be included.
Distribution:
Regulatory file
Be sure to include all appropriate people. Some folks are extra-sensitive about being left off the list!

IND, investigational new drug; NDA, new drug application.

What are the key concepts? These are specificity and objectivity. For this type of document, your opinion should not be reflected. Accuracy and getting specific information is most important. This might take work either before or after the call, but it allows a reader (and a reader who looks at this either 2000 miles away or 20 years later) the ability to put the contact in perspective.

SUBMISSIONS

Submissions to regulatory authorities are the ultimate "product" created by a regulatory department, and they also, in terms of content, format, and quality, *represent* the company and product. Often voluminous and spanning multiple technical areas, regulatory submissions are complex documents in every sense—from an editorial, scientific, and paper management perspective. At the same time, these documents represent the ideal opportunity for a regulatory professional to shine—not just in the quality of the final product but in the way the document is brought together.

WHO WRITES THESE DOCUMENTS, ANYWAY?

The two absolute answers to this question are both, in my opinion, wrong. At one end of the spectrum are those folks who believe the regulatory department is completely responsible for writing all content of all submissions to regulatory authorities. At the other end are those who would believe all that regulatory does is place a postage stamp on a document written completely by the technical departments. The answer is, of course, somewhere in the middle. I will always believe that the best discussion and presentation of the data will come from those closest to the data. This means the scientists, engineers, and technicians who produce the data, do the experiments, etc. At times, it can be difficult convincing these folks why regulatory submissions need to be a priority. It is worthwhile reminding them that the regulators are the gatekeepers to further development of their projects and that the regulatory process is a necessary one (even if viewed by some to be a necessary evil). Another way to encourage inclusion into the regulatory writing process is to point out that regulatory writing and key contributions to significant regulatory milestone submissions (e.g., IND and NDA) are career-enhancing and help broaden professional development.

In terms of the writing process, keys to success include recognizing the following:

1. Submission writing is an iterative process.
2. Submission writing is a back-and-forth process.
3. You need to lower your expectations.

Figure 11.3 illustrates the first two points. Scientists have usually gained expertise at writing scientific documents such as papers, abstracts, and technical reports. They want (and require) guidance as to what specific data need to be in a submission. Regulatory needs to point to specific regulations and guidelines that provide justification for the work, and guidance as to specific content and format. Expectations as to the quality of the work (e.g., print-ready manuscript vs. handwritten notes) and the timing of drafts are very important to resolve, and to resolve early. At the same time, the technical counterparts to regulatory have their own responsibilities. They need to hit deadlines, be engaged in the process, and deliver quality work. The easiest way to achieve this is to assure *ownership* of the submission or parts of the submission. Ownership implies an individual with responsibility and accountability for the section. This person may not do any writing, but he or she is the one who must be sure that things are delivered, and delivered on time. As with expectations

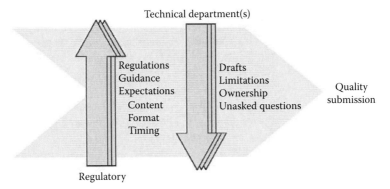

FIGURE 11.3 The process of writing submissions.

from regulatory, gaining concurrence on owners of sections and concurrence on their responsibilities is important to establish early and to communicate upward through management.

The technical owners of submission writing are also responsible (or should be) for making sure unasked questions get asked. That is to say, if key data are not requested by regulatory, or an important issue seems to go unaddressed in the document, the technical group has to mention it. The goal of every regulatory person is to never let this happen, but no one is perfect, and the submission needs to be a true collaboration. In addition, it may not be possible for a regulatory person to anticipate the unasked. It may be so specific, or technical, as to require the scientific expertise of a technical owner.

The multiple arrows going up and down in Figure 11.3 represent the multiple drafts that need to be exchanged as the process continues. These cycles of review, comment, feedback, and rewrite need to be on strict timetables, and regulatory needs to avoid being on the critical path.

The last concept on the list regarding writing was partially facetious: lowering expectations. This refers to reality and the old adage that "the perfect is the enemy of the good." Odds are you are not going to train a technical group to produce "submission-ready" output during the writing process for one submission. The amount of effort this would take leads to diminishing returns when the "polish" on a document can be done within the RA group. I have worked with brilliant scientists who write wonderfully, if they were writing for *Nature* or *Science*. Initially, I attempted to alter some of their styles when it came to summary paragraphs that sought to raise more questions instead of simply presenting data. I learned that this was just the way they wrote. It was how they knew to write for journals and editorials and one lowly regulatory submission was not going to alter that. Instead of focusing on style, I focused on content and data, knowing that the stylistic issues were easy enough to correct in the regulatory edits and reviews and in the writing process that the regulatory group owned.

REGULATORY REVIEW: CONTINUITY AND CONNECTIONS

Most large regulatory submissions involve multiple technical sections that are written by separate technical groups. As the overall "owner" of the submission, regulatory is responsible to assure the overall quality. This can usually be broken down into the concepts of continuity and connectivity.

Earlier it was implied that regulatory should avoid writing a submission—when it comes to continuity, regulatory must take the lead in writing. Sections of the document need to flow into each other, so the document appears at some level to have one voice. This is particularly important when concepts and data from multiple sections are brought together, as in introductory sections, synopses, and summary conclusions—cut and paste does not cut it. The language needs to be fluid, and the order of data logical.

Connectivity is a concept that is seldom recognized overtly by the regulatory community, but is in fact one of our most important responsibilities when it comes to submissions. As the owner of a submission, regulatory is really the only "person" who sees the entire document, and the document is not a linearly attached series of sections—it has multiple internal cross-references and connections. For example, data on preclinical safety connect to clinical protocols in terms of dose ranges and duration of dosing. This same connection is dependent on CMC data showing that the material used for preclinical safety data is truly supportive of the material intended for use clinically. The connections within even a relatively simple document, such as an IND exemption, are multifold and may differ from product to product. Who else is going to check and maintain these connections? Regulatory should have this responsibility and is ideally positioned to manage them. A systematic approach to reviewing these connections is important, and templates for review and cross-check are useful tools.

PRESENTING DATA IN SUBMISSIONS

With the advent of electronic submission production (e.g., Word, Excel, multiple graphics packages), we far too often resort to a quick "cut-and-paste job" when it comes to presenting data. I would suggest that rather than blindly including graphs and tables of data, it is the regulatory's job to look at these data presentations and make sure that the message behind them is clear and that the presentation is suited to the message. If an upward trend in the data is what you want a reviewer to see, a graph is better than a table, for example. Having a y-axis that has a maximum value of 100 when all your data skirt between 0 and 10 may not make sense (of course, if your message is that the data are all well below some threshold, let us say 30, it might make sense!). Edward Tufte has written several books on the inherent value of how data are presented, and I strongly encourage you to read his books and see his lectures.[10] One of the most important concepts is to make sure the data speaks as loudly as possible, and that it speaks the right message without being lost in the noise of the presentation. Bold colors and three- or four-dimensional artwork mean little if a reader cannot grasp the data or the experiments behind the data. A classic example is when multiple experimental points (e.g., subjects in a clinical trial) are compressed into a small number of data points. The goal was clarity, but power is lost—a reader may assume only a few experiments (or a small number of subjects) produced the data. The power of the data is thus diminished.

THE ART OF HANDLING LARGE DOCUMENTS

Never underestimate the difficulty of handling large volumes of paper or even electronic files. The electronic publishing era is maturing, but the concept still holds. One of the key lessons here is to keep sections of large documents separate, until they really need to be together. This "patience with paper" avoids the need to recollate or edit multiple volumes when only a few pages are in need of work. This concept is at odds with another notion—give yourself enough time to go through the mechanics of printing and copying. When you must move ahead and some small sections (I prefer to restrict them to individual pages) are not ready, use placeholders (colored paper works well) so the pages that need last-minute replacement or fixing can be identified.

Never underestimate the value of individuals who support the handling of large documents (be they electronic or paper). All the content in the world is useless if you cannot get 10 copies on a reviewer's desk by Tuesday. Expertise in this area comes from experience. Forget to paginate a document before copying it once, and you remember it forever.[11]

CONCLUSIONS

This was a disjointed roller coaster ride through RA. While not purposeful, this ride is a perfect microcosm of RA: many topics of varying technical detail, connecting the seemingly unrelated, with moments of panic, moments of boredom, but never a moment exactly like another. It is this complexity that makes the profession interesting, and it is the position of regulatory at the juncture of so many technical, managerial, and legal disciplines that makes it so vital to the industry. As professionals, we need to go beyond documenting regulations and guidelines, and document how we think, why we do things one way or another, and what has worked. This chapter and this book are intended to be a small start in the right direction.

NOTES

1. This is because an RA professional should not be writing CMC material from scratch! He or she should be a conduit for this information, an editor, and a reviewer. Later sections will expand on this method.
2. He has a Web page at http://www.math.temple.edu/~paulos/. One example from his book *Innumeracy* discusses diagnostics and specificity. It is a fascinatingly simple study, and I have gotten a lot of mileage out of it in discussions and presentations.
3. He has a great Website at http://www.gladwell.com/ that includes all of his writings.

4. *New Yorker*, August 21–28, 2000.
5. Please do not look this reference up. I made it up in its entirety and any resemblance to regulations past, present, or future is purely coincidental.
6. I first wrote about this analogy when interviewed in the RA column of *Biotechnology* magazine in the fall of 2000. I honestly cannot recall the first time I heard of it.
7. If you have not scoured every square inch of www.fda.gov, do so. It is a treasure trove of information. If it did not update so frequently, a book could be written about it.
8. Every few years, the Tufts Center for the Study of Drug Development (http://csdd.tufts.edu/) does a survey that says how much and how long a typical drug takes to develop—the 2002 numbers were seven years (on average) and ~US$900 million. Take these numbers with a few grains of salt—they are based on a limited sample size. At a minimum, they give some sense of scale for the biggest and longest projects.
9. I have been in drug development for 21 years, and am on my sixth job. This seems excessive, but is becoming a more common trend both in biotechnology and the economy as a whole.
10. Tufte's Website at http://www.edwardtufte.com is almost as good as his books, which are not only educational but also works of art. Get them and read them.
11. Imagine three copies of a document that has to be paginated. Imagine someone (certainly not me!) paginating all three copies and the last page of copy one is 340, while copy two ends at 337 and copy three finishes at 341. A lesson never forgotten.

12 FDA Advisory Committees

Christina A. McCarthy and David S. Mantus

CONTENTS

Introduction 327
Historical and Regulatory Framework 328
Structure 330
Composition 330
    Membership Type 330
    Qualification Requirements 332
    Appointment Process 332
    Membership Training 333
Conflicts of Interest 333
    Conflict of Interest Waivers 334
    FDA-Initiated Conflict of Interest Studies 335
Operation 338
Industry Perspective 340
    Affecting FDA's Decision to Call a Meeting 340
    Preparing for the Advisory Committee Meetings 341
        Success Criteria: Dedicated Team 341
        Success Criteria: External Expertise 342
        Success Criteria: Rehearsal 342
    After the Advisory Committee Meetings 343
Conclusion 344
References 344

INTRODUCTION

An integral part of the Food and Drug Administration (FDA)'s mission is to protect the public health by assuring the safety, efficacy, and security of human drugs, biological products, and medical devices. Because of the great magnitude and implication of this mission, the FDA occasionally calls upon external experts for advice and counsel. One of the ways in which the FDA can access external scientific expertise is through the use of the FDA advisory committee system. Advisory committees have been under considerable scrutiny over the past decades and the system has been the focus of congressional oversight. According to 21 Code of Federal Regulations (CFR) 14.5(a), "An advisory committee is utilized to conduct public hearings on matters of importance that come before FDA, to review the issues involved, and to provide advice and recommendations to the Commissioner." Utilizing advisory committees, the FDA can seek advice and input from scientific and medical experts, consumers, and patient advocacy groups on a number of issues ranging from approvals of new medical products to providing guidance on appropriate methods of clinical investigation. The advisory committee system is most frequently employed in reviewing products or topics that are controversial or novel; however, there are no concrete guidelines as to when to expect using an advisory committee. In addition, the FDA is not bound to the recommendations of the advisory committee.

According to Dr. Linda A. Suydam, DPA, former senior associate commissioner of the FDA:

> The FDA advisory committee system was established to provide independent expertise technical assistance related to the development and evaluation of products regulated by the FDA; to lend credibility to the product review process; to speed the review of products by providing a visible sharing of the responsibility for evaluation and judgment; to provide a forum for public discussion on matters of significant public interest; to allow sponsors and consumer to stay current with trends in the product development and review process and changes in regulations and guidelines related to FDA-regulated industries; and to provide external review of FDA intramural research programs.[1]

HISTORICAL AND REGULATORY FRAMEWORK

Informally, the FDA's practice of seeking external scientific and consumer advice began in 1964.[2] An early example of this practice includes a series of meetings held by the FDA, in which it consulted manufacturing and nutritional experts, as well as the general population, regarding what types of ingredients should be included in white bread.[3] Formal implementation of the system as we know it today, however, did not occur until the enactment of the Federal Advisory Committee Act (FACA) of 1972.

FACA was passed by Congress in 1972 to provide federal agencies with a formal mechanism for seeking external expertise and advice. The purpose of the federal advisory committees is to "provide independent, expert, and objective advice on policy, the funding of research, and other issues."[4,5] This act was passed in part because of concern by some legislators that there were too many informal and nonpublic advisory committees within sectors of the federal government.[5] FACA created a formal system that allowed government agencies to seek external advice, ensuring appropriate checks and balances. FACA also defines the operation of the federal advisory committees and emphasizes the importance of public involvement in the advisory committee system.

In 1997, the FDA Modernization Act (FDAMA) amended the Food, Drug, and Cosmetic Act [21 United States Code (USC) 355], directing the FDA to establish or use panels of experts to provide advice on the research and approval of drugs.[6] FDAMA focused on Advisory Board membership, including improving training and defining conflicts of interest, as well as stressing the need for timely considerations and notifications of committee deliberations.[7] Inclusion of provisions for use of advisory committees in FDAMA reinforced the importance of the advisory committee system within the FDA.

The FDA Amendments Act (FDAAA) of 2007 included additional provisions for the advisory committee system. Most notably, FDAAA stipulates that before a new drug [specifically, "a drug no active ingredient (including any ester or salt of the active ingredient) of which has been approved in any other application"] is approved, the Secretary must refer the new drug for review to an advisory committee. If the drug is not referred to an advisory committee for review prior to approval, the action letter must include the reasons stating why.[8] The effect of this stipulation on the advisory committee system, as well as on industry as a whole, has been significant. Between 50 and 60 advisory committee meetings are held each year, and their increased impact on the new drug review and approval process has led to the continued evolution of regulation and guidance.

The FDA Safety and Innovation Act (FDASIA) of 2012 included multiple provisions affecting the FDA advisory committee system.[9] From the perspective of process, FDA is required to provide sponsors of applications for new molecular entities (NMEs) with a review timeline letter 74 days after submission of a new drug application (NDA) or biologics license application (BLA). This letter will contain a discussion on whether or not an advisory committee meeting is necessary for the FDA to complete its review. This is the first time a formalized, prior notice of advisory committee usage has been included in the regulations. FDASIA eliminated several provisions of the FDAAA, as they related to conflicts of interest determinations for the FDA advisory committee members. This was prompted by concerns expressed by the FDA that it was difficult recruiting members with the necessary expertise to provide assistance. FDASIA aligns the FDA advisory committee conflict of interest rules with those of other federal committees. As a result of these changes, the FDA is to issue new guidance related to conflict determinations. FDASIA also modified the FDA's review goals, lengthening them for reviews requiring the advisory committee input, and it permanently authorized the Pediatric Advisory Committee.

More than 50 years of policy development and rulemaking has made the advisory committee system highly regulated. 21 CFR 14 describes almost every operational aspect of the advisory committee system. The detailed regulation provides a platform for uniform application of the advisory committee system throughout the government. Since the enactment of FACA, the FDA has also issued a number of guidance documents on the advisory committee system, ranging from guidance detailing the impact of FDAMA on the advisory committee system to multiple documents discussing disclosure of confidential information and management of conflicts of interest.

The FDA advisory committee Website includes general information, such as guidance documents, frequently asked questions (FAQs), meeting schedules, and transcripts from previous meetings. Detailed information on each committee, including financial status and membership rosters, are available for public review on the federal advisory committee database Website.[10]

STRUCTURE

As detailed in 21 CFR 14.100, the FDA has 35 standing advisory committees. In addition to the standing advisory committees, the FDA uses policy and technical advisory committees. While all of the centers at the FDA each have at least one advisory committee, 90% of standing FDA advisory committees can be found within the FDA's three major review centers handling human medicinal products [i.e., Center for Biologics Evaluation and Research (CBER), Center for Drug Evaluation and Research (CDER), and Center for Devices and Radiological Health (CDRH)]. While there are only 35 standing advisory committees detailed in the regulations, review of the FDA advisory committee Website indicates that there are additional active advisory committees.[7]

Those FDA advisory committees that are not mandated by law are created at will by the Department of Health and Human Services (DHHS). Committees are chartered for two years; at the end of the two-year term, the FDA, in conjunction with DHHS, must determine whether the committee should continue its service. If the committee is no longer useful to the FDA, it is dissolved.

Each FDA center is responsible for the general administration of its advisory committees. However, because there may be variability between the centers in regard to how advisory committees are managed, there is a central office, which is responsible for the general administration of the FDA advisory committee system. The Advisory Committee Oversight and Management Staff is an office within the FDA, whose primary function is to ensure that the advisory committee system runs smoothly. This office handles member training as well as administrative aspects such as organizing travel and reimbursement for committee members.

COMPOSITION

The FDA advisory committees are made up of various individuals who range from practicing clinicians to individual patients and other interested citizens. The membership roster includes those individuals who are considered standing members, but the law also provides the opportunity to call upon other individuals on a temporary basis. At certain types of meetings, depending on the type of input that the FDA is seeking, a vote may be necessary; however, not all members are voting members. Regulations explicitly define which members can vote; voting status is dependent upon the type of committee (standing, policy, or technical) on which members serve (21 CFR 14.80). Voting status may also be dependent on documented conflicts of interest.

FACA requires that all advisory committee members be appropriately trained. As such, the Advisory Committee Oversight and Management Staff provide training to all advisory committee members.

Membership Type

In general, the role of an FDA advisory committee member is to provide independent advice and expertise to questions asked by the FDA on a particular topic or product.

FACA mandates diverse committee membership representing the general public. The FDA takes measures to ensure the diversity of its advisory committees with regard to demography as well as professional/scientific expertise. All outside members who are hired by the FDA for their input on advisory committees, except for industry representatives, are considered special government employees (SGEs).[11] Members are paid per day of committee meeting service and are reimbursed for travel, food, and lodging expenses. The payment, however, is minimal. In 1995, the FDA reported that members receive US$150 a day during meeting service.[2]

For each advisory committee, there is an executive secretary and a committee chairperson. The executive secretary is an FDA employee who is assigned to oversee the general administration of the advisory committee. The executive secretary does not participate as an advisory committee member, but rather as a liaison between the FDA and the committee. The executive secretary ensures that all regulations are followed in committee conduct. In contrast, the committee chairperson, mandated by 21 CFR 14.30, is a committee member and is most often one of the more experienced members, and has the authority to conduct hearings and meetings.

With the implementation of FDAMA, advisory committee membership is divided between core membership and *ad hoc* membership. Core members are those individuals who are appointed by the commissioner on the basis of scientific expertise, while *ad hoc* members are those who are asked to serve on committees when needed. The standard term of service is usually four years, and membership extensions are rarely given.[12]

Members of the FDA advisory committees include scientists/academicians/practitioners, consumers, patients, and industry representatives. Each of these types of members plays a different and important role on the advisory committee panel. According to a 2001 FDA survey, 80% of the advisory committee members are scientists, followed by consumers, industry representatives, and then patient representatives.[13] The academicians/practitioners are frequently employed as chairperson because of their expertise in the field. The role of an industry representative is to provide advice and address concerns from an industry standpoint. While any industry member will have individual ties to a particular company, their role is not to represent that specific company, but rather to represent the industry as a whole.[14] Industry representatives are permitted membership on advisory committees, but because of conflict of interest concerns, industry members are by law always nonvoting.[7]

One of the most important aspects of the FDA advisory committee system is public involvement in important and/or controversial issues, products, and policies that the FDA is considering. The public is formally involved in the advisory committees through consumer representative membership. The FDA is careful to ensure that the consumer representative is well qualified to handle the scientific nature of the discussions, as well as be a true representative of the public, and not simply provide an individual opinion.[15]

Patient representatives are intended to bring a unique and humanistic viewpoint to the advisory committees. Patient representation allows for the input of those individuals who are directly affected by the issue or product being discussed. Historically, it was the HIV/AIDS patient advocacy groups that lobbied the FDA for representation in the decisions being made regarding drug approvals. Committee members were initially resistant to the inclusion of patient representatives; however, patient members provided valuable input, and today their representation is highly respected.[3] Currently, patient representatives are predominantly used on the advisory committees handling HIV/AIDS and oncology issues; however, the FDA requests patient representation on other advisory committees, discussing serious and/or life-threatening illnesses on an *ad hoc* basis.[16] Patient representatives usually have had direct experience, individually or through a family member, with the disease being discussed, and can articulate how the disease affects quality of life. Oftentimes, the patients are well informed and have formal affiliations with advocacy groups.[17] While patient representatives can be both voting and nonvoting, patient representatives who are members on the advisory committees that review oncology products/therapies are voting members, and those representatives serving on most other nononcology advisory committees are nonvoting. The FDA maintains a comprehensive Website on patient representation for those individuals interested in participating in the program.[16]

QUALIFICATION REQUIREMENTS

21 CFR 14.80 details membership qualifications; these are dependent on the type of committee represented, and the regulation mandates that members have diverse interests, education, training, and experience. However, technical expertise, unless as a member on a technical committee, is not a formal requirement.

APPOINTMENT PROCESS

The nomination and selection process of committee advisory members is highly regulated. For voting members of the standing advisory committees, the process begins with the commissioner publishing a notice in the *Federal Register* requesting nominations. Nominations are then screened by the appropriate product divisions within the Centers to ensure that the nominee possesses the required expertise, and to screen for potential conflicts of interest.[12] Persons nominated and selected in this manner serve on the committee as an individual and not as an advocate for a larger organization (e.g., consumer representative).

For consumer and industry representatives, a request for nomination is published in the *Federal Register*. For consumer representatives, the regulations urge that nominations be filtered through consumer advocacy groups. For industry representatives, regulations state that the industry organizations with corresponding member nominees are to decide among themselves who is to be the representative. If no decision is made, then the commissioner selects the industry representative.

Anyone can nominate a candidate to serve as a patient representative.[16] Nominations are sent to the FDA patient representative program where the selection process is vetted.

MEMBERSHIP TRAINING

The FDA is required by law to provide training to every advisory committee member prior to participation in a committee meeting. All members undergo a comprehensive training program run by the FDA staff. The FDA advisory committee new member training program is available for public review on the FDA Website.[12] In addition to the formal member training, patient representatives are oriented by the Office of Special Health Issues. According to the FDA's patient representative Website, newly selected patient members receive training on an individual basis, which includes observing an advisory committee meeting and discussions with previous patient representatives.

CONFLICTS OF INTEREST

No aspect of the FDA's advisory committee system has been more scrutinized, than member conflict of interest. Because many of the advisory committee members are also experts in their respective fields, they are often closely involved with cutting-edge research, and such research is often funded from a variety of sources. It is because of their direct knowledge of new research and/or products that the FDA seeks their advice. Sometimes, however, members' involvement in research is closely linked to the development of specific products, which can become an issue when it is on those products that the FDA is seeking advice. While there can be many types of conflicts of interest, the primary conflict with which the FDA is concerned is financial. Prior to every advisory meeting, committee members are sent a confidential questionnaire (Form FDA 3410, "Conflict of Interest Disclosure Report for SGEs"), which asks about financial interests in regard to the product or the product's sponsor to be discussed at the meeting. Along with the form, a list of all products and sponsors associated with the meeting is sent. Using this information, the member is asked to determine if they have financial interest in anything that is being reviewed. Financial interests can range from direct investments in a particular company to receiving research grants from a sponsor whose product is under review.[1]

Once the FDA receives the completed financial disclosure form, it is then determined whether, and to what extent, a member has a conflict of interest. If there is a conflict, the FDA staff determines if the conflict qualifies for a waiver, or, if it is too significant, the conflicted member should be excluded from the meeting. Most often, this decision is made by the FDA official from the division or office, requesting the advisory committee's assistance.[1] The Director of Advisory Committee Oversight and Management is also closely involved in the determination of conflicts of interest.[18] Further, if there is any question about any waivers that are granted, an independent review by the FDA Ethics Office is conducted.

The advisory committee members are subject to two conflict of interest laws, under which criminal prosecution is possible. The Criminal Conflict of Interest statute regulates conflict of interest for all federal government employees, including SGEs; since voting advisory committee members are considered SGEs, these members are subject to the law (18 USC 208). FDAMA included provisions for conflict of interest management, FDAAA expanded on those provisions in an attempt to make the process simpler and more transparent, and FDASIA realigned some provisions.[6,8,9]

CONFLICT OF INTEREST WAIVERS

According to the Criminal Conflict of Interest statute, if an SGE has a financial conflict of interest, then they are not allowed to participate in related advisory committee meetings unless a waiver of exclusion is granted.[19] The Criminal Conflict of Interest statute, FDAMA, FDAAA, and FDASIA all include provisions allowing for waivers of conflicts of interest.

18 USC 208 (b) allows for three types of waivers to be granted to the advisory committee members with conflicts of interest. The first type of waiver is for federal employees serving on an advisory committee and experts who are performing tasks other than serving on an advisory committee. This waiver is granted when the financial interest is determined not to be significant enough to affect outcomes. The second type of waiver is for committee members participating in meetings; this waiver is slightly more lax than the first waiver. This waiver is granted when it is determined that "the need for the individual's services outweighs the potential for a conflict of interest." The third type of waiver is for financial interests that are determined by regulation to be minimal by the Office of Government Ethics.

FDAMA included a provision for a waiver [which only applies to the advisory committee members for CDER and CBER panels], which allowed committee members to vote in a particular matter, from which they or an immediate family member may receive financial gain, if their expertise is essential (i.e., no one else has the needed expertise) to the committee.[5]

With the implementation of FDAAA, additional, more detailed requirements detailing how the FDA should address conflicts of interest for the advisory committees were enacted. Members, or their immediate family members, with financial interests having the potential to affect the meeting, are prohibited from participating in the FDA advisory committee meetings. And while FDAAA maintains that the FDA has the authority to grant a waiver if the member's expertise is essential to the advisory committee, it does limit how many waivers the FDA can grant annually, and requires the publication of an annual report. Further, FDAAA requires that the FDA review potential conflicts of interest when initially considering new members for the advisory committees. In addition, FDAAA requires that all members of the advisory committees disclose to the public any conflicts of interest in regard to the meeting's subject matter. In August 2008, final guidance, the FDA states that prior to an advisory committee meeting each member must disclose the type, nature, and magnitude of any financial conflicts of interest, and stipulates that members cannot participate until this disclosure is made. The FDA will post the disclosure, along with any waiver granted by the Agency, on the FDA Website.[20]

FDA Advisory Committees

In 1994, the FDA issued a guidance document titled "FDA Waiver Criteria," which outlined how, when, to whom, and under which conditions waivers may be granted for committee members with conflicts of interest. That document was updated in 2000 (titled "Waiver Criteria 2000"),[21] and more recently in 2007.[18] Because of the scrutiny of its financial conflict of interest policies, the FDA issued a final guidance document in August 2008, which attempts to simplify the process by which the FDA assesses conflicts of interests and determines meeting participation, including the granting of waivers. In part, the guidance was intended to "enhance public trust" in the advisory committee function by better describing the algorithm for determining when waivers for conflicts of interest are granted.[18] This algorithm is presented in Figure 12.1.

The August 2008 guidance sets forth the following stipulations:

- Members should not participate in the advisory committee meetings, regardless of need for expertise, if their disqualifying financial interest exceeds US$50,000.
- When the disqualifying financial interest is less than US$50,000, members can only participate when the need for the member's service outweighs the potential conflict, and can only participate as nonvoting members.
- The FDA can limit participation when there may be a perceived conflict of interest, even if none have been determined under law.

The FDA focuses primarily on individual financial conflicts of interest. While there are other types of conflicts, like previous involvement with a particular product/industry sponsor or an SGE's institutional potential financial gain, the FDA is often not required to grant a waiver in those broader situations; instead, only a public disclosure is made.[19]

FDA-Initiated Conflict of Interest Studies

Periodically, the FDA conducts surveys in which conflicts of interest in the advisory committee system are explored. In 2001, the FDA sent a survey out to 400 advisory committee members. The survey asked general questions about members' attitudes toward conflicts of interest, particularly regarding public disclosure.[22] The FDA received answers from 73% of the members polled.[22] Member attitudes toward public disclosures is important because anytime a waiver is granted for a conflicted member, public disclosure is legally required. Of note, 65% of the respondents indicated that additional disclosure did not add credibility to the process, whereas 33% said it did.[23] Also notable were members' response to the question, "If FDA asked for more disclosure, I would ...." To this question, 58% answered that they would "act dependent on what was required" and 36% answered that they would "do what was asked."[23] Interestingly, 5% of the respondents answered that they would consider resignation, and one committee member responded that he or she would resign if the FDA asked for more disclosure.[23] Again, this is an important information for the Agency to know and

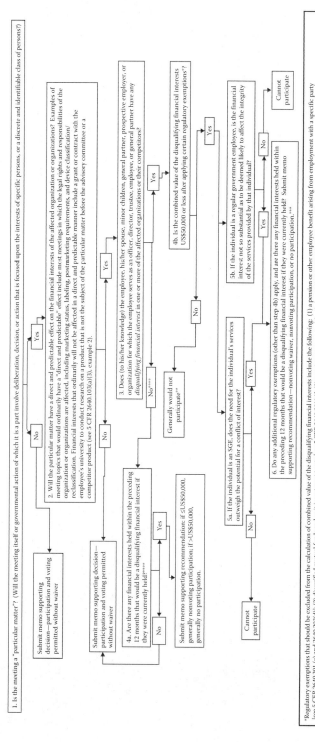

FIGURE 12.1 Algorithm for considering advisory committee member participation. (From http://www.fda.gov/oc/advisory/waiver/ACAlgorithm.pdf. With permission.)

consider when making policies on how far public disclosure should be taken. The FDA concluded from the survey results that most members would be willing to publicly disclose more of their financial information; however, they also concluded that about half of the members indicated that continuation of membership was dependent on the severity of disclosure procedures.[24] While it is important to abide by the conflict of interest public disclosure laws, if the disclosures are so strict that the committee members refuse to participate, then the system would be defunct.

In 2003, the FDA conducted a second survey titled "Conflicts of Interest and FDA Advisory Committee Meetings: A Study of Public Attitudes and Opinions."[25] The two-part study consisted of the FDA surveying attendees and the advisory committee members from 11 advisory committee meetings throughout the spring of 2003. According to the study summary report, "The study's intent was to examine the perceived fairness and credibility of FDA advisory committee meetings related to FDA's management of real or potential conflicts of interest among advisory committee members."[25] The FDA was also interested in finding out how much audience members knew about the FDA's procedures to manage conflicts of interest, as well as how satisfied they were with the FDA's current conflict of interest procedures. Further, the survey asked what aspect of the advisory committee meetings was most important to audience members and inquired about general satisfaction with the FDA.[24] The study summary stated that there was a 21% overall response rate by audience members polled and 66% response rate by the committee members who were polled.[25]

A percentage (44.8%) of the audience members indicated familiarity with the FDA's procedures for reviewing conflicts of interest of the advisory committee members, but only 33.8% of the audience respondents considered themselves knowledgeable about how the FDA monitors conflicts of interest among its advisory committee members.[25] Further, 75.1% of the audience indicated disagreement with the statement, "The FDA should not allow members with conflicts of interest to participate in any capacity at advisory committee meetings." Of note, 83.5% of audience respondents considered it "unreasonable to expect that advisory committee members won't have some conflicts of interest."[25] In regard to the most important aspects of the advisory committee meetings, audience respondents stated that "Fairness of decision or outcomes" and "Committee members are top experts" were equally important aspects and of top importance. It is worth noting that 82% of the audience respondents were paid by an employer or organization to attend the meeting, and therefore it appears that the responses were industry heavy, and thus subject to bias.[25]

In regard to the advisory committee member survey, while it was much shorter than the audience version, the overwhelming response was that the FDA's policies and procedures on conflicts of interest for the advisory committee system are impartial and are fair to both committee members and to the general public.

It is not evident what specific impact the two surveys have had on the FDA policies or decision making. Following the first survey, the FDA published a draft guidance titled "Draft Guidance on Disclosure of Conflicts of Interest for Special Government Employees Participating in FDA Product Specific Advisory

Committees; Availability." It is important to note, however, that the FDA has not indicated that the guidance was a direct result of the SGE financial disclosure survey.

The FDA contracted Eastern Research Group, Inc., to assess the "relationship between expertise and financial conflicts of interest of the FDA advisory committee members."[26] In October 2007, data on member expertise and conflict of interest were published. The study was based on a sample of the advisory committee meetings held between December 2005 and October 2006. The study found that members with higher levels of expertise were more likely to have been granted waivers for conflicts of interest.[26] Of note, however, was that in the cases where expert members had financial conflicts of interest, alternative members with equal expertise, while easy to find, would also have similar conflicts of interest. For members granted waivers, it was reported that median total dollar value of financial interest was US$14,500. The report concluded,

> Overall we judge the ability to create alternative conflict-free advisory panels to be speculative. If possible, it would represent an uncertain and potentially substantial additional burden on the cost and the timeliness of advisory committee operations. Further, the FDA might not always be able to match the specialized expertise of some existing advisory committees.[26]

The removal of some of the more restrictive provisions of FDAAA by FDASIA is likely a response to both real and perceived difficulties in recruiting members of the advisory committees. The issue of conflicts of interest within the advisory committee system will continue to be scrutinized and criticized by constituents. It can be speculated that with the new provisions and simpler algorithms provided by the FDA, the manner in which member conflict of interest is handled will become more transparent.

OPERATION

The primary function of the FDA advisory committee system is to provide the FDA with expertise on products and/or topics for which the Agency is seeking external guidance. These recommendations are provided at formal meetings that are highly regulated. An integral part of the laws and regulations governing the advisory committee system is the inclusion of public involvement. As a result, the advisory committee meetings are generally open to the public; however, regulations do permit full and/or partial closure of meetings under certain conditions.

Before the advisory committee can be held, it is important for members to be debriefed on the subject matter under deliberation. In cases where a specific product is to be reviewed, the FDA may request that the sponsor submit a written information summary (briefing document) detailing safety, efficacy, or other pertinent data. If written information is requested by the FDA, the sponsor (the company whose product is under review) is required to submit the summary at least three weeks before the meeting. The Agency then distributes the sponsor's summary, in addition to its own summary of pertinent information, to the committee members. In addition to the briefing documents, the committee members are also given a list of questions that the FDA would like addressed. These questions are usually posted in the *Federal Register* and on the FDA's Website for public review (not public comment) in advance of the meeting (often no more than a couple of business days before the meeting). In 2008, the FDA issued a comprehensive guidance document for sponsors on preparation and public availability of material given to the advisory committee members.[27]

The format in which the FDA advisory committee meetings are to be conducted is detailed in 21 CFR 14.25. Regulations stipulate that the advisory committee meetings may have four portions: open public hearing, open committee discussion, closed presentation of data, and closed committee deliberations. Open meetings are those in which the public may attend and present information, whereas closed meetings are those in which the meeting is only open to the advisory committee members and associated support staff. It is important to note that not all committee meetings require all of the above four elements. In meetings at which topics of a general nature are discussed, there may be no need to close any portion of the meeting. As per regulation, all the FDA advisory committee meetings are published in the *Federal Register* at least 15 days before the meeting's date, thereby allowing anyone interested in the opportunity to attend and speak at the meeting (21 CFR 14.20). Federal regulations also detail the required elements to be published in the *Federal Register*; items such as the meeting agenda, nature of topics to be discussed, and contact information for the FDA. People who wish to use the public forum are not required to contact the FDA in advance of the meeting, but they are urged to register beforehand. Preregistration of public speakers allows the FDA advisory committee oversight office to plan accordingly for speakers. According to 21 CFR 14.29, at least 60 minutes of each advisory committee meeting must be allotted for open public comment. However, if the public comment portion of the meeting does not take 60 minutes, then that time may be decreased. Alternatively, if the topic under discussion is of great public interest and 60 minutes is insufficient to hear all public comments, then the advisory committee chairperson is allowed to increase the time allotted.

In addition to presentations made by the public, during advisory committee meetings at which specific products are under review, the sponsor presents pertinent data. The sponsor's address to the committee is usually the most formal and well-rehearsed presentation. The preparation required for the sponsor presentation will be explored in further detail in the section "Industry Perspective."

The purpose of the advisory committees is for advisory members to provide the FDA with guidance on selected topics. As mentioned earlier, prior to the meeting the Agency sends committee members a list of questions, which are to be addressed at the meeting. Therefore, an integral part of the advisory committee meetings is the committee discussion of these questions. Frequently, committee deliberation of the FDA's questions occurs during the open portion of the meeting. From review of transcripts from previous advisory committee meetings, it is typical for the chairperson to go through each question and elicit a response from the members; however, depending on whether the Agency is seeking product-specific or general policy guidance, the methodology for how the questions are answered varies. If a vote is required, the chairperson asks for the members' vote and then often summarizes the votes after all members have answered the question.

The FDA can ask the advisory committee members to deliberate on a variety of topics, ranging from issues of general subject matter to advice regarding specific marketing applications. Depending on the type of advice that the FDA is seeking, the advisory committee can issue a number of different types of recommendations.

General topics reviewed typically include request for guidance on policy making, advice regarding clinical study design for certain disease indications or conditions, or input on safety of certain classes of products. When the FDA calls upon an advisory committee to review a particular marketing application, several recommendations are possible and are dependent on the type of advice the FDA is seeking. When the FDA is asking whether the committee concurs that the data presented are sufficient for marketing approval, the committee may concur and recommend approval, or they may issue recommendations for additional information prior to final approval. These recommendations may include additional studies, more safety or efficacy data, population/age restrictions, and changes to the proposed indication/labeling.

The FDA is not bound to the recommendations made by the advisory committee members; however, recommendations are taken under serious consideration. It is difficult to predict whether the FDA will concur with an advisory committee recommendation; however, data tend to suggest that there is a positive correlation between the determinations; or at a minimum, the FDA takes the recommendations into consideration in their own deliberations. While one must consider that there are a multitude of other factors affecting the FDA determination of final action, the impact of an advisory committee determination cannot be discounted.

INDUSTRY PERSPECTIVE

The very public and important nature of the advisory committee meetings and their content makes them one of the most resource-intensive interactions between sponsor companies and the FDA. The future of a product's development, a company, and even an industry may be impacted by the recommendations of a particular advisory committee. A sponsor company can attempt to manage and optimize the advisory committee process in three key areas: (1) influencing when advisory committees are used, (2) preparing for the advisory committee meetings, and (3) once a committee has met and provided guidance to the FDA, a sponsor company must then manage the impact of the meeting.

Affecting FDA's Decision to Call a Meeting

Since the FDA calls on its advisory committee system for guidance when the internal expertise of the Agency is insufficient to either decide on an issue or provide more general guidance for product development, the back-and-forth between the FDA and a sponsor company can increase the likelihood of an advisory committee consultation. For example, if the FDA and a sponsor company cannot reach agreement on a particular issue, including the factors that influence market approval of a product, the sponsor may ask the FDA to bring the issue to an advisory committee. Even without such a request, the FDA may bring an issue to the committee because discussions with a sponsor are no longer progressing and the opinion of an external and expert panel may aid in reaching mutual understanding. Even if there are no contentious issues between the FDA and a sponsor company, a particular product development path, or an entire class of products, may be so novel as to require the assistance of an advisory committee to provide guidance on paths forward to both the FDA and the industry.

When a sponsor company has not asked for an advisory committee meeting, it is common for the FDA to provide advance notice to a company before publicly announcing an advisory committee meeting. While this notice provides a company with the theoretical opportunity to influence the FDA's decision, in general the FDA's initial decision to convene a committee meeting is final.

PREPARING FOR THE ADVISORY COMMITTEE MEETINGS

A sponsor company may have as long as one year or as little as two months to prepare for an FDA advisory committee meeting. The amount of time depends on whether or not the guidance needed from the committee is part of a long-term development program (or a group of programs) that is years away from approval, or a group of approved products or product classes, or whether committee input is needed urgently on a time-sensitive subject, such as safety, or as part of a user-fee driven review timeline. The least amount of preparation time for advisory committee meetings are usually the result of meetings called as part of priority product reviews, in which the FDA targets total review times of six months or less. FDASIA now requires the FDA to discuss the possibility of the advisory committee involvement in a letter to the applicant for NMEs, 74 days into the review. Even with this notification, most advisory committees will (1) have very short lead times for preparations and (2) occur only weeks before action dates (approval, approvable, or not approved decisions).

Successful preparation for the advisory committee meetings is not achieved via a single approach, and each meeting is unique. In general, the following principles are critical to success, and the approaches described have proven successful in some instances.

Success Criteria: Dedicated Team

A sponsor should form a team of employees and consultants who will be dedicated to planning and carrying out the advisory committee meeting functions. It is important that the team working on the advisory committee meeting be experienced and extremely knowledgeable of the product under review. In general, this means that the team is the same team that has led product development to date, or a subteam, but additional team members and leadership may be required.

Proactive team leadership is vital to success, as preparations are required on a variety of fronts, from reanalyses of data to logistics.

Each speaker on behalf of the sponsor company should have a dedicated support team to aid in presentation development and in response to questions from the committee. These support teams should have as intimate a knowledge of the data as the speaker. Each speaker should also have an assigned backup, in case of last-minute emergencies that prevent their participation in the meeting.

Success Criteria: External Expertise

All advisory committee meetings include challenges of content and logistics. It is unlikely that any company has all the content experts required for successful presentation of data, and perspective on that data, to the committee. Experts with a level of technical accomplishment and global recognition at the same level as the committee members (or beyond) are usually present at these meetings. This is not to say that such experts present all of the data. Normally, they are used to present either key findings or help place findings in the perspective of medical standard of care and medical need. Using such experts brings the communication of data to the peer-to-peer level with the committee.

In addition to outside content experts, most companies facing the challenge of an advisory committee meeting employ experts in an array of logistical and style challenges of presenting vast quantities of data in a short time period (often 90 minutes or less), to a relatively unfamiliar audience, and respond to questions from the committee. There are a number of firms that specialize in the preparation, cataloging, and presentation of slides for the advisory committee meetings. These companies provide staff expert at presenting data, managing thousands of slides, and retrieving slides in a manner of seconds for display in answering questions. It is important for the sponsor company to develop a strong working relationship with the presentation vendor as early as possible and to demand a dedicated staff presence on the committee meeting team. Additional outside experts can be called upon to help hone the presentation skills of speakers, the overall flow and messaging of the data presentations, as well as the logistics of practices (see section "Success Criteria: Rehearsal") and last-minute preparations.

Success Criteria: Rehearsal

Rehearsing for the advisory committee meetings can be highly analogous to practices in team sports. Early rehearsals are loosely organized with the intent of working out timing, general flow, and allowing the speakers to refine their presentations. Each presentation should be scripted, so that eventually, every word is documented to allow for last-minute replacements in case of emergency. Every rehearsal should have an audience. Early on, these can be other team members, but eventually they should be external experts who play the role of committee members. As practices continue, a log should be maintained of all questions, and brainstorming sessions to consider all possible questions should be held. Special rehearsal of individual presentations, or just focused on one set of technical issues, may also occur. Eventually, mock committee meetings should be held. Some companies go to the extreme of simulating the room in which the meeting is scheduled, so as to acclimate the speakers and support staff to the logistical realities of the day of the meeting. Some rehearsals are staged with all the aspects of a real meeting, including assigned roles for the audience, an agenda that matches the day, presentations by the mock FDA staff, and rules that no one breaks until the practice is concluded. As in sport, all of these practices are intended to make the team comfortable with all the possible twists of the day. Effective practices lead to high-quality presentations of data, and quick and accurate responses to committee questions.

Briefing document and slides. As mentioned earlier, sponsors are usually required to submit summary data of pertinent information, typically referred to as a briefing document. The 2008 final guidance provides details on preparation and public availability of material given to the advisory committee members, and applies to the briefing document.[27] The summary is a comprehensive compilation of all information and data that are relevant for the committee member review. Typically, the briefing document is about 20 to 30 pages long. It is important to remember that the document is used by the committee as a reference in considering their guidance to the FDA, and that often the committee only has a few weeks to prepare for the meeting. This heightens the need for clear, concise, and accurate summaries of data, and a flow to the document that anticipates, to the degree possible, potential questions committee members might have and potential points of contention with the FDA. It is also important to remember that this document is posted publicly, and will be in the public domain forever. The sponsor is also given the opportunity to review the briefing information created by the Agency and to request changes, if errors are noted.

The slides accompanying the oral presentations given by the sponsor company are also critical documentation, and will be presented publicly, and be archived as a public record of the meeting. Slides are carefully prepared to present data that are integral to the committee members' deliberations, specifically in regard to those questions that the FDA has asked the committee to answer. A key component to the sponsor presentation is the ability to anticipate what questions committee members or agency representatives may ask. Backup slides, as many as several thousand, are created as backup for potential questions. The speakers must learn every backup slide and be able to recall appropriate data on the spot to address questions at the meeting.

AFTER THE ADVISORY COMMITTEE MEETINGS

The outcome of an advisory committee meeting is usually some form of guidance to the FDA, often in the form of a vote on key questions. As has been stated, these recommendations are nonbinding; but as a matter of public record, they can have significant impact on a company, or even an industry. In terms of regulatory follow-up, it is important for a sponsor company to quickly digest the questions raised at the meeting, committee's recommendations, and the potential impact on the FDA's ongoing review(s) of product applications. In some instances, the positive recommendations of the committee can be leveraged to improve the chance of a successful review. However, this leverage is unlikely to come from simply quoting a positive tally of the committee votes. Arguments should be based on the justifications presented by the committee for the vote, new insight from the committee on the data, and the broader perspective the committee often brings to the discussion. In the case of a negative vote, a sponsor company should continue to work with the FDA to resolve issues and present, or re-present, their data to support their arguments. Whether the cumulative response from the committee is positive or negative, their feedback on the data, and their revelation of potential previously unaddressed issues, should be considered. The goal should be seamless inclusion of the advisory committee meeting into the review process.

CONCLUSION

The FDA advisory committee system is a significant part of the FDA's decision-making process and can be a highly charged component of the process. As a public forum, the advisory committees have the potential to impact the political and commercial environment of entire industries. Even when considering a single product application, a committee can have the power to halt development and change the course of a company's future. These factors make it imperative that sponsor companies fully understand the advisory committee process and are well prepared to be part of that process. This chapter has attempted to outline the process and provide guidance on how to maneuver successfully within the process. A typical regulatory professional may only have a few opportunities in a career to have projects reviewed at an advisory committee, and therefore any and all references, including this chapter and the counsel of experienced consultants, should be used in preparations. A well-prepared company, with a well-rehearsed and scientifically strong presentation of the issues, should not only receive a fair and balanced review by an advisory committee but also present an image of the company to the public as a diligent, compassionate, and rigorously data-centric player in the public health industry.

REFERENCES

1. Department of Health and Human Services Website. Assistant Secretary for Legislation (ASL). Testimony on vaccine advisory committees by Linda A. Suydam, DPA, senior associate commissioner, the Food and Drug Administration, Department of Health and Human Services, before the Committee on Government Reform, US House of Representatives. June 15, 2000. Available at http://www.hhs.gov/asl/testify/t000615a.html, accessed September 19, 2012.
2. Farley, D. Getting outside advice for close calls. *FDA Consumer Magazine*. March 1995, pp. 23–26.
3. Lewis, C. Advisory committees: FDA's primary stakeholders have a say. *FDA Consumer Magazine*. September–October 2000, pp. 30–34.
4. Federal Advisory Committee Act. 5 USC App. 1 Public Law 92-463. Available at http://www.accessreports.com/statutes/FACA.htm, accessed September 19, 2012.

5. Steinbrook, R. Science, politics, and federal advisory committees. *New England Journal of Medicine*, 2004, 350(14): 1454–1460. Available at http://www.nejm.org/doi/full/10.1056/NEJMp058108, accessed September 19, 2012.
6. FDA Modernization Act of 1997. Public Law 105-115. November 1997. S. 830.
7. US Food and Drug Administration Website. FDA advisory committees. Available at http://www.fda.gov/AdvisoryCommittees/default.htm, accessed September 19, 2012.
8. Food and Drug Administration Amendments Act of 2007. Available at http://www.fda.gov/RegulatoryInformation/Legislation/FederalFoodDrugandCosmeticAct FDCAct/SignificantAmendmentstotheFDCAct/FoodandDrugAdministration AmendmentsActof2007/ucm2005652.htm, accessed September 19, 2012.
9. Food and Drug Administration Safety and Innovation Act (FDASIA). Available at http://www.fda.gov/RegulatoryInformation/Legislation/FederalFoodDrugandCosmetic ActFDCAct/SignificantAmendmentstotheFDCAct/FDASIA/ucm20027187.htm, accessed September 19, 2012.
10. Federal Advisory Committee Database Website. Available at http://www.fido.gov/facadatabase/, accessed September 19, 2012.
11. Sherman, L. A. Looking through a window of the Food and Drug Administration: FDA's Advisory Committee system. *Preclinica* 2004; 2(2): 99–102.
12. US Food and Drug Administration Website. Available at http://www.fda.gov/AdvisoryCommittees/AboutAdvisoryCommittees/CommitteeMembership/ucm117646.htm, accessed September 19, 2012.
13. US Food and Drug Administration Website. Available at http://www.fda.gov/oc/advisory/conflictofinterest/2001Survey/COISurveyRslts2001Q2Q6.pdf, accessed September 19, 2012.
14. Rados, C. Advisory committees: Critical to the FDA's product review process. *FDA Consumer Magazine*. January–February 2004, pp. 17–19.
15. US Food and Drug Administration Website. Advisory committee consumer representatives. Available at http://www.fda.gov/AdvisoryCommittees/AboutAdvisory Committees/CommitteeMembership/ucm231782.htm, accessed September 19, 2012.
16. US Food and Drug Administration Website. FDA patient representative program. Available at http://www.fda.gov/ForConsumers/ByAudience/ForPatientAdvocates/PatientInvolvement/ucm123858.htm, accessed September 19, 2012.
17. Meadows, M. Bringing real life to the table: Patient reps help FDA review products. *FDA Consumer Magazine*. January–February 2002, pp. 10–11.
18. US Food and Drug Administration Website. Draft guidance for the public, FDA advisory committee members, and FDA staff on procedures for determining conflict of interest and eligibility for participation in FDA advisory committees. Available at http://www.fda.gov/ForConsumers/ConsumerUpdates/ucm125637.htm, accessed September 19, 2012.
19. US Food and Drug Administration Website. Guidance for FDA advisory committee members and other special government employees on conflict of interest 2000. Available at http://www.fda.gov/oc/advisory/conflictofinterest/waiver.html, accessed September 19, 2012.
20. US Food and Drug Administration Website. Guidance for the Public, FDA Advisory Committee Members, and FDA Staff on Procedures for Determining Conflict of Interest and Eligibility for Participation in FDA Advisory Committees. Available at http://www.fda.gov/downloads/RegulatoryInformation/Guidances/UCM125646.pdf, accessed September 19, 2012.
21. US Food and Drug Administration Website. Conflict of interest. Available at http://www.fda.gov/oc/advisory/conflictofinterest/guidance.html, accessed September 19, 2012.

22. US Food and Drug Administration Website. Brief report: SGE financial disclosure survey. Available at http://www.fda.gov/oc/advisory/conflictofinterest/2001Survey/COISurveyRslts2001Q1Intro.pdf, accessed September 19, 2012.
23. US Food and Drug Administration Website. Brief report: SGE financial disclosure survey. Q7. Available at http://www.fda.gov/oc/advisory/conflictofinterest/2001Survey/COISurveyRslts2001Q7ToQ11.pdf, accessed September 19, 2012.
24. US Food and Drug Administration Website. Comparing answers for all types of financial interest. Available at http://www.fda.gov/oc/advisory/conflictofinterest/2001Survey/COISurveyRslts2001AnlysCncl.pdf, accessed September 19, 2012.
25. US Food and Drug Administration Website. Conflicts of interest and FDA advisory committee meetings. Available at http://www.fda.gov/oc/advisory/acstudy0904/JIFSANresearch.html, accessed September 19, 2012.
26. US Food and Drug Administration Website. Measuring conflict of interest and expertise on FDA advisory committees ERG, Inc., Submitted to Nancy Gieser Office of Policy, Planning and Preparedness US Food and Drug Administration; Submitted by Nyssa Ackerley, John Eyraud, Marisa Mazzotta, Eastern Research Group, Inc., October 27, 2007. Available at http://www.fda.gov/oc/advisory/ergcoireport.pdf, accessed September 19, 2012.
27. US Food and Drug Administration Website. Guidance for Industry Advisory Committee Meetings—Preparation and Public Availability of Information Given to Advisory Committee Members. Available at http://www.fda.gov/downloads/RegulatoryInformation/Guidances/UCM125650.pdf, accessed September 19, 2012.

13 Biologics

Florence Kaltovich

CONTENTS

Introduction 347
Current FDA Oversight 348
Biologics Development 350
    Design 350
    Process Development and Manufacturing 351
    Analytical Development 354
    Nonclinical Evaluation 356
    Clinical Development 358
Biologics Approval Process 359
References 360

INTRODUCTION

With the threats of global infectious disease outbreak and biological weapon terrorism, development of biologics to prevent and treat pathogens requires innovation, production speed, and capacity. The technology needed to monitor and respond to these threats is highly sophisticated and still evolving both in capability and reliability. Advances in biotechnology have resulted in novel vaccines to accelerate production or improve immunogenicity (Table 13.1).[1] The use of biologics to treat noninfectious disease is also expanding, with applications in such areas as oncology and inflammation. In this chapter, we will review regulatory oversight of biologics by the Food and Drug Administration (FDA) from design, through development and licensure, and provide an overview of the regulatory processes that govern biologics entry to interstate commerce in the United States.

In July 2002, the Center for Biologics Evaluation and Review (CBER) celebrated the 100-year anniversary of the passage of the 1902 Biologics Control Act, which gave the public Hygienic Laboratory (a precursor of the FDA/CBER) the authority to regulate biological products and ensure their safety for the American public. To commemorate the occasion, CBER published an eloquent history titled Science and the Regulation of Biological Products (available at http://www.fda.gov/AboutFDA/WhatWeDo/History/ProductRegulation/100YearsofBiologicsRegulation/default.htm). Readers of this chapter are encouraged to read more about CBER's unique history, which includes the amendment to the Food, Drug, and Cosmetic Act (FD&C Act) to regulate biological products and control communicable diseases in 1944 (Public Health Service Act or PHS Act); the transfer of the regulation of biologics from the National Institutes of Health (NIH) to the FDA in 1972; and enactment

TABLE 13.1
Vaccine Technologies

Class	US Licensed Product in Class
Reverse vaccinology (sequence based)—prior to culture	
Immunogen identification technology—cross-reactive epitopes	
Reassortants, reverse genetics	√
Cell culture—scalability, potential use of contract facility	√
Live attenuated—rapid, broad antigenicity, antibodies, and cytotoxic T lymphocytes, but safety	√
Viral/bacterial vectored	
DNA (poor human responses) and "prime/boost": antibodies and cytotoxic T lymphocytes	
Recombinant protein(s): mono- or multiantigen	√
–Insect/animal cells for glycosylation	
–Plant, edible: dosing, environmental issues	
Synthetic or natural peptide	√
Virosome/pseudovirus/liposome—antibodies and cytotoxic T lymphocytes	

of the Public Health Security and Bioterrorism Preparedness and Response Act of 2002, implemented in the wake of the events of September 11, 2001, was to improve the country's ability to prevent and respond to public health emergencies. Provisions include a requirement that the FDA issue regulations to enhance controls over imported and domestically produced commodities. Later, the Project Bioshield Act of 2004 authorized the FDA to expedite its review procedures to enable rapid distribution of treatments as countermeasures to chemical, biological, and nuclear agents that may be used in a terrorist attack against the United States, among other provisions.

CURRENT FDA OVERSIGHT

The primary FDA center for review of biologics and related products is the CBER. Biologics, in contrast to drugs that are chemically synthesized, are derived from living sources (such as humans, animals, and microorganisms), are not easily identified or characterized, and many are manufactured using biotechnology.[2] On June, 30, 2003,[3] the FDA transferred some of the therapeutic biological products from CBER to the Center for Drug Evaluation and Research (CDER), which now has regulatory responsibility and oversight of the transferred products (Table 13.2). The current definition for a biological product, as provided by the FDA,[4] is as follows:

> Biological products, like other drugs, are used for the treatment, prevention or cure of disease in humans. In contrast to chemically synthesized small molecular weight drugs, which have a well-defined structure and can be thoroughly characterized, biological products are generally derived from living material—human, animal, or microorganism—are complex in structure, and thus are usually not fully characterized.

TABLE 13.2
Therapeutic Products Regulated by CBER and CDER

Therapeutic Biological Products Regulated by CBER

- Cellular products, including products composed of human, bacterial, or animal cells (such as pancreatic islet cells for transplantation), or from physical parts of those cells (such as whole cells, cell fragments, or other components intended for use as preventative or therapeutic vaccines)
- Gene therapy products. Human gene therapy/gene transfer is the administration of nucleic acids, viruses, or genetically engineered microorganisms that mediate their effect by transcription and/or translation of the transferred genetic material, and/or by integrating into the host genome Cells may be modified in these ways *ex vivo* for subsequent administration to the recipient, or altered *in vivo* by gene therapy products administered directly to the recipient
- Vaccines (products intended to induce or increase an antigen-specific immune response for prophylactic or therapeutic immunization, regardless of the composition or method of manufacture)
- Allergenic extracts used for the diagnosis and treatment of allergic diseases and allergen patch tests
- Antitoxins, antivenins, and venoms
- Blood, blood components, plasma-derived products (e.g., albumin, immunoglobulins, clotting factors, fibrin sealants, proteinase inhibitors), including recombinant and transgenic versions of plasma derivatives (e.g., clotting factors), blood substitutes, plasma volume expanders, human or animal polyclonal antibody preparations including radiolabeled or conjugated forms, and certain fibrinolytics, such as plasma-derived plasmin, and red cell reagents

Therapeutic Biologic Products Regulated by CDER

- Monoclonal antibodies for *in vivo* use
- Proteins intended for therapeutic use, including cytokines (e.g., interferons), enzymes (e.g., thrombolytics), and other novel proteins, except for those that are specifically assigned to CBER (e.g., vaccines and blood products). This category includes therapeutic proteins derived from plants, animals, or microorganisms and recombinant versions of these products
- Immunomodulators (nonvaccine and nonallergenic products intended to treat disease by inhibiting or modifying a preexisting immune response)
- Growth factors, cytokines, and monoclonal antibodies intended to mobilize, stimulate, decrease, or otherwise alter the production of hematopoietic cells *in vivo*

Section 351 of the PHS Act defines a "biological product as a 'virus, therapeutic serum, toxin, antitoxin, vaccine, blood, blood component or derivative, allergenic product, or analogous product, … applicable to the prevention, treatment, or cure of a disease or condition of human beings.' FDA regulations and policies have established that biological products include blood-derived products, vaccines, *in vivo* diagnostic allergenic products, immunoglobulin products, products containing cells or microorganisms, and most protein products. Biological products subject to the PHS Act also meet the definition of *drugs* under the FD&C Act. Note that hormones such as insulin, glucagon, and human growth hormone are regulated as drugs under the FD&C Act, but not as biological products under the PHS Act."

From the point of view of an organization engaged in biologics development, initial laboratory and animal testing is required to show that investigational use in humans is reasonably safe. Biological products, such as other drugs, can be studied in clinical trials in humans under an investigational new drug (IND) application in accordance with the regulations at 21 CFR 312. Hence, this holds true for products regulated by both centers, CBER and CDER. Regulatory differences arise when the data generated by advanced clinical studies demonstrate that the product is safe and effective for its intended use. A new drug application (NDA) is used for drugs subject to the drug approval provisions of the FD&C Act, while a biologics license application (BLA) is required for biological products subject to licensure under the PHS Act. A biological license is issued to distribute a biologic into interstate commerce.

The remainder of this chapter focuses on the generalities needed to plan and then develop a biomanufacturing process to produce a biological product of high quality, for example, a vaccine, under CBER regulatory authority, or a monoclonal antibody, under CDER regulatory authority. All other biologic products, including combination products (a biologic/device or a biologic/drug combination), will not be discussed, except to note that the choice of centers for regulation is product dependent, and relies on the identification of primary mode of action.

BIOLOGICS DEVELOPMENT

Design

The objective of biomanufacturing development is to design a product that has attributes[5]—strength, identity, purity, potency, and safety—for its intended use and can be consistently manufactured throughout the product's life cycle. Each biomanufactured or biotechnology product is unique and is, or will be, produced using well-known commercial processes and specific analytical methods developed for the particular product or class of products. There are three stages of biomanufacture, namely, (1) upstream processing; (2) downstream processing; and (3) formulation, fill, and finish. In upstream processing, the product is produced from raw materials using process technologies, such as cell culture, fermentation, and synthesis. Downstream processing involves purification of the desired product by separation from impurities and contaminants. The output of downstream processing is the bulk drug substance (DS) and is the active pharmaceutical ingredient (API), the entity with the therapeutic activity. The output of the formulation, fill, and finish stage is the final drug product (DP) in a container/closure (vial or syringe) ready for human administration.

The International Committee on Harmonization (ICH) published a pharmaceutical development guidance (Q8 Pharmaceutical Development) in May 2006 and, more recently, revised the guidance, "Q8(R1) Pharmaceutical Development," to describe the principles of quality by design (QbD) and to further clarify the key concepts of QbD. The Q8(R1) guidance gives examples

Biologics

of a systematic process development plan (PDP) and definitions including the following elements:

A. Quality target product profile—forms the basis of design for the development of a product
B. Critical quality attributes (CQAs)—a physical, chemical, biological, or microbiological property or characteristic that should be within an appropriate limit, range, or distribution to ensure the desired product quality
C. Risk assessment—process used in quality risk management[6] that can aid in identifying which material attributes and process parameters potentially have an effect on product CQAs
D. Design space—the relationship between the process inputs and the CQAs
E. Control strategy—ensures that a product of required quality will be produced consistently
F. Product life cycle management and continual improvement—evaluation of innovative approaches to improve product quality throughout the product life cycle[7]

CBER and CDER adopted this guidance in June 2009 and have strongly encouraged biologic product developers to use these elements in the PDPs and during the product life cycle. Further clarification to facilitate understanding and use of Q8, Q9, and Q10 was published by the FDA in July 2012 in "Guidance for Industry, Q8, Q9, & Q10 Questions and Answers."

PROCESS DEVELOPMENT AND MANUFACTURING

Biomanufacture involves various processes including gene cloning, development of an expression vector, and production of cell banks. The first step is to isolate and characterize the gene, clone the gene into an expression vector, transform the host cell with the vector, clone the host cell, select and expend the cell to research seed bank, and then produce the master cell bank (MCB). The manufacturer will need to submit the following information to the FDA:

- Source of DNA and gene isolated, sequence, and molecular weight
- Description of the methods used to identify and clone the gene of interest
- A detailed description of the vector into which the gene is inserted. This will include a map of the plasmid used, identifying other genes carried; other active sequences (e.g., promoters and enhancers); and selection markers employed such as a gene conferring antibiotic resistance
- The sequence and plasmid map of the cloned vector
- Description, history, and purity of the host cell to be used
- A description of the techniques used to transfect the plasmid into the host (production) cell. Methods for amplification of the copy number of

the plasmid, stability, expression gene purity, and a description of how the clones were selected
- Description of the cloned host cell, purity, copy number, and gene expression
- Description of the research cell bank, purity, copy number, gene expression, viability, and sterility

From the research cell bank, the manufacturer will establish an MCB, from which working cell banks (WCBs) will ultimately be produced. The banks are typically kept frozen in ampoules under defined conditions (e.g., in the vapor phase of liquid nitrogen). Each ampoule contains a standard cell density from the single clone. The history of the cells laid down for the MCB should be described, including the copy number or generation number of the cells when frozen, as well as any selection markers used. While there are a number of cell lines suitable for production, the manufacturer should carefully consider the needs of the protein product and the regulatory history of the host line when selecting a cell-line candidate. For example, bacteria, notably *Escherichia coli*, and yeasts, such as *Pichia pastoris,* are common choices. Mammalian cell lines include Chinese hamster ovary (CHO) cells (used for over 50 years), Vero, MDCK, human embryonic kidney (HEK-293), baby hamster kidney (BHK) Per C6, and NSO.

The MCB will need to be tested for incorporation and maintenance of the expression system. In broad terms, the expression system may be either incorporated to the host cell genome or maintained extrachromosomally. In either case, the stability of the transfectant will need to be demonstrated, as well as the fidelity of the copy to the original construct. It is possible that the newly incorporated gene can mutate in the same manner as any other gene. The manufacturer will therefore need to either reclone or isolate the gene from the MCB, sequence it, and compare the results with those from the original construct used in transfection. The nucleic acid sequences should be identical. Finally, the limits of *in vitro* age must be established to define an acceptable generation number for production use. This is verified by comparison of end-of-production cells to the original MCB, as noted earlier.

Having established the production cell line, the manufacturer will need to describe the conditions under which the cells are grown for production. As with a small molecule, the detail to which this is described will increase with the product's proceeding through the development process. The media used to grow the cells will need to be described fully, and it is advisable for the manufacturer to use fully defined media where possible. Cell lines of mammalian origin may harbor adventitious viruses that may be harmful to humans. For purposes of safety, it is advisable to limit this theoretical exposure to the cell line itself. Any human- or animal-derived proteins or components used in the growth media—indeed, any such components used in any stage of the production process—should be fully described and their sources verified. In practice, the use of such components should be minimized to provide the highest theoretical assurance of safety from adventitious viruses or other potentially infective agents. To support the many

Biologics

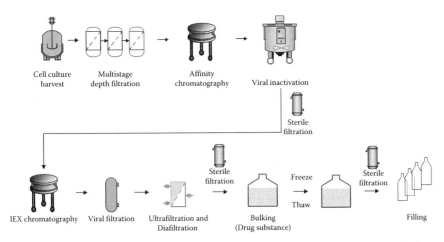

FIGURE 13.1 Downstream purification equipment and processes routinely used during recombinant product purification resulting in filled drug product.

types of vaccines in development, the FDA has published specific guidance, such as "Considerations for Plasmid DNA Vaccines for Infectious Disease Indications and Characterization and Qualification of Cell Substrates and Other Biological Materials Used in the Production of Viral Vaccines" as well as others.

The removal of adventitious viruses from recombinant products is complex, and viral safety must be assessed early in process development. A conference on detection and mitigation strategies was held in December 2010, and proceedings were published.[8] The manufacturer will need to show that the purification techniques used would effectively remove viruses that may be present in the cell line used and the raw materials employed or introduced to the production from an external source (e.g., production personnel). Validation of viral removal will need to be performed prior to licensure, but should be considered early in the program in selecting or developing purification processes. Figure 13.1 illustrates downstream purification equipment and processes routinely used during recombinant product purification.

An important aspect of the purification process is the removal of nonproduct proteins and other biomolecules. Proteins of host cell origin may be copurified with the protein of interest, and acceptable limits for these must be established. The composition of host cell proteins will be specific to a given production cell line and the conditions under which it is grown. Therefore, specific methods will be developed over time to characterize and quantify host cell protein in a given product. Moreover, residual DNA from the production cell line may be present in the final product; specific purification steps are therefore employed to minimize the presence of this contaminant.

We have so far been concerned with what is not in the final product—viruses, host cell contaminants—but we cannot lose sight of what must be in the final product, that is, a fully active biologic of sufficient purity and potency. Biologics tend to be large, complex molecules with significant secondary, tertiary, and often quaternary structures. The process must yield a final material having the correct conformational state and any modifications necessary to its biological activity. Protein denatured or otherwise inactivated in processing may be copurified with fully active material, and the presence of such impurities must be understood. Additional purification methods may be included to exclude such material or current methods may be refined to limit them. In all cases, the process will need to be characterized using CQAs at each stage of production to understand the status of the active component.

Select Guidance Documents:

> International Conference on Harmonization (ICH) Q5B: Quality of Biotechnological Products: Analysis of the Expression Construct in Cells Used for Production of rDNA Derived Protein Products, November 30, 1995.
> ICH Q5D: Derivation and Characterisation of Cell Substrates Used for Production of Biotechnological Products, July 16, 1997.
> FDA: Content and Format of Chemistry, Manufacturing, and Controls Information and Establishment Description Information for a Vaccine or Related Product, January 1999.
> FDA: Compliance Program Guidance Manual Chapter 45—Biological Drug Product Development, December 1, 2004.
> FDA: For the Submission of Chemistry, Manufacturing, and Controls and Establishment Description Information for Human Plasma-Derived Biological Products, Animal Plasma or Serum-Derived Products, February 1999.

ANALYTICAL DEVELOPMENT

Analytical methods measure the product's attributes. For biologics, multiple methods may need to be employed to characterize a given attribute of the product. Purity may be defined by several methods, each of which contributes information on a different molecular characteristic of the product. Size-exclusion high-performance liquid chromatography (HPLC) may provide data on the molecular weight of the protein, and will demonstrate the presence of higher molecular weight species (such as aggregates) or lower weight species (such as fragments). Similarly, reverse-phase HPLC and ion-exchange HPLC will expose variations in hydrophobicity and charge, respectively, and may demonstrate the presence of isoforms or degradants in the product. No single method will capture the full heterogeneity of the compound, and therefore several may be employed to fully characterize a given attribute of the protein. The data will be assessed as a whole, and will provide a composite picture of the product in development. The rationale for method selection—what information is provided and the relevance thereof—will be submitted to the FDA.

TABLE 13.3
Example of Methods for Recombinant Protein Drug Substance Release

Attribute	Analytical Method	Reference
Appearance	Visual inspection of DS in clear glass tube	SOP
Safety	Microbial limits test	USP <61>
Safety	pH	SOP
Identity	N-terminal sequence	SOP
Identity	SDS-PAGE	SOP
Safety	Endotoxin, Gel Clot LAL	USP <85>
Purity	SDS-PAGE	SOP
Purity	HPLC	SOP
Purity	Aggregates by size exclusion chromotography	SOP
Purity	Peptide map	SOP
Purity	Host cell protein, ELISA	SOP
Purity	Host cell DNA	SOP
Purity	Aggregated protein, light scatter	SOP
Strength	BCA, concentration and amount of protein	SOP
Potency	Receptor binding	SOP
Potency	Viability of cultured cells	SOP

HPLC, high-performance liquid chromatography; SOP, standard operating procedure; USP, US Pharmacopeia

Further complicating the analytical picture for a biologic is the inherent variability of the methods used. As potency is often a function of biologic response in a living system, cell-based assays may be developed. Acceptable specifications may therefore be in the range of 50% to 150% of a standard response, a function of the inherent variability of living systems. This argues further for the use of orthogonal methodology to describe specific attributes under consideration. Table 13.3 outlines a number of methods used in the characterization of a recombinant protein DS.

Once the DS has been fully characterized and released, the bulk material is formulated, filled into the final container, such as a vial or syringe, packaged, and labeled. Unique to vaccines is the addition of adjuvant to the bulk material. The attributes of the final DP are assessed, similar to that performed on the DS, with a focus on safety and intended use. Most important is that the final product is sterile (USP <71>) and free of endotoxin or other toxic substances. In the United States, the General Safety Test, as described in 21 CFR 610.11, is required for release. Additional methods that may be used for characterization and release testing include the following:

- Osmolality
- Electrophoretic methods, such as immunoelectrophoresis
- Western blot

- Host cell RNA
- Carbohydrate
- Isoform characterization
- Amino acid composition
- Mass spectroscopy—time of flight
- Electrospray mass spectroscopy
- Intrinsic fluorescence
- Liquid chromatography—mass spectroscopy
- Neutralization assays

To demonstrate potency, *in vivo* and *in vitro* biological testing, or bioassays, must be described and performed on relevant lots of DS. Due to the various types of these bioassays, the sensitivity, specificity, and variability of the assay should be evaluated to set the acceptance limits for the assay.

Select Guidance Documents:

FDA: Analytical Procedures and Methods Validation. Chemistry, Manufacturing, and Controls Documentation, August 2000.

ICH Q6B: Specifications: Test Procedures and Acceptance Criteria for Biotechnological/Biological Products.

ICH Q5C: Quality of Biotechnological Products: Stability Testing of Biotechnological/Biological Products, September 1995.

NONCLINICAL EVALUATION

The requirements for nonclinical evaluation of drugs naturally extend to biologics. Before human testing may be performed, adequate safety in the predicted dosing range and duration must be demonstrated in model species. However, there exist unique challenges in the design and execution of nonclinical studies for a biologic. These challenges result from both the variety of molecules that are considered biologics (a gene therapy nonclinical program will by its nature differ from that for a peptide or a vaccine) and the variety of biological effects possible in multiple species. No single toxicology program can be applied to all drugs, and this is magnified in the case of biologics.

ICH guideline S6[9] offers the following as elements to consider in the design of nonclinical studies:

- Selection of the relevant animal species
- Study design—dose, route of administration, and treatment regimen
- Immunogenicity
- Stability of the test material under the conditions of use

Primary consideration must be given to the nature of the biological effect and the validity or availability of relevant model systems. For many viral diseases and certain malignancies, the only relevant species for a human therapeutic is humans. Animal models of disease may be incomplete or may not offer direct comparison to the human. For example, certain cytokines may have profound effects on murine models of cancer, which have not yet been shown to be fully predictive of human response.[9] Basic proof of concept data may be provided by *in vitro* systems alone, with little or no whole animal experience that may be predictive of a human response. Some molecules, such as cytokines, trigger a complex series of physiologic events, and these events may differ in their specifics among species. Cell culture data are extremely useful in elucidating the molecular events at the level of a single cell type, but may not be predictive of a whole animal response. It is possible to clone and produce the animal analogue of the protein of interest and use this in pharmacology or mechanistic studies; however, consideration must be given to both the homology of the protein and the similarity of biological effect in the human and animal species. Alternatively, cell lines from the target nonclinical species may be screened for biological activity of the protein to support or discount the validity of the species chosen. The effects of age on the physiological state of the animal species being tested will also be considered as they might impact the usefulness of the species chosen. Differences may be seen in the pharmacokinetics of a particular molecule as a function of the developmental status of the animal to which the biologic is given. The approach taken for these studies and the rationale behind the approach should be described in the IND application.

Safety assessments are similarly clouded. As always, consideration must be given to the product being used and its impurity profile at the given stage of development; however, impurities are often product related and may be active. Of great concern is the potential for a given product to elicit an immune response in humans, a response that cannot be predicted confidently through nonhuman studies. By definition, a human protein administered to a nonhuman is a foreign protein, and would be expected to elicit an immune response (here, the assumption is <100% homology among the species). Moreover, a human protein produced in nonhuman mammalian cells may carry posttranslational modifications that differ from the endogenously produced protein because of inherent differences in cellular processing machinery. Depending on the extent to which the molecule is understood in terms of its activity, and depending on the nature and extent of anticipated human dosing, it may not be possible to maintain test animals without an immune response that renders null the data from long studies. Often complicating matters is the long lead time associated with assay development for quantifying antibody responses in animals. Another factor impacting the relative immunogenicity of a compound will be the route of administration. In general, intravenous delivery is associated with less immunogenic potential than subcutaneous or intramuscular delivery. As we have seen, the effects of formulation excipients or the presence of contaminants may also play a role in the development or absence of an immune response as well.

Specific nonclinical safety studies may be required as a function of the mechanism of action of the product studied. Monoclonal antibodies, by their nature, bind a specific epitope on a target. However, antibodies may bind to other epitopes that offer a similar conformational or structural presentation. It will therefore be necessary to perform tissue cross-reactivity studies *in vitro* to demonstrate that the antibody does not bind to tissues unrelated to the pharmacological effect of the antibody.

Carcinogenicity studies are not typically required for biologics, but this must be justified on the basis of knowledge of the specific function of the molecule under development. If the product is associated with hyperplasia, it may be necessary to perform such studies. In general, long-term studies are difficult because of the likelihood of an immune response. Any product likely to be given to women of childbearing potential will require some degree of reproductive toxicology studies; however, this will vary on the basis of both the likely extent of dosing and the specific disease being treated.

Finally, the stability of the test article for use in nonclinical studies must be described. Early in development, test article will often be supplied as a nonformulated product in solution. Studies should be performed to demonstrate that the product is stable and active in the period of use, and dosing materials should be analyzed to confirm that the animals receive the appropriate exposure. As the program matures, and as the formulation of the biologic is refined, these data will provide a foundation upon which to assess the nonclinical effect of the material over time.

In May 2002, the "Animal Rule" was published which allows appropriate studies in animals in certain cases to provide substantial evidence of effectiveness of new drug and biological products used to reduce or prevent the toxicity of chemical, biological, radiological, or nuclear substances (21 CFR Part 601, Subpart H). This rule applies when definitive human efficacy studies are not ethical or feasible. Postmarketing studies of the biological are generally required after licensure.

Select Guidance Documents:

ICH S6 Preclinical Safety Evaluation of Biotechnology-Derived Pharmaceutical.
Guidance for Industry: S6 Addendum to Preclinical Safety Evaluation of Biotechnology-Derived Pharmaceutical, FDA, May 2012.
Final Rule, New Drug and Biological Drug Products: Evidence Needed to Demonstrate Effectiveness of New Drugs When Human Efficacy Studies Are Not Ethical or Feasible, 67 FR 37988, May 31, 2002.

CLINICAL DEVELOPMENT

The clinical indication for which a biologic is developed will drive the design and execution of the clinical trials aimed at demonstrating safety and efficacy. The sponsor will need to demonstrate the safety and efficacy of the biologic, and will need to submit the proposed clinical plan, with statistical considerations where appropriate, to the FDA. The biological activity of the product, the limits on measurement and assessing efficacy (as through the use of surrogate markers), and the availability of patients suitable for clinical trials must all be weighed in designing these trials.

The potential for immunogenicity for a biologic can be both desirable and of great concern. Baseline, preexposure samples must be drawn from patients exposed to the biologic and later, after administration, immunogenic responses toward the targeted disease evaluated. The potential for anaphylaxis or anaphylactoid reactions must be considered, and appropriate steps to ameliorate these reactions must be taken.

Of particular concern is the unpredictability of a human response to a biologic despite nonclinical studies that may predict no adverse effect level doses in man. The potential for unforeseen side effects by biologics must be given careful consideration, and changes to routine approaches to initial dosing, dose escalation, and patient versus volunteer selection may be indicated. The rationale for such selection will be submitted as part of an initial IND and subsequent amendments.

Select Guidance Documents:

> Guidance for Industry: E8 General Considerations for Clinical Trials, FDA, December 1997.
> Guidance for Industry: Providing Clinical Evidence of Effectiveness for Human Drug and Biological Products, FDA, May 1998.

BIOLOGICS APPROVAL PROCESS

For products of biological origin defined for regulatory purposes as a biologic (e.g., gene therapy and vaccines), the sponsor will submit a BLA and submit the Form FDA 354h; the same form used to submit an NDA. The FDA has moved to a common technical document (CTD) format, and mandating that new submissions be filed electronically (eCTD). Review timelines by the FDA are dictated by the most recent amendment of the Prescription Drug User Fee Act (PDUFA, now PDUFA V), which are described elsewhere in this book. The extent to which these timelines are met is dependent on the adequacy and completeness of the data submitted to the FDA. An early and open dialog with the FDA, coupled with excellence in execution of agreed-upon development, will ensure that a safe and efficacious biological is brought to market in a timely manner.

Under the FDA Amendments Act (FDAAA) of 2007, Section 524 was added to the FD&C Act enabling the FDA to publish guidance documents on regulation of vaccines for global infectious diseases and accelerating approval of these vaccines: General Principles for Development of Vaccines to Protect Against Global Infectious Diseases (December 2011), and, Tropical Disease Priority Review Vouchers (October 2008). The "General Principles" guidance clarified that FDA can license vaccines for diseases that are not endemic to the United States using the same regulatory pathway as other vaccines and that sponsors may submit data from clinical trials conducted outside the United States to support licensure. The FDA can grant priority review of these applications for treatment and prevention of specified tropical diseases including malaria, tuberculosis, cholera, and "any other infectious disease for which there is no significant market in developed nations and that disproportionally affects poor and marginalized populations, designated by the Secretary." To incentivize development and licensure of these vaccines, the priority review voucher (PRV) [Section 524(a)(2)] would be granted to the sponsor of a "tropical disease" product application at the time of approval; the PRV entitles the holder to qualify for a six-month priority FDA review of another application that would otherwise be reviewed under the FDA's standard 10-month review clock. Economic models have predicted that this faster time to market could be worth between US$50 million and US$300 million.[9] To date, the FDA has granted only a single tropical disease PRV—in connection with the April 2009 approval of Novartis' NDA No. 022268 for Coartem (artemether; lumefantrine) for the treatment of acute, uncomplicated malarial infections in adults and children weighing at least 5 kg. Novartis redeemed the PRV in connection with the submission of an application for Ilaris (canakinumab) for the treatment of gouty arthritis attacks in certain patients. After the six-month priority review, the FDA ultimately denied approval of Ilaris due to insufficient data to evaluate the overall safety profile.

REFERENCES

1. Goodman, J. L. How fast can a new vaccine for an emerging respiratory virus be developed and available for use? *CDC ICEID*, March 22, 2006.
2. http://www.fda.gov/AboutFDA/CentersOffices/OfficeofMedicalProductsandTobacco/CBER/ucm123340.htm
3. http://www.fda.gov/AboutFDA/CentersOffices/OfficeofMedicalProductsandTobacco/CBER/ucm133463.htm
4. http://www.fda.gov/Drugs/DevelopmentApprovalProcess/HowDrugsareDevelopedandApproved/ApprovalApplications/TherapeuticBiologicApplications/ucm113522.htm
5. ICH Q6A: Specifications: Test Procedures and Acceptance Criteria for New Drug Substances and New Drug Products: Chemical Substances.
6. ICH Q9: Quality Risk Management.
7. ICH Q10: Pharmaceutical Quality System.
8. Khan, A. S., Hughes, P., and Wiebe, M. Introduction to workshop: Adventitious viruses in biologics—detection and mitigation strategies. *PDA Journal of Pharmaceutical Science and Technology*, 2011, 65(6): 544–546.
9. Noor, W. Placing value on FDA's priority review vouchers. *In Vivo: The Business Medicine Report*, 2009, 27(8): 1–8.

14 Regulation of Combination Products in the United States

John Barlow Weiner, Esq.

CONTENTS

Introduction 361
Combination Product Classification and Assignment 362
    Classification: Is the Product a Combination Product? 362
    Assignment: Which Center Has the Lead for a Combination Product? 364
    Obtaining an Agency Classification or Assignment Determination 366
Regulation of Combination Products 367
    Premarket Regulation: Marketing Authorization Requirements and Processes 367
        Investigational and Marketing Submissions 368
        Marketing Authorization Standards for Combination Products 369
        Intercenter Coordination and Sponsor–FDA Interaction for Premarket Review of Combination Products 370
    Postmarket Regulation of Combination Products 372
What Is OCP and What Does It Do? 373
Recent and Upcoming Developments 374
Acknowledgments 374
Notes 374

INTRODUCTION

In the United States, combination products comprise two or more different types of medical products, either a biological product (or biologic) and device, biologic and drug, device and drug, or all three combined.[1] These different types of medical products are regulated under distinct regulatory programs by different components of the US Food and Drug Administration (FDA). As a consequence, combination products raise particular regulatory questions and challenges, for example, with regard to regulatory standards and procedures, as well as coordination within the Agency and between the Agency and sponsors.

As this chapter explains, combination products are a separate class of medical products in the United States, distinct from drugs, devices, and biological products. The regulatory requirements, processes, and expectations for combination products reflect the intent to address the issues raised by each constituent part (drug, device, or biologic), including those issues arising from their combined use in the combination product. The essential goal is to ensure the safety and effectiveness of the combination product in light of the constituent parts of which it is comprised and how they interrelate and may interact with one another.

A particular challenge is that the statutory authorities implemented by the FDA expressly state whether they apply to drugs, devices, and/or biological products, but do not state whether or how they apply to combination products. Further, to date the FDA has promulgated only two regulations expressly stating how these various statutory provisions apply to combination products, and is developing one other such rule as discussed later. The Agency has published some guidance expressly for combination products and additional guidance for combination products is being developed and finalized, as discussed later.

Looking to the statutory authorities, regulations, and guidance that currently exist, this chapter discusses the following:

- Combination product classification and assignment
- Regulation of combination products
- The role of the Office of Combination Products (OCP)
- Recent and upcoming developments

In discussing these issues, the chapter addresses some practical considerations to help manage regulatory challenges for this class of products.[2]

For further information on combination products, the OCP's Website offers a wealth of reference materials (http://www.fda.gov/CombinationProducts/default.htm). In addition, as discussed later, the OCP serves as a contact point to obtain assistance from the Agency as needed to address regulatory issues for combination products.

COMBINATION PRODUCT CLASSIFICATION AND ASSIGNMENT

This section addresses considerations for determining whether a product is a combination product; how the Agency component with primary jurisdiction for a combination product is determined; and how to obtain an agency determination of the classification or assignment of a product.

CLASSIFICATION: IS THE PRODUCT A COMBINATION PRODUCT?

An initial question that may arise in assessing the regulatory pathway for a product is whether it is a combination product. The Federal Food, Drug, and Cosmetic (FD&C) Act does not define the term "combination product," but agency regulations do. Essentially, they state that combination products include the following three categories of products:

1. Products that physically or chemically combine different types of medical products (single-entity combination products), such as prefilled syringes and drug-eluting stents
2. Products in which different types of medical products are copackaged (copackaged combination products), such as a first aid kit or surgical procedure kit that includes both drugs and devices
3. Certain products comprised of different types of medical products that are separately marketed but related through their labeling (cross-labeled combination products), as might be the case for a drug and a laser that activates it[3]

Products falling within the first two categories (single-entity or copackaged combination product) generally are fairly readily identified. For example, they would include products comprised of one article that is clearly a drug or biologic, and another that serves merely to deliver the first (i.e., a device) that is either filled with the drug or biologic (e.g., a prefilled syringe) or copackaged with it as a kit (e.g., a first aid or surgical kit containing both drugs and devices). However, if a product consists of a liquid or gel that includes an active ingredient, for example, the answer may not be as obvious. Are the liquids or gels devices, or are they merely inactive drug ingredients?[4] Similarly, even if it is clear that the product includes two distinct medical products, it may not be clear how one or the other of those products should be classified (e.g., is one a drug or a biologic; is the other a device or a biologic?).

Analysis can involve consideration of the composition of the articles in a product and the mechanism(s) by which the product accomplishes its intended effect(s), as well as parsing of the statutory definitions for biologic, device, and/or drug.[5] It can also involve comparing the product to existing products regulated by the FDA and reviewing prior agency classification determinations. Comparing products to existing ones and reviewing prior agency classification determinations can be challenging, however. Product classification determinations are made with regard to investigational products. As a result, these determinations may remain confidential unless the product receives marketing authorization. In addition, classification often depends on product-specific characteristics that may not be readily apparent and may vary among otherwise similar products.

The Agency has published jurisdictional updates as an aid to determining the classification and assignment of products (available at http://www.fda.gov/CombinationProducts/JurisdictionalInformation/JurisdictionalUpdates/default.htm). It has also published intercenter agreements that address these issues (available at http://www.fda.gov/CombinationProducts/JurisdictionalInformation/IntercenterAgreements/default.htm). The Agency has indicated, however, that these agreements are incomplete and/or out of date and should not be read in isolation. The Agency has noted that new laws, regulations, and guidance now apply; that new products have come into existence; and that the specific characteristics of products can lead (and have led) to classifications and assignments that differ from those that the agreements might indicate apply to a class of products, of which that specific product might appear to be a member.[6] Similar considerations may apply to the jurisdictional updates.[7]

Members of the third category of combination products, cross-labeled combination products, can be difficult to identify because of the complexity of the regulatory language, which includes both objective elements (Are both products needed for the intended use?) and subjective ones (Is one product intended to be used only with the other?). The FDA has announced its intention to develop guidance to clarify under what circumstances separately marketed products constitute cross-labeled combination products under the Agency's regulations.

ASSIGNMENT: WHICH CENTER HAS THE LEAD FOR A COMBINATION PRODUCT?

Medical products for humans are regulated by three "centers" within the FDA: the Center for Biologics Evaluation and Research (CBER), Center for Devices and Radiological Health (CDRH), and the Center for Drugs Evaluation and Research (CDER). The center with primary regulatory responsibility (the "lead") for a combination product is determined based upon the "primary mode of action" (PMOA) of that product.[8]

There are three types of modes of action: (1) biologic, (2) device, and (3) drug. These modes of action are contributed, respectively, by the biologic(s), device(s), or drug(s) included in the combination product.[9] "Primary mode of action" is not defined by statute. The FDA promulgated regulations in 2005 to define the term and address how to determine the PMOA of a combination product.[10] In accordance with that rule, since that date, PMOA has been defined as the mode of action that provides the "greatest contribution to the overall therapeutic effects" of the combination product, with "therapeutic" effect or action defined to include any effect or action that is "intended to diagnose, cure, mitigate, treat, or prevent disease, or affect the structure or any function of the body."[11] Accordingly, CBER would generally have the lead for a combination product if it has a biologic PMOA, CDRH if it has a device PMOA, and CDER if it has a drug PMOA.[12]

For some types of combination products, the constituent part that contributes the PMOA is well established. For example, if the combination product consists of a drug or biologic and a device that either delivers or activates the drug/biologic and does not otherwise contribute to the therapeutic effect, the Agency consistently ascribes PMOA to the drug/biologic because that constituent part is the one that treats the disease or condition once delivered or activated. Accordingly, the drug or biologic in a prefilled syringe would be considered to provide the PMOA, for example.

For other types of combination products, a case-by-case analysis may be needed to determine PMOA. PMOA can vary among similar combination products. For example, one drug–device combination product indicated to accelerate wound healing might include a higher strength of a drug or a drug that is more powerful than that included in another combination product that has the same intended use. As a result, the device may provide the PMOA for the combination product with the weaker drug, while the drug might provide the PMOA for the combination product that includes the stronger drug. Similarly, two combination products that both include the same or similar drug and device constituent parts may have different indications, and the respective contributions of those constituent parts may differ depending on the indication. One indication may be for a shorter-term use during which structural support provided by the device might be primary, for example, while for the other use, longer-term therapeutic effects of the drug might be more significant. As a result, the device may provide the PMOA for one indication, and the drug may provide it for the other.

If possible, PMOA is determined directly if the Agency can determine with reasonable certainty which constituent part appears to contribute the most to the product's intended therapeutic effects.[13] In some cases, however, sufficient data may not exist, may not be generated by the sponsor, or may not be possible to generate (e.g., because the product has two distinct indications with one constituent part achieving one of these effects, and the second achieving the other). In such instances, which center will have the lead is determined through a two-step algorithm established by regulation. The first step is to determine whether one of the centers is already regulating a combination product raising similar questions of safety and effectiveness. If so, the product would be assigned to that center. If not, the second step asks which center has the greatest expertise with respect to the most significant questions of safety or effectiveness raised by the combination product, and that center would have the lead for it.[14]

In short, under the 2005 PMOA rule, assignment of combination products is based, if possible, on determination of which constituent part contributes most to the product's overall effects if that constituent part can be identified, regardless of which center may have the lead for similar combination products or which are the most challenging regulatory issues raised by the product. This means that past assignments for combination products may not be a reliable indicator of the assignment, and similar combination products may be assigned to different centers. For example, scientific understanding of how products achieve their intended effects may shift over time, and this may result in a different PMOA determination than was made for a similar product in the past. Also, the Agency may have applied a different standard in determining the PMOA for a product prior to promulgation of the 2005 rule. This also means that the center assigned the product may not be the center with the most expertise with respect to the most challenging regulatory issues raised by that combination product. As discussed later, however, the centers and other agency components work together to regulate combination products to ensure appropriate consideration of all regulatory issues, including those associated with each constituent part and how the constituent parts interact and interrelate. Sponsors can also help ensure that appropriate offices and personnel from the nonlead center are playing an appropriate, timely role in the review and regulation of the combination product, as is also discussed later.

Obtaining an Agency Classification or Assignment Determination

In many cases, it may be clear that a product is a combination product and what its center assignment is. If the classification or assignment is unclear or in dispute, however, a sponsor can seek an informal view or submit a request for designation (RFD) to the OCP for a formal determination. The merits of each approach are discussed later.

Submitting a marketing application or clearance notification without seeking agency input poses some risk if the classification or assignment for the product is unclear. It can mean incurring potentially substantial costs only to have the Agency conclude that the product has a different classification or center assignment, and a different type of submission should be made to another center as a result. Or the Agency may be unsure of the classification or assignment and determine that it needs to be vetted with the OCP. Further, if the Agency is unsure as to the appropriate classification or assignment of the product, the submission can be put on hold while the issue is resolved, delaying product development or review.[15] Seeking input from the center to which the sponsor believes the product should be assigned poses less risk but can lead to delay if the center disagrees with the sponsor or is uncertain of the classification or assignment for the product. In both cases, the center can be expected to contact the OCP or direct the sponsor to contact the OCP regarding the classification or assignment.

Contacting the OCP is generally the most efficient, cost-effective means to determine the classification and/or assignment for a product if either is unclear. Consistent with its statutory duty to classify and assign combination products, the OCP also is the office delegated authority to classify products as drugs, devices, or biological products, as well as to classify combination products, and assign such products to centers.[16] If the OCP has a ready answer, an informal inquiry can be the fastest way to confirm Agency thinking. This might be the case, for example, if the OCP has had occasion to consider the same or similar scientific, technical, or legal issues in addressing the classification or assignment of a similar product or products. If the OCP does not have a ready answer, an informal query still can help the requester to determine what information to provide in an RFD to ensure an informed determination by the OCP.

As indicated earlier, the RFD process is a mechanism to obtain a formal classification or assignment determination from the Agency. An FDA response to an RFD is binding and can only be changed with the consent of the RFD submitter or for public health reasons based on scientific evidence. Also, FDA must respond to an RFD within 60 days or the classification or assignment recommended in the RFD applies with the same binding effect.[17]

Agency regulations stipulate the required elements of an RFD and establish a 15-page size limit.[18] In addition, the OCP has published a guidance titled "How to Write an RFD," which elaborates upon the information that should be provided in an RFD. Important elements include a detailed description of the product, its intended use(s), the mechanisms by which it achieves its intended use(s), and a recommendation for the lead center assignment based on the PMOA for the product or the algorithm if PMOA cannot be demonstrated directly. This guidance is available on the OCP's Website (http://www.fda.gov/RegulatoryInformation/Guidances/ucm126053.htm).

OCP classification and assignment determinations reflect not only the judgment of the OCP but also the expertise and views of the interested centers. For example, when RFDs are submitted, current practice is for the OCP to circulate them to the jurisdiction officers for the interested centers (e.g., to CDER and CDRH for a product that may be a drug–device combination product) for center input. The jurisdiction officers share the RFDs with expert staff in those centers, and the centers offer their recommendations for how the product should be classified and assigned, to inform the OCP's determination. In some cases, the OCP also seeks the input of FDA's Office of the Chief Counsel before issuing a determination.

If the RFD submitter disagrees with the OCP's determination, it may submit a request for reconsideration of the RFD to the OCP within 15 days, to which the OCP must reply within 15 days.[19] The submitter may also appeal the OCP's determination in accordance with 21 CFR 10.75. Both of these are mechanisms for the Agency to reconsider its decision in light of the information before it at the time it made the decision. However, if an RFD submitter believes that the OCP would have made a different decision if it had had additional information before it, the submitter can submit a new RFD presenting that information.

REGULATION OF COMBINATION PRODUCTS

This section discusses regulatory standards and procedures for combination products, both in the premarket and postmarket settings. As discussed later, the marketing authorization pathways, regulatory standards, and procedures for combination products are essentially those for drugs, devices, and biological products. The principal issues for combination products concern how to ensure that all of the regulatory issues raised by a combination product are appropriately addressed, regardless of the regulatory pathway by which it may enter the Agency. Associated challenges include coordination across centers and between sponsors and the Agency. This section focuses on regulatory issues for combination products in particular. It does not delve into the details of drug, device, and biological product marketing authorization pathways, review standards, or procedures, except as needed to elucidate considerations for combination products.

PREMARKET REGULATION: MARKETING AUTHORIZATION REQUIREMENTS AND PROCESSES

The PMOA standard determines which center will have the lead for regulation of a combination product. However, it does not expressly address what types of investigational and marketing authorization submissions should be made for the product, nor does it expressly address what review standards or data requirements should apply for combination products or whether these standards should vary based upon which center has the lead. It also does not establish precisely how the lead and nonlead centers should coordinate or how sponsors should interact with either. However, as will be discussed further, statutory language and agency policies, statements, and practice offer some insight.

Investigational and Marketing Submissions

The PMOA standard expressly addresses only center assignment. However, CBER, CDER, and CDRH do not currently have the delegated authority to review all marketing authorization types. Specifically, CDER has the authority to review biologics licensing applications (BLAs), new drug applications (NDAs), Abbreviated NDAs (ANDAs), and investigational new drug applications (INDs); CDRH has the authority to review PMAs, 510(k)s, Humanitarian Device Exceptions (HDEs), and IDEs, and CBER has the authority to review all of these types of submissions.[20] Further, while the FDA has not stated that the submission types associated with the constituent part that provides the PMOA must or may always be used, they typically have been.

Accordingly, key questions to consider in evaluating what investigational and marketing authorization submissions to make for combination products are (1) which constituent part (drug, device, biological product) provides the PMOA and (2) which submissions type(s) associated with that constituent part are available for the combination product. For example, if a drug–device combination product has a device PMOA, and it includes a drug that is a new molecular entity or the combination product has an indication for which the drug has not been approved, a PMA could be submitted but not a 510(k) because safety and efficacy would need to be demonstrated. If a drug–device combination product with a drug PMOA includes the same drug but a different device than a previously approved combination product for the same indication, an NDA could be submitted, but submission of an ANDA would be possible only if the device is sufficiently similar in performance and design characteristics to the one included in the previously approved combination product.[21]

If two marketing authorizations are permitted or required[22] (e.g., for each constituent part of a cross-labeled combination product), the marketing authorization for each constituent part would be of a type normally associated with that kind of product [e.g., an NDA or ANDA for a drug constituent part or a PMA or 510(k) for a device constituent part, as appropriate], and these submissions would be made to the center normally responsible for that type of product (e.g., an NDA for the drug constituent part generally would be submitted to CDER and a PMA for a device constituent part to CDRH). The centers still coordinate as appropriate on the review of the product as discussed later even though each would receive its own submissions to act upon. To the extent it is relevant to review of each constituent part, the same data could be presented and relied upon for both marketing authorizations, though additional data might also be needed to support either authorization, or both, individually.

While the formal submission type may have limited significance for the data needed to support marketing authorization for a combination product as discussed later, the type of submission(s) available may have other implications relevant to business judgments and product development planning. For example, although waivers and reduced fees may be available, user fees vary considerably for marketing submissions, with standard fees for NDAs currently ranging from ~US$1 to US$2 million, for PMAs being nearly US$250,000, for ANDAs being over US$50,000, and for 510(k)s being nearly US$5,000.[23] Combination products reviewed under a single marketing authorization should be subject to the fee associated with that type of authorization; if two authorizations are reviewed, the fee associated with each applies.[24] Waivers may be available under the same standards applicable to noncombination products.

In addition, some marketing submission types offer protections from competition while others do not. If a product is being marketed under an NDA or BLA, provisions would apply, for example, that facilitate protection of patent rights and grant periods of marketing exclusivity during which the Agency cannot approve follow-on products that seek to rely on the Agency's prior approval of the same or a similar product. However, such abbreviated marketing authorization mechanisms would be available to enable such follow-on applicants to come to market once such exclusivities expire.[25] If instead a product is marketed under 510(k), no marketing exclusivity applies, so a follow-on product could be cleared at any time. For products marketed under PMA, no follow-on pathway has been made available to date.[26] So, subsequent applicants for similar combination products would have to generate their own, stand-alone sets of data to support PMA approval of their products.

Marketing Authorization Standards for Combination Products

The Agency has not published general guidance on what substantive requirements must be met to obtain marketing authorization for combination products. However, the Agency has stated that the constituent parts of a combination product retain their legal status as a drug, device, or biologic.[27] In keeping with this position, the Agency has indicated in guidance that the review considerations raised by each constituent part should be addressed in keeping with standard approaches for such products.

For example, the Agency has issued guidance on injectors that reflects an expectation that review considerations that would be addressed for injectors if marketed under a device pathway (e.g., performance and human factors) should be included among those addressed for an injector if reviewed as part of a combination product under NDA or BLA.[28] Similarly, draft guidance for drug-eluting stents marketed under PMA reflects an expectation that study of the drug constituent part address issues relevant to its use with the stent (e.g., clinical pharmacology and tolerance for an as-yet unstudied drug), in a manner consistent with how those issues would be addressed for a drug reviewed under an NDA or ANDA pathway.[29] Agency guidance on early development considerations also reflects the FDA's view that marketing authorizations for combination products must address the various questions associated with the constituent parts of the combination product, as would be the case if the constituent parts were to be marketed independently, as well as the safety and effectiveness issues raised by their being combined.[30] In addition, these FDA publications indicate that sponsors should look up to the standards and guidance applicable to the constituent parts as an aid in determining how to satisfy review requirements for the combination product.

In assessing what sort of data may be needed to support marketing authorization for a combination product, an important question (as for any FDA-regulated product) is what kind of safety and effectiveness issues the combination product raises. These issues may relate to the product as a whole, individual constituent parts, and the interaction of the constituent parts. For example, in assessing the appropriateness of a syringe as a delivery device for a particular drug, relevant considerations would include whether the syringe's materials interact with the drug, whether it will deliver the correct drug dosage, and whether the syringe will maintain its integrity in accordance with the combination product's shelf life. Similarly, in the case of a drug-eluting stent, for example, considerations for the drug constituent part would include (in addition to its effectiveness to achieve its intended therapeutic purpose) such factors as whether the formulation of the drug is appropriate in light of the need to control the elution rate and resist flaking from the stent.

The more novel the combination product, the more issues may need to be addressed in supporting data.[31] For example, a combination product that includes a new molecular entity, or a device relying on a novel or highly complex technology, would be expected to raise more questions and require more extensive safety and effectiveness data than one that includes a previously approved drug for the same indication and an off-the-shelf, low-risk device having marketing authorization for the same intended use. Because combination products are not merely their constituent parts however, development challenges can arise, for example, due to new scientific or technical issues raised by the combining or combined use of the constituent parts. As discussed later, timely coordination with the Agency can be helpful in formulating an appropriate product development process and study designs.

Intercenter Coordination and Sponsor–FDA Interaction for Premarket Review of Combination Products

Coordination within the Agency and between the Agency and sponsors can be particularly important and complex for combination products, as can relations between sponsors and other regulated entities. As explained briefly later, standard operating procedures (SOPs) and other established procedures and mechanisms facilitate intercenter coordination, agency–sponsor interaction, and coordination between sponsors and third parties important to the development of the product. Mindful engagement by sponsors may also be helpful to ensuring efficient premarket assessment of combination products.

Premarket review systems for combination products provide for coordination between the lead center and the center(s) that typically regulate the other constituent part(s) included in the combination product. One set of SOPs, for example, includes a formalized process for enabling the lead center to seek input from the secondary center(s). This process provides both for the secondary center(s) to advise the lead center and also to have final decision-making authority on review questions as appropriate.[32] OCP's annual reports to Congress include data tracking the number of consults between centers. These data reflect the substantial amount and increasing frequency of coordination among the centers.[33]

Sponsors coordinate with the Agency through the lead center as a general matter. Sponsors may work through the lead center to confirm participation in meetings and timely review of submissions by relevant offices and staff from the other center(s) with expertise and regulatory interest for the product. In addition, as discussed later, the OCP is available as a resource to facilitate scheduling of meetings and other coordination with the Agency, and also to help resolve disputes regarding product review.

A related consideration for combination products is the importance of strong business relationships. Such relationships can be important in general for FDA-regulated products, for example, between sponsors and contract manufacturers. However, these relationships may be particularly significant for combination products because these products include two or more different types of regulated articles, and sponsors familiar with one type of regulated article and the regulatory regime associated with it may be looking to third parties for data and expertise relevant to the other constituent part and the regulatory issues relating to its use in the combination product. Development of combination products can often involve coordination between sponsors and manufacturers of different types of articles, such as a drug sponsor and device manufacturer, for example. This coordination can include providing rights of access for the Agency to rely on master files, investigational or marketing authorization files, and other sources of data needed to support the development and marketing authorization of the combination product. It can also involve sharing of resources and expertise to support product study and development.

Cooperation between product sponsors and manufacturers can be important not only to support development and pursuit of initial marketing authorizations for combination products, but also to enable agency consideration of postmarketing changes to the combination product. For example, postmarket changes to one constituent part of a combination product may affect its interaction with the other constituent part, and thereby affect the safety and effectiveness of the combination product. Therefore, ongoing willingness to provide access to data or otherwise support new versions and uses of the combination product may be critical to seeking and obtaining marketing authorizations for postmarket changes. With respect to cross-labeled combination products, such considerations may inform decisions regarding whether it makes better sense to seek a single marketing authorization for the combination product or to seek separate marketing authorizations for each of the separately marketed constituent parts. Separate applications may facilitate further development of these constituent parts for independent uses not involving the other constituent part. However, reliance upon separate applications may also pose challenges, for example, with respect to coordination of postmarket modifications to either constituent part, particularly if the applications are (or become) held by different sponsors.

Postmarket Regulation of Combination Products

As discussed above in relation to premarket regulation, the Agency has stated that constituent parts retain their legal status and, therefore, combination products are subject to regulatory requirements applicable to each of their constituent parts. As detailed briefly later, a final rule on current good manufacturing practices (cGMPs), a proposed rule on postmarketing safety reporting (PSR) for combination products, and a final rule on unique identification for devices as well as agency statements concerning product listing requirements, reflect a consistent analytical and policy approach to address these legal duties.[34] In each case, the Agency has attempted to simplify and streamline compliance with the regulatory requirements applicable to the constituent parts of a combination product, while ensuring that the regulatory needs to ensure that the safety and effectiveness of combination products are met.

In developing the cGMP and PSR rules for combination products, the OCP in conjunction with expert staff from the centers and other agency components has reviewed the underlying regulations applicable to drugs, devices, and biological products, to determine how best to ensure that the regulatory needs served by these regulations could be met for combination products while minimizing any unnecessary overlap or redundancy in their application to these products. Based on such analysis, the cGMP rule, for example, offers a streamlined compliance option under which compliance with both the cGMP regulations for drugs and quality system regulations for devices can be demonstrated by fully implementing either of these two sets of regulations and implementing only specified provisions from the other. The Agency's final rule on unique identifiers (UDIs) for devices reflects a similar policy objective. It establishes UDI duties for combination products and devices that are limited so as to ensure that a UDI is associated with all marketed devices through duties that apply either to the combination product or device constituent part, but not both.[35]

Agency guidance issued to date on registration and listing for combination products addresses how to list combination products that have a device PMOA and states that facilities should list the product with the lead center for it, but also can choose to list with the secondary center(s) for the product.[36] The Agency has enabled this approach by updating electronic listing systems for devices so that combination product manufacturers can identify their products as combination products rather than devices, to enable the Agency to track and coordinate regulatory activities for these products even if the product is not listed with the secondary center(s) as well.

Coordination across centers and other agency components is an important feature of postmarket regulatory activity for combination products, as it is for premarket activities as discussed earlier. For example, centers and the Office of

Regulatory Affairs may coordinate on manufacturing facility inspection activities or on evaluation and response to postmarket safety reports.

WHAT IS OCP AND WHAT DOES IT DO?

OCP is a statutorily mandated office, established in 2002 in response to industry calls to establish an office within the Office of the Commissioner (i.e., outside the centers) to coordinate and oversee the regulation of combination products. Under this statutory mandate, the OCP is tasked with ensuring timely, effective premarket review of combination products and consistent and appropriate postmarket regulation for these products.[37] In accordance with its statutory mandate, the OCP pursues a variety of activities, including many designed to assist stakeholders to understand Agency policies and practices and to facilitate their interactions with the Agency.

Specifically, the OCP:

- Develops guidance and regulations for combination products with assistance from the centers and other relevant Agency components.
- Participates in the development of other guidance and regulation that may have implications for combination products.
- Makes medical product classification and assignment determinations both informally and in response to RFDs as discussed earlier.
- Responds to public inquiries about the regulation of combination products.
- Assists regulated entities in identifying appropriate agency resources and components to address regulatory issues.
- Coordinates and participates in meetings of centers and other Agency components with product sponsors, to identify and resolve regulatory issues.
- Resolves disputes between sponsors and centers and between centers.[38]
- Enhances transparency, including through postings and links on its Web page and through information technology enhancement efforts.
- Conducts outreach and seeks input and feedback through public presentations, publications, and meetings with stakeholder groups.

OCP is of particular note to stakeholders having questions or needing assistance navigating the FDA with regard to combination products. OCP is an unusual resource that facilitates access to Agency staff and expedites Agency feedback on combination products. It can sometimes be challenging to schedule meetings and obtain expedited input for FDA-regulated products due to limited Agency resources. OCP can enable timely scheduling of meetings and other measures to expedite resolution of challenging regulatory issues including differences of opinion between the Agency components and between the Agency and sponsors.

The value of the OCP to stakeholders is reflected in the volume of requests it receives and addresses. As its annual reports to Congress show, the OCP typically addresses several hundred inquiries and requests for assistance each year.[39]

RECENT AND UPCOMING DEVELOPMENTS

As noted earlier, the Agency continues to pursue development of regulations and guidance on various issues relating to combination products. Topics the Agency has announced it is addressing include the following: cross-labeled combination product status; when one or two marketing submissions may be appropriate for a combination product; postmarket changes to combination products, classification standards for drugs and devices; and finalization and development of additional guidance on specific classes of products, including injectors, drug-eluting stents, imaging products, and companion diagnostics.

In addition to guidance and regulations, the Agency continues to address procedural and systems-related issues for combination products. As noted above, for example, the Agency has recently updated its listing system for device-led combination products. In addition, it has updated the search features for the Agency's 510(k) and PMA online databases to enable the use of "combination product" as a search term. Additional information technology projects are being pursued, including enhancing postmarket safety reporting quality and efficiency, and argumenting search capabilities on the Agency's Website for information on combination products.

Efforts also continue to encourage coordination and convergence internationally on regulation of combination products. These efforts depend in part on the success of underlying efforts to promote such developments for drugs, devices, and biological products. However, the OCP and the FDA more broadly continue to work with foreign counterparts and to offer technical assistance to them both through government-to-government channels and through workshops and other events in conjunction with stakeholders.

ACKNOWLEDGMENTS

Mr Weiner is the associate director for Policy for the United States Food and Drug Administration (US FDA)'s Office of Combination Products. The views expressed in this chapter do not necessarily reflect those of the US FDA. He thanks Thinh X. Nguyen, Director, OCP, for his review and comment on this chapter, and also many others at the FDA and in counterpart agencies overseas, and the stakeholders, who have informed his thinking on the regulation of combination products, including his colleagues in OCP: Michael Berman, Leigh Hayes, Kristi Lauritsen, Patricia Love, and Joseph Milone.

NOTES

1. See Section 503(g) of the Federal Food, Drug, and Cosmetic (FD&C) Act [21 USC 353(g)]; Title 21 of the Code of Federal Regulations (CFR), part 3, section 3.2(e) [21 CFR 3.2(e)].
2. This chapter does not address the application of the combination product concept to medical products intended for use in animals. While the concept might be applied to such products, FDA regulations and guidance currently do not expressly address this topic.

3. Specifically, 21 CFR 3.2(e) states that combination products include the following:
    i. A product comprised of two or more regulated components, that is, drug/device, biologic/device, drug/biologic, or drug/device/biologic, that are physically, chemically, or otherwise combined or mixed and produced as a single entity
    ii. Two or more separate products packaged together in a single package or as a unit and comprised of drug and device products, device and biological products, or biological and drug products
    iii. A drug, device, or biological product packaged separately that according to its investigational plan or proposed labeling is intended for use only with an approved individually specified drug, device, or biological product where both are required to achieve the intended use, indication, or effect and where upon approval of the proposed product the labeling of the approved product would need to be changed, for example, to reflect a change in intended use, dosage form, strength, route of administration, or significant change in dose
    iv. Any investigational drug, device, or biological product packaged separately that according to its proposed labeling is for use only with another individually specified investigational drug, device, or biological product where both are required to achieve the intended use, indication, or effect.
4. See, for example, Heparin Catheter Lock-Flush Solutions; Transfer of Primary Responsibility from Center for Drug Evaluation and Research to Center for Devices and Radiological Health, 71 Fed. Reg. 47,499 (August 17, 2006) [stating that heparin catheter lock-flush solutions intended to maintain catheter patency are combination products comprised of sterile saline or water that physically occupies space and exerts pressure (the device) and heparin (the drug) that contributes an anticoagulant effect].
5. Section 201(g) of the FD&C Act [21 USC 321(g)] provides that the term "drug" means the following:
    (1) Articles recognized in the official US Pharmacopoeia, official Homoeopathic Pharmacopoeia of the United States, or official National Formulary, or any supplement to any of them; (2) articles intended for use in the diagnosis, cure, mitigation, treatment, or prevention of disease in man or other animals; (3) articles (other than food) intended to affect the structure or any function of the body of man or other animals; and (4) articles intended for use as a component of any articles specified in clause (1), (2), or (3) …
    Section 201(h) of the FD&C Act [21 USC 321(h)] provides that the term "device" means:
    … an instrument, apparatus, implement, machine, contrivance, implant, in vitro reagent, or other similar or related article, including any component, part, or accessory, which is
    i. recognized in the official National Formulary, or the US Pharmacopeia, or any supplement to them;
    ii. intended for use in the diagnosis of disease or other conditions, or in the cure, mitigation, treatment, or prevention of disease, in man or other animals; and
    iii. intended to affect the structure or any function of the body of man or other animals, and which does not achieve its primary intended purposes through chemical action within or on the body of man or other animals and which is not dependent upon being metabolized for the achievement of its primary intended purposes.
    Section 351(i) of the Public Health Services Act 42 USC 262(i) states that the term "biological product" means a virus, therapeutic serum, toxin, antitoxin, vaccine, blood, blood component or derivative, allergenic product, protein (except any chemically synthesized polypeptide), or analogous product, or arsphenamine or derivative of arsphenamine (or any other trivalent organic arsenic compound),

applicable to the prevention, treatment, or cure of a disease or condition of human beings.
6. See Review of Agreements, Guidances, and Practices Specific to Assignment of Combination Products in Compliance with the Medical Device User Fee and Modernization Act of 2002; Request for Comments, 71 Fed. Reg. 56,988, 56,989 (September 28, 2006); see also 21 CFR 3.5.
7. The agency has also published draft guidance focused on how to determine whether an article is a drug or device, which, when finalized, would represent the first time the agency has published general guidance on this topic. See "Classification of Products as Drugs and Devices and Additional Product Classification Issues" and "Interpretation of the Term 'Chemical Action' in the Definition of Device Under Section 201(h) of the FD&C Act" (http://www.fda.gov/RegulatoryInformation/Guidances/ucm122047.htm). A critical issue is often whether the article "achieve[s] its primary intended purposes through chemical action." See 21 USC 321(h). If it does, it cannot be a device. The agency is also developing guidance addressing some issues in determining whether a product should be classified as a biologic, including what is a "protein" for purposes of the biological product definition. See, for example, Biosimilars: Questions and Answers Regarding Implementation of the Biologics Price Competition and Innovation Act of 2009 (http://www.fda.gov/downloads/Drugs/GuidanceComplianceRegulatoryInformation/Guidances/UCM273001.pdf).
8. See 21 USC 353(g)(1). These guidances should be helpful to classification of potential combination products where the classification of the articles of which the product is comprised is not obvious.
9. See 21 CFR 3.2(k).
10. See Definition of Primary Mode of Action of a Combination Product, 70 Fed. Reg. 49,848, 49,852 (August 25, 2005) (final rule) (amending 21 CFR Part 3).
11. See 21 CFR 3.2(k), (m).
12. Certain therapeutic biologics are regulated by CDER (see Transfer of Therapeutic Biological Products to the Center for Drug Evaluation and Research (http://www.fda.gov/CombinationProducts/JurisdictionalInformation/ucm136265.htm), and certain drugs and devices used in relation to blood are regulated by CBER. It follows that combination products that include a biologic, drug, or device falling into these categories could be assigned to the Center that regulates such biologics (CDER) or drugs and devices (CBER) if that constituent part provides the PMOA of the combination product.
13. The agency offered an interpretation of "reasonable certainty" in the preamble to its 2005 amendments to 21 CFR Part 3 establishing the definition for primary mode of action. The agency states that "In general, it would be possible to determine the PMOA of a combination product with 'reasonable certainty' when the PMOA is not in doubt among knowledgeable experts, and can be resolved to an acceptable level in the minds of those experts based on the data and information available to the FDA at the time an assignment is made." See Definition of Primary Mode of Action of a Combination Product, 70 Fed. Reg. at 49,852.
14. See 21 CFR 3.4.
15. See 21 CFR 3.10.
16. See FDA Staff Manual Guides, Volume II—Delegation of Authority Regulatory—Product Designation Authority Relating to Determination of Product Primary Jurisdiction (SMG 1410.701) (June 4, 2010) (http://www.fda.gov/AboutFDA/ReportsManualsForms/StaffManualGuides/ucm052536.htm).
17. See 21 USC 360bbb-2.
18. See 21 CFR 3.7(c).

19. See 21 CFR 3.8.
20. See FDA Staff Manual Guide Volume II—Delegations of Authority, 1410.104, 1410.204, 1410.408 (http://www.fda.gov/AboutFDA/ReportsManualsForms/StaffManualGuides/ucm136380.htm).
21. See, for example, Janet Woodcock to Sunil Mehra (Dey Pharma, L. P.) re docket no. FDA-2009-P-0578 (May 27, 2010); Janet Woodcock to Thomas K. Rogers (King Pharmaceuticals, Inc.) re docket nos. FDA-2007-P-0128 and FDA-2009-P-0040 (July 29, 2009) (affirming the availability of the ANDA pathway for a follow-on auto-injector that has some design differences so long as these differences "do not significantly alter product performance or operating principles and do not result in impermissible differences in labeling").
22. See 21 CFR 3.4(c).
23. See user fee information posted to FDA's website (http://www.fda.gov/ForIndustry/UserFees/default.htm).
24. See Application User Fees for Combination Products (April 2005) (http://www.fda.gov/RegulatoryInformation/Guidances/ucm126017.htm).
25. See, for example, 21 USC 505(c), (j); 42 USC 262(k), (l).
26. See Guidance for Industry and for FDA Reviewers: Guidance on Section 216 of the Food and Drug Administration Modernization Act of 1997 (August 9, 2000) (http://www.fda.gov/medicaldevices/deviceregulationandguidance/guidancedocuments/ucm073707.htm); 21 USC 520(h)(4).
27. See, for example, Current Good Manufacturing Practice Requirements for Combination Products, 78 Fed. Reg. 4307 (January 22, 2013) (Proposed Combination Product CGMP Rule) (https://www.federalregister.gov/articles/2013/01/22/2013-01068/current-good-manufacturing-practice-requirements-for-combination-products); Postmarketing Safety Reporting for Combination Products, 74 Fed. Reg. 50,744, 50,751 (October 1, 2009) (Proposed Combination Product PSR Rule) (http://www.gpo.gov/fdsys/pkg/FR-2009-10-01/pdf/E9-23519.pdf).
28. See Technical Considerations for Pen, Jet, and Related Injectors Intended for Use with Drugs and Biological Products (Draft Guidance) (April 2009) (http://www.fda.gov/downloads/Regulatory Information/Guidances/UCM147095.pdf); "Glass Syringes for Delivering Drug and Biological Products: Technical Information to Supplement International Organization for Standardization (ISO) Standard 11040-4" (Draft) (April, 2013) (http://www.fda.gov/RegulatoryInformation/Guidances/UCM122047.htm).
29. See Coronary Drug-Eluting Stents—Nonclinical and Clinical Studies (Draft Guidance) (March 2008) (http://www.fda.gov/downloads/Drugs/GuidanceComplianceRegulatoryInformation/Guidances/UCM072193.pdf).
30. See Early Development Considerations for Innovative Combination Products (September 2006) (Early Development Guidance) (http://www.fda.gov/RegulatoryInformation/Guidances/ucm126050.htm).
31. See, for example, Early Development Guidance at 6-10.
32. FDA Staff Manual Guides, Volume IV—Agency Program Directions Combination Products Intercenter Consultative/Collaborative Review Process (June 2004) (SMG 4101) (http://www.fda.gov/downloads/AboutFDA/ReportsManualsForms/StaffManualGuides/UCM283569.pdf).
33. See, for example, FY 2012 Performance Report to Congress for the Office of Combination Products (reporting a rise from nearly 400 consult requests in 2008 to over 650 in 2012, even as the number of combination product submissions remained steady or fell over the period; multiple consults may be associated with the same premarket review) (http://www.fda.gov/downloads/AboutFDA/ReportsManualsForms/Reports/PerformanceReports/CombinationProducts/UCM365878.htm).

34. See Combination Product CGMP Rule; Proposed Combination Product PSR Rule; Frequently Asked Questions about the New Device Registration and Listing Requirements, Questions 10, 11 (Frequently Asked Questions) (http://www.fda.gov/MedicalDevices/DeviceRegulationandGuidance/HowtoMarketYourDevice/RegistrationandListing/ucm318796.htm).
35. See Unique Device Identification System (Final Rule) 78 Fed. Reg. 58786 (September 24, 2013) (https://www.federalregister.gov/articles/2013/09/24/2013-23059/unique-device-identification-system)
36. See Frequently Asked Questions, no. 11; How to Register and List (Listing for Combination Products) (http://www.fda.gov/MedicalDevices/DeviceRegulationandGuidance/HowtoMarketYourDevice/RegistrationandListing/ucm053185.htm#listing).
37. See 21 USC 353(g)(4).
38. See, for example, Submission and Resolution of Formal Disputes Regarding the Timeliness of Premarket Review of a Combination Product (http://www.fda.gov/RegulatoryInformation/Guidances/ucm126006.htm).
39. See, for example, FY 2011 Performance Report to Congress for the Office of Combination Products (reporting nearly 500 sponsor contacts with OCP to address premarket review issues, and nearly 60 to address postmarket regulatory issues) (http://www.fda.gov/AboutFDA/ReportsManualsForms/Reports/PerformanceReports/CombinationProducts/ucm317884.htm).

Index

Note: Locator "*f*" and "*t*" denote figures and tables in the text

A

Abbreviated NDA (ANDA), 84, 174, 368–369
Absorption, distribution, metabolism, and excretion (ADME), 60
Accelerated approval, 25, 86
Accreditation Council for Continuing Medical Education (ACCME), 298
AC meetings, preparing for
  success criteria
    dedicated team, 341–342
    external expertise, 342
    rehearsal, 342–343
Active moiety, 194, 196
Adverse drug experience (ADE), 101
Adverse drug reports, 26
Adverse event (AE), 2
  summary information, 71–72
Advisory Committees (AC), FDA, 107–108, 119–120, 330–333
  composition, 330–333
    appointment process, 332–333
    membership training, 333
    membership type, 330–332
    qualification requirements, 332
  conflict
    of interest studies, 335–338
    of interest waivers, 334–335
  historical and regulatory framework, 328–329
  industry perspective
    FDA's decision, affecting, 340–341
    meetings, outcome, 343–344
    meetings, preparing for, 341–343
  operation, 338–340
  structure, 330
Agency guidance, 369, 372
Alert report, 26
Amendment, 182
ANDA. *see* Abbreviated NDA (ANDA)
Approvable letter, 24
Approval letter, 24

B

Batch analysis, 205–206
Best Pharmaceuticals for Children Act, 10
Biocompatability assessment, 152

Biological product development (BPD), 108–109
Biologics
  approval process, 359–360
  biosimilar, 31
  core concept, 30
  definition, 29
  development
    analytical, 354–356
    clinical, 358–359
    design, 350–351
    nonclinical evaluation, 356–358
    process and manufacturing, 351–354
  FDA oversight, current, 348–350
  generic, 31
  innovator/brand, 31
  potency, 30
  purity, 30
  safety, 30
Biologics/biosimilars law, finding, 183–184
Biologics Control Act. *see* Virus–Serum–Toxin Act
Biologics license application (BLA), 350, 359, 368
  pre-NDA/BLA meeting, 86
  review process, 97–100, 98*f*
Biologics Price Competition and Innovation Act (BPCIA), 13–14
Bioresearch Monitoring Program (BIMO), 100, 246–247, 275, 277
Biosimilar, 31
Biosimilar Initial Advisory Meeting, 108, 122
BLA. *see* Biologics license application (BLA)
BPCI. *see* Biologics Price Competition and Innovation Act (BPCIA)
Breakthrough therapy, 16–17, 25
Briefing document
  DAVP, 112
  end-of-phase II meetings, 118
  structure, 112–113

C

Center for Biologics Evaluation and Research (CBER), 17, 30, 63, 301, 330, 347–348, 350, 364, 368

Center for Devices and Radiological Health (CDRH), 17, 32, 302–303, 330, 364, 368
Center for Drug Evaluation and Research (CDER), 17, 30, 45, 287, 301–303, 330, 348, 351, 364, 368
CFR. *see* Code of Federal Regulation (CFR)
cGMPs. *see* Current Good Manufacturing Practices (cGMPs)
Chemistry, manufacturing, and controls (CMC), 200, 211, 218–219, 222, 226–227, 230, 311
  information
    drug product, 57–58
    drug substance, 57
    later-phase, 56
    phase 1 and 2, 54–56
    placebo, 58
  introduction, 56
  meetings with FDA
    EOP2 meeting, 227
    pre-IND meeting, 226–227
    pre-NDA meeting, 227–228
    type C, 228
  safety concerns, 55–56
Citation, 179
Citizens United v. Federal Election Commission, 9
Claims, new and now-available, 297
Classification, combination products medical products
  copackaged, 363
  cross-labeled, 363
  single-entity, 363
Clinical hold, 42
Clinical outcome, 25
Clinical protocol
  elements, 53
  later-phase, 52–53
  phase 1, 52
Clinical Research Associate (CRA), 170, 271–273
Clinical trials
  phases, 21–22
  surrogate endpoint, 25
CMC. *see* Chemistry, manufacturing, and controls (CMC)
Code of Federal Regulation (CFR), 44, 170, 185–186, 238, 258, 274, 289, 291, 294, 301–302, 327, 329–332, 336*t*, 339
  clinical data, medical device
    evidence, 157
    international, 158
  NDA, 80, 93–94

Combination products
  agency classification, obtaining, 366–367
  assignment, 364–365
  classification, 362–364
  regulation, 367–373
Common Technical Document (CTD), 200–201, 359
  electronic (eCTD), 65, 81
  FDA format, 87
  IND, 46, 65, 65*t*
  NDA, 86–97, 96*t*–97*t*
Compliance program guidance manuals (CPGMs), 246, 247*t*
Continuing medical education (CME) events, 298, 302–303
Contract research organizations (CROs), 246, 275, 277
Controlled Substances Act (CSA), 6
Corporate Integrity Agreements (CIAs), 294–295
Corrective and preventive action (CAPA), 159, 161, 161*f*
CPGMs. *see* Compliance program guidance manuals (CPGMs)
CRA. *see* Clinical Research Associate (CRA)
Critical Path Initiative (CPI), 11
Critical process parameter (CPP), 229
Critical quality attribute (CQA), 229, 351, 354
CRO. *see* Contract research organizations (CROs)
CTD. *see* Common Technical Document (CTD)
Current Good Manufacturing Practices (cGMPs), 56, 372

D

Database classification, 147
Debarment certification, 92
Department of Health and Human Services (HHS), 171, 184, 192, 330
Developments, recent, 374
Device history record (DHR), 163
Device master record (DMR), 163
Devices
  classification, 32
  definition, 32
  marketing, 33
  promotion
    CDER and CDRH policies, similarities, 303
    FDA enforcement, 304–306
    investigational device, advertising, 303
    medical devices, DTC promotion, 303
    PE promotion, 303
  regulatory standards, 32–33

Index

Direct-to-consumer (DTC), 288
   advertising, 299
   promotion, 303
Disease, rare, 26
Division of Antiviral Product (DAVP), 112
DP. *see* Drug product (DP)
Draft Guidance for Industry and FDA Staff, 154, 156
Drug
   definition, 19
   development process, 83*f*
   innovator, 31
   new
      approval, 19–20
      definition, 19
      investigational, 20–21
      labeling, 94
      preclinical investigation, 20
   orphan, 26
   product (DP), 350, 355
   radioactive, 62
   regulation, 33
Drug Efficacy Amendments, 5
Drug Efficacy Study Implementation (DESI), 5
Drug Importation Act, 3–4
Drug master file (DMF), 63–64
Drug Price Competition and Patent Restoration Act, 6–7
Drug product (DP)
   components
      drug substance, 208
      excipients, 208
   control, 211–214
      degradation products, 213–214
      drug product impurities, miscellaneous, 214
      residual solvents, 214
      specifications, 211–214
   development
      compatibility, 209–210
      container closure system, 209
      formulation development, 208
      manufacturing process development, 208–209
      microbiological attributes, 209
      overages, 208
Drugs and devices, policy differences, 302–303
Drug substance (DS), 350, 355

E

electronic CTD (eCTD), 65
End-of-review conference, 121–122

Environmental Protection Agency (EPA), 238–239
Establishment inspection report (EIR), 255, 277

F

FACA. *see* Federal Advisory Committee Act (FACA)
Fast track, 24, 82, 86
FDA. *see* Food and Drug Administration (FDA)
FDA Amendments Act (FDAAA), 329, 334, 338, 359
FDA Modernization Act (FDAMA), 8–9, 192, 293, 296, 328, 329, 334
FDA Safety and Innovation Act (FDASIA), 8, 10, 14–17, 179, 182, 329, 338, 341
FD&C Act. *see* Food, Drug, and Cosmetic (FD&C) Act
Federal Advisory Committee Act (FACA), 328, 330
Federal Register, 185–186, 192, 332, 339
Federal Trade Commission (FTC), 299, 302
Financial disclosure, 62–63, 92–93
510(k) premarket notification
   accuracy statement, 153
   class II devices, 130
   clinical data, 153
   components, 150–153
   database, 133*f*–135*f*, 147
   electronic, 154
   goal, 147
   guidance documents, 154
   labeling, 151
   paradigm, 149*f*
   postsubmission, 155
   predicate device, 147–148
   substantial equivalence statement, 151
   technological characteristics, 147–148
   types
      abbreviated, 135–136, 148–150
      de novo, 150
      special, 150
      traditional, 135, 148
FOI. *see* Freedom of information (FOI)
Food and Drug Administration (FDA), 170, 171, 173–176, 184, 186, 190–193, 200, 211, 220, 222–226, 236–239, 242, 245–259, 260*t*, 264, 267–278, 315, 317, 327–346, 347–348, 351, 354, 356, 359
   advertising and promotions, regulations approvals relationship, 293
   biologic products, 301

Food and Drug Administration (*Continued*)
  CDER and CDRH policies,
    similarities, 303
  FD&C Act, enforcement/violation,
    304–306
  medical devices, 302
  prescription drugs, 287
  enforcement
    primary enforcement tools, 304
    untitled letter (NOV), 304
    warning letter, 304–306
  evolution, 3–17
  Form 1571, 46–49, 47–48*f*
  Form 1572, 53, 54–55*f*, 145
  Form 3455, 93
  Form 3514, 150
  Form 3654, 151
  Form 3500A, 101
  Form 356h, 88–94. 89–91*f*
  history
    1848–1979, 3–6
    1980–2012, 6–13
  IND, 42–75
  initiative and impact
    control strategy, 229
    CPP. *see* Critical process parameter (CPP)
    CQA. *see* Critical quality attribute (CQA)
    design space, 229
    lifecycle, 229
    QbD. *see* Quality by design (QbD)
    QTPP. *see* Quality target product profile (QTPP)
  meetings
    AC, 107–108, 119–120
    ad hoc technical, 107
    BPD, 108–109, 122
    briefing document, 112–113, 118
    conduct, 114–115
    do's/don'ts, 116
    end-of-phase II, 107, 117–118
    error, 115
    expectations, 109–111
    guidance, 109, 110*t*
    key factors, 106
    labeling, 108, 121
    late cycle, 108, 120
    letter, request, 111–112
    milestone, 109
    objectives, 116–122
    pitfalls avoidance, 115–116
    PMR/PMC and REMS, 108, 121
    postaction, 108, 121–122
    pre-IND, 45, 106–107, 117
    preliminary responses, 113–114
    pre-NDA/BLA, 107, 119
    preparation, 111–114
    purpose, 106
    schedule, 109
    SPA, 107, 118–119
    type A, B, and C, 109, 111
    type 1, 2, 3, and 4, 122
  mission statement, 2–3
  NDA, 78–102
  organization, 17–19
  protocol assessment, 119
  review divisions, 106, 111–113, 114, 118
  violation and enforcement, 33–34
Food and Drug Administration Amendments Act (FDAAA), 11
Food, Drug, and Cosmetic (FD&C) Act, 3, 42, 170, 173–176, 181–183, 301, 304, 347, 349, 359, 362
  Medical Device Amendments, 128
  and USC law, difference, 181
Freedom of Information (FOI), 154, 317

G

Generally recognized as safe and effective (GRASE), 19, 29
Generic biologic, 31
Generic Drug Enforcement Act, 8
Genotoxicity, 59
Good Clinical Practice (GCP), 192, 236, 256–278
Good Laboratory Practice (GLP), 20, 61, 236, 237–247
Good Manufacturing Practice (GMP), 236, 248–256
Guidance, 186–187

H

Hatch–Waxman Act. *see* Drug Price Competition and Patent Restoration Act
Health-care professionals, promotion
  broadcast and print media, 299–300
  direct-to-consumer (DTC) advertising, 299
  drug detailing, 297–298
  financial community
    materials, 300
    press releases, 300
    video news releases (VNRs), 300
  industry organizations, 301
  medical conferences and exhibits, 298
  scientific information, exchange, 298
  single-sponsor publications, 299
  spokespersons, use, 299

Index

Health Insurance Portability and
 Accountability Act, 259
Help-seeking advertisements, 290–291
HHS. *see* Department of Health and Human
 Services (HHS)
High performance liquid chromatography
 (HPLC), 354, 355*t*
Human factors evaluation, 138–141
Humanitarian Device Exemption (HDE), 9
Hygienic Table, 4

I

IB. *see* Investigator's brochure (IB)
IND. *see* Investigational new drug (IND)
Information amendments, 71
Information, communicating
 critical, 318
 noncritical, 317
Inspections and consequences, GMPs,
 255–256
Institutional Review Board (IRB), 21,
 43–44, 142, 192, 246, 263–264,
 266, 269
International Conference on Harmonization
 of Technical Requirements for
 Registration of Pharmaceuticals
 for Human Use (ICH), 200, 207,
 212–213, 215, 217, 222, 350, 354,
 356, 358
 CTD, 88*f*
 goal, 87
 mission, 87
Investigational device exemption (IDE),
 141–142
Investigational new drug (IND), 9, 20–22,
 192–193, 200, 201, 218–219,
 226–227, 237, 239, 254, 350, 359
 additional information, 74
 amendments, 46, 49, 66, 69–71
 annual report, 46, 50, 71–72
 application, 42
 charging, 74
 clinical protocol, 52–53, 117
 clinical trial, 43
 CMC information, 53–59
 content and format, 46–66
 cover sheet, 46–49
 electronic CTD (eCTD), 46, 65, 65*t*
 FDA review, 66
 Form FDA 1571, 46–49, 47*f*–48*f*
 Form FDA 1572, 53, 54*f*–55*f*
 human experiences, 61–62
 introductory statement, 50–51
 investigator's brochure (IB), 51–52
 pharmacology and toxicology, 59–61
 phases, 22
 pre-IND meeting, 45–46
 promotion, 73
 regulation exemptions, 43–44
 regulation needs, 42–43
 safety report, 46, 50, 67–69
 submission information, 62–65
 table of contents (TOC), 50
 types, 72–73
 Websites, 74–75
Investigations Operation Manual (IOM), 18
Investigator's brochure (IB), 263, 267
 information, 51–52
IRB. *see* Institutional Review Board (IRB)

K

Kefauver–Harris Act, 5

L

Labeling, product, 121
Law, drug/device
 availability, 176–178
 biologics/biosimilars, finding, 183–184
 current, finding, 178–179
 definition, 170–171
 enforcement, 171–172
 FDCA and USC, difference, 181
 federal and state, difference, 172
 federal/state, importance, 172–175
 interpretation, 171
 making, 171
 numbers, meaning, 179–180
 publication, 181
 and regulation, difference, 184–185
 USC and public, difference, 181
Law/regulation, importance, 185
Lithotripter, 129

M

Master cell bank (MCB), 351–352
Medical device
 approval process, 135–137
 classification
  class I, II, and III, 129–130
  database search, 130*f*
  determination, 130–132
 clinical research
  exempt, 142
  IDE, 144
  ISO 14155, 144
  NSR and SR, 142–144, 143*t*
  regulation, 141–142, 141*f*
  unique aspects, 144–146

Medical device (*Continued*)
  definition, 128
  design control
    components, 138–141, 139*t*–140*t*
    input, 138
    QSR, 137
    review, 138
    risk analysis, 138–141
  effect reporting, 146*f*
  reclassification, 132–133
  test protocol, 152
  types, 129, 129*t*
Medical Device Amendments, 128
Medical device reporting (MDR)
  decision tree, 162*f*
  key definitions, 166
  reports, 165
  requirement, 166
Medical Device User Fee and Modernization Act (MDUFMA), 8, 9–10
Medical need, unmet, 24
Medical products, control, 175–176
Medical science liaison (MSL), 298
MedWatch, 101, 165
Meetings with FDA
  CMC, 45
  conduct, 114–115
  do's/don'ts, 116
  end-of-phase II, 107, 117–118
  error, 115
  expectations, 109–111
  guidance, 109, 110*t*
  key factors, 106
  labeling, 108, 121
  late cycle, 108, 120
  letter, request, 111–112
  mid-cycle, 100
  milestone, 109
  objectives, 116–122
  pitfall avoidance, 115–116
  preliminary responses, 113–114
  preparation, 111–114
  purpose, 106
  schedule, 109
  type(s)
    A, B, and C, 45, 109, 111
    AC, 107–108, 119–120
    ad hoc technical, 107
    BPD, 108–109, 122
    Briefing Document, 112–113, 118
    PMR/PMC and REMS, 108, 121
    postaction, 108, 121–122
    pre-IND, 45, 106–107, 117
    pre-NDA/BLA, 86, 98, 107, 119
    SPA, 107, 118–119

N

National Environmental Policy Act (NEPA), 58–59
National Institutes of Health (NIH), 8, 11
National Organization for Rare Disease (NORD), 85
NDA. *see* New drug application (NDA)
New drug application (NDA), 172, 200, 201, 218–219, 222–224, 227, 238, 277, 368–369
  annual report, 101
  content and format, 86–97
  CTD
    administrative information, 87–94
    clinical study, 95
    ICH, 88*f*
    mapping, 95, 96*t*–97*t*
    nonclinical study, 95
    quality, 95
    summary, 94–95
  development, 83–86
  FD&C Act, 79–80
  goal, 22–23
  guidance, 80–81
  maintenance, 101
  PDUFA, 81–82, 86, 97, 98, 100
  pre-NDA/BLA meeting, 86, 98
  presubmission, 97–98
  RDP, 83–86
  regulation, 80
  requirements, 79
  review
    process, 23, 97–100, 98*f*
    results, 24
  submission to FDA, 78
  type, 84–85
New molecular entity (NME) NDA
  AC meetings, 119
  late cycle meetings, 108, 120
  PDUFA, 14–15, 81
Nonapprovable letter, 24
Nonsignificant risk (NSR)
  SR comparison, 143*t*
  studies, 142
Nutrition Labeling and Education Act, 7

O

Obamacare, 178
Office of Combination Products (OCP), 362, 366–367, 373
Office of Device Evaluation (ODE), 130
Office of Orphan Products Development (OOPD), 193, 196

Index

Office of Prescription Drug Promotion
  (OPDP), 18, 287–288, 292,
  296, 304
Office of Regulatory Affairs (ORA), 18
Off-label promotion, 293–295
  private request/response, 295
  public request/response, 295
Orange Book, 27–28
Organization for Economic Cooperation
  and Development (OECD), 238,
  241–242, 246
Orphan
  status, 82, 86
  subsets, 194
Orphan drug
  clinical investigations, 190, 192–193
  designation, 193–197
    clinical superiority, 196
    experience in the United States, 196–197
    medically plausible, 194
    nonrare disease/condition, orphan
      subsets, 194
    patient population determination,
      193–194
    sameness, determination, 194–195
  development, incentives
    clinical development, tax credit,
      191–192
    exclusive approval, 191
    marketing application fee exemptions,
      192–193
    orphan products grants, 192
    written recommendations, 193
  marketing approval, 191–194, 196–197
Orphan Drug Act, 6, 26, 82
  amendments, 190–191
  incentives
    financial, 190
    regulatory, 190, 193
Orphan Drug Regulations, 191, 194–195
Over-the-counter (OTC)
  drug review, 5–6
  monographs, 5, 29
  regulation, 28–29

P

Pacemaker, 129
Patent
  certification, 94
  information, 93
Patient population, 193–194
Patient-reported outcome (PRO), 296
PDP. *see* Process development plan (PDP)
PDR. *see* Physician's Desk Reference (PDR)

Pediatric Research Equity Act, 10–11
Pharmaceutical quality
  biotechnology products, information,
    215–217
  for protein products, 216–217
  viral adventitious and TSE agents,
    215–216
  characterization, 204–205
  combination products, information, 217
  container closure system, 206–207,
    214–215
  drug product
    components, 208
    control, 211–214
    description and composition, 207
    development, 208–210
    pharmaceutical development, 207
  drug substance
    control, 205–206
    general properties, 202
    nomenclature, 201
    structure, 201–202
  excipients, control, 211
    human/animal origin excipients, 211
    non-USP/NF compendial, 211
    novel excipients, 211
    USP/NF compendial excipients, 211
  manufacturer(s), 202–203, 210–211
    material control, 203–204
    method, 210–211
    process description, 203
    reference standards/materials, 206, 214
    stability, 207, 215
Pharmaceutical Research and Manufacturers
  Association (PhRMA), 295, 301
Pharmacoeconomic (PE)
  claims, 296
  promotion, 303
Pharmacology and toxicology
  elements, 60
  integrated summary, 60–61
  safety studies, 59
Physician's Desk Reference (PDR), 317
Poison Squad, 4
Postmarketing issues
  advertising and promotion, 167
  medical device
    510(k), modifications, 164, 164*t*
    PMA, modifications, 164–165, 165*t*
    registration and listing, 163
    reporting, 165–166
    unique device identification, 166–167
Postmarketing requirement/commitment
  (PMR/PMC), 108
Postmarketing safety reporting (PSR), 372

Postmarketing surveillance, 26
Postmarket report, 101
Preclinical investigation, 20
Predicate device
    comparative information, 152
    definition, 147
Preemption, 172–173
Premarket approval (PMA), 32, 175
    advisory panel, 156–157
    annual report, 165
    class III devices, 130
    clinical data, 157
    components, 158
    modular, 156
    process, 155–156
    shell, 156
    supplements, 127
    user fee, 156
Premarket notification, 32
Prescription Drug Marketing Act (PDMA), 7
Prescription drugs, promoting policies, 288–301
    advertising, requirements
        directed to physicians, 290
        exemptions, 290
    approved products, unapproved products/uses, 292
    claims
        comparative and superiority, 295–296
        new and now-available, 297
        pharmacoeconomic (PE), 296
        product, 292
        quality-of-life (QOL), 296–297
    product name and placement, 291
    promotion
        health-care professionals, 299
        see also Health-care professionals, promotion
        off-label; Off-label promotion
    promotional materials
        fair balance requirements, 289–290
        preclearance requirements, 288–289
        prior approval, 288
        submission to OPDP, 292
Prescription Drug User Fee Act (PDUFA), 8, 179, 182, 192, 359
Primary mode of action (PMOA), 364–365, 367–368
    biologic, 364
    device, 364
    drug, 364
Priority review, 25, 86
Priority review voucher (PRV), 360
Process development plan (PDP), 351
    control strategy, 351
    CQAs, 351
    design space, 351
    product life cycle management, 351
    quality target product profile, 351
    risk assessment, 351
Product jurisdiction, 128–129
Project Bioshield Act, 11
Protocol amendments, 69–70
PRV. *see* Priority review voucher (PRV)
PSR. *see* Postmarketing safety reporting (PSR)
Public Health Security and Bioterrorism Preparedness and Response Act, 11
Public Health Service (PHS) Act, 301, 304, 347, 349

Q

Quality assurance unit (QAU), 238, 242–244, 246
Quality by design (QbD), 228, 350
Quality-of-life (QOL) claims, 296–297
Quality system inspection technique (QSIT), 159
Quality system regulation (QSR)
    audits, 160–161
    class II, 130
    design control, 137–141, 139t–140t, 160
    exempt, 129
    goal, 159, 163
    management control, 160–161
    preamble, 137
    production and process control, 163
    subsystem, 160f
Quality target product profile (QTPP), 228
Quincy, M.E., 190

R

RAs. *see* Regulatory affairs (RAs)
Reagan, Ronald, 190
Reasoned proposals, 116
Recall, 34
Regulation, 184, 185–186
    current, finding, 185–186
    older, finding, 186
    postmarket, 372–373
    premarket
        intercenter coordination, 370–372
        investigational/marketing submissions, 368–369
        marketing authorization standards, 369–370
Regulatory affairs (RAs), 309–311, 313, 320
    background and training, 311–315
    attitude and approach, 313
    degree programs, 311–312

Index

regulatory, 313–314
self-education, importance, 312
zealotry, 314–315
documentation
  documents, managing, 320
  FDA contact, documenting, 320–321, 320*t*
  memo, 319–320
  information
    communicating, 317–319
    gathering, 316–317, 316*t*
    inherent limitations, 315–316
  spectrum, 311*f*
  submissions
    data presentation, 324–325
    documents, writing, 322–323, 323*f*
    large documents, handling, 325
    regulatory review, 324
Regulatory compliance, CMC
  managing changes and maintaining, 218–219
    during IND Stages, 219–223, 221*t*
    during postapproval stages, 223–226, 225*t*
Regulatory development plan (RDP), 83–84
Regulatory patent, 136
Regulatory project manager (RPM), 99
Reminder advertisements, 290–291, 301
Reproduction toxicity, 59
Request for applications (RFAs), 192
Request for designation (RFD), 366–367
Risk evaluation and mitigation strategy (REMS), 12–13

S

Safe harbors, 93
Safe Medical Devices Act, 7
Safety pharmacology, 59
Section 510(k) in USC, finding, 183
Securities and Exchange Commission (SEC), 300
SGEs. *see* Special government employees (SGEs)
Shelf life, 153
Sherley Amendment, 4
Significant risk (SR), 144
Source document/data verification (SDV), 273

Special government employees (SGEs), 331, 334
Special Protocol Assessment (SPA), 107, 118–119
Specification(s), justification, 214
Split predicates, 148
Standard operating procedure (SOP), 137, 243–244, 248, 269–271, 370–371
Standard review, 25
Sterile devices, 147
Sulfa, 4

T

Toxicity, 152
Tufts Center, 106

U

UDI. *see* Unique identifier (UDI)
Unique Device Identifier (UDI), 167
Unique identifier (UDI), 372
United States Code (USC), 170, 179, 181, 185, 328, 334
  and CFR, difference, 185
United States Pure Food and Drug Act (PFDA), 4
Untitled letters (NOV), 301, 304
USC and public law, difference, 181
US Pharmacopoeia (USP), 3

V

Video news releases (VNRs), 300
Virus–Serum–Toxin Act, 4

W

Warning letter, 34, 301, 304–306
  remedies, 305–306
Washington Legal Foundation (WLF), 293–294
Wiley Act. *see* United States Pure Food and Drug Act (PFDA)
Wiley, Harvey W., 4
Writing process
  back-and-forth process, 322–323, 323*t*
  iterative process, 322–323
  lowering expectations, 323